THE ORIGINS OF THE NEW

The Origins of the New

NOVELTY AND INNOVATION
IN THE HISTORY OF LIFE,
CULTURE, AND TECHNOLOGY

DOUGLAS H. ERWIN

PRINCETON UNIVERSITY PRESS
PRINCETON & OXFORD

Published by Princeton University Press
41 William Street, Princeton, New Jersey 08540
99 Banbury Road, Oxford OX2 6JX

press.princeton.edu

GPSR Authorized Representative: Easy Access System Europe - Mustamäe tee 50, 10621 Tallinn, Estonia, gpsr.requests@easproject.com

All Rights Reserved

ISBN 978-0-691-17894-3
ISBN (epub) 978-0-691-28753-9
ISBN (PDF) 978-0-691-27292-4
Library of Congress Control Number: 2025947267

British Library Cataloging-in-Publication Data is available

Editorial: Alison Kalett, Hallie Schaeffer, and Laura Lassen
Production Editorial: Jenny Wolkowicki
Jacket design: Katie Osborne
Production: Danielle Amatucci
Publicity: Matthew Taylor
Copyeditor: Maia Vaswani

Jacket images: Patrick Guenette / Book Worm / INTERFOTO / Alamy Stock Photo

This book has been composed in Arno

Printed in the United States of America

10 9 8 7 6 5 4 3 2 1

CONTENTS

1

Introduction

OPABINIA WAS only a few inches long, with about 15 paired, undulating flaps on the sides of the body, a head with five bulbous eyes, and a long, flexible proboscis ending in a pair of spiny claws. Long one of the weird wonders from the 505-million-year-old rocks of the Burgess Shale in western Canada, the few specimens of *Opabinia* are now known as part of an evolutionary burst leading to arthropods in the Cambrian Explosion of animal life. Finding a home for *Opabinia* among the arthropods makes it no less remarkable. If anything, it sharpens our questions about the processes that generated such remarkable evolutionary novelties.

Evolutionary novelties abound. Consider grasses. Grasses have been supremely successful, with grasslands carpeting the temperate interiors of North and South America, Asia, and Africa for the last 15–20 million years. Grasslands changed the structure of terrestrial communities, modifying regional climate, and stabilizing soils against erosion, probably even reducing the sediment flowing down rivers to the oceans. Studies of the fossil record of minute silica particles (known as phytoliths) in grass stems by paleobotanist Caroline Strömberg, coupled with molecular evidence, show that grasses originated 55 million years ago, or perhaps earlier. From their widespread success one might assume that grasses spread soon after they first appeared. Yet grasses were ecologically insignificant for tens of millions of years before changing climate accelerated a pervasive change in terrestrial ecosystems. So, grasses originated and diversified into their major clades long before they became ecologically or evolutionarily successful.[1]

Similar lags have long been recognized between invention and the impact of a new technology. Although patent records are often used to study new technologies, most patents have little economic impact, and there is often a lag between discovery of a new technology and the onset of a significant role for it in the economy. The complexity of technological innovation is illustrated by the early history of digital computers. Over the winter of 1937–1938 John Vincent Atanasoff, a physicist at the Iowa State College, built the first digital computer to speed up laborious calculations. His computer was not programmable, but like modern digital computers it solved complex equations using Boolean logic and binary numbers. Although Atanasoff is often credited as the inventor of the digital computer (and eventually won recognition in a patent suit), the spread of digital computers was due to the development of ENIAC (Electronic Numerical Integrator and Computer) in 1943–1944 at the University of Pennsylvania (some of Atanasoff's ideas may have found their way into ENIAC, but that is a story for another day). As ENIAC was being built, British codebreakers at Bletchley Park were building Colossus, a fully digital computer, to help solve German codes. Atanasoff's computer was discarded as scrap at Iowa State, and the existence of Colossus would remain secret for decades, leaving ENIAC as the first fully digital, programmable computer. More importantly, some of the builders of ENIAC founded a company that became part of the Sperry-Rand Corporation and the beginning of the digital revolution.[2]

The importance of such lags is best illustrated through a brief look at three approaches among biologists since Darwin published *The Origin of Species* in 1859.

Darwin argued for a continuity of evolutionary processes from small-scale changes observed in living plants and animals to the longer-term patterns revealed by the fossil record. By denying the discontinuities suggested by some earlier natural scientists, Darwin emphasized the power of natural selection to explain the diversity of life. Although most scientists quickly accepted Darwin's views of descent with modification, natural selection was only one of many explanations debated in the interval between 1860 and the 1930s, in part because of controversies over how organismal attributes were passed to the next generation. These controversies were seeming resolved by geneticists and evolutionary biologists in the Modern Synthesis of evolutionary biology (1920s–1950s). Many contributors to the Modern Synthesis adopted Darwin's extrapolationist approach, updated with the recent discoveries in genetics. Paleontologist George Gaylord Simpson and evolutionary biologist Ernst Mayr focused on the origin of larger groups as a proxy for evolutionary innovation. For example, from this perspective the critical issue in the origin of turtles was the vertebrate order Testudines, which encompassed their morphological novelties. Mayr viewed evolution as opportunistic, capitalizing on useful morphological novelty, while Simpson beautifully articulated the variety of evolutionary patterns documented by the fossil record (his 1944 book *Tempo and Mode in Evolution* remains an intellectual touchstone for paleontologists). By the 1950s Simpson had adopted the reigning views of the Synthesis, that natural selection could, over time, craft the full panoply of the history of life. This extrapolationist view, that selection at the level of individuals within populations via small changes in genes, is, over time, responsible for remarkable evolutionary novelties, has long been the default assumption by evolutionary biologists and viewed as immune to deeper inquiry.[3]

A second approach to novelty extends the extrapolationist view. Confronted by new opportunities a species may diversify into a range of new species, each specialized for different opportunities—food type, habitat, and so on. Simpson explored these adaptive radiations from the 1940s to the 1960s, extrapolated from living plants and animals, often on islands, to more complex diversifications such as the spread of placental mammals and even the Cambrian Explosion of animals. In 1960 Simpson wrote: "On a broader scale, we now see, even more clearly than Darwin did, that every marked expansion of a group, whether it be a genus or a phylum or the whole animal kingdom is an adaptive radiation."[4]

Subsequent generations of evolutionary biologists have examined adaptive radiations, including Darwin's finches in the Galapagos Islands, repeated invasions of small marine fish into lakes in British Columbia, and the remarkable expansion of the *Anolis* lizards across the Caribbean, establishing them as important contributors to diversity. But the theory of adaptive radiation assumes that the generation of new morphologies is a sufficiently regular occurrence that we do not need to worry about the supply side of the equation. The hinge point is the ecological opportunities that facilitate the radiation, whether migration to a largely unoccupied island, the aftermath of an extinction, or the acquisition of a "key innovation" that allows access to new resources. But the significance of adaptive radiations for morphological novelties seems limited, since few are associated with adaptive radiations. In later chapters evidence for lags between the origin and success of novelties will support the distinction between novelty and innovation. Moreover, these chapters reveal that not all novelties are a response to an opportunity or need.

Extrapolationist views invoke the principle that explanations with the fewest assumptions should be favored over more complex scenarios. Known as Ockham's razor, this principle of parsimony is attributed to William of Ockham, a medieval English monk. Isaac Newton made similar arguments, writing in his *Principia*: "Nature does nothing in vain, and more causes are in vain when fewer suffice. For nature is simple and does not indulge in the luxury of superfluous causes." Darwin had enough challenges arguing for the evolutionary unity of life and for the power of natural selection in crafting biological diversity without multiplying the range of processes. Similarly, advocates of the Modern Synthesis were trying to consolidate evolutionary biology through the incorporation of field natural history, genetics, and the fossil record, which also favored extrapolating from selection within populations. Thus, neither extrapolationist nor adaptive radiation scenarios for the origins of novelty and innovation can be rejected out of hand. The assumptions made by Simpson, Mayr, and their colleagues were reasonable, particularly as they were working to produce a more scientific view of evolutionary change. In fact, elements of them may make important contributions to a broader understanding of these problems. But nonetheless they seem to fall short of a full and comprehensive understanding of the phenomena of novelty and innovation.[5]

The third, more recent, approach reflects growing experimental evidence that new morphologies often reflect different sources of genetic variation from those fueling adaptation. New molecular methods of interrogating the genes and gene networks responsible for development from an embryo to an adult have revealed a remarkable and unexpected conservation of genes across many kinds of animals. Hox genes, involved in animal patterning, and *Pax6*, responsible for eye development in vertebrates, have become broadly known, but there are many more. From the extensive conservation of these genes among living animals we can infer that their last common ancestors also possessed these genes, even if the genes may have had somewhat different roles. These discoveries have led to the rise of comparative evolutionary developmental biology, or "evo-devo." The origins of novel attributes of animals (and some plants) have attracted considerable attention and have suggested the need for an expanded view of the origins of novelty.

While the classic model of an adaptive radiation involving a "key" morphological innovation assumes a close connection between the origins of novelties and their evolutionary success, there is no logical or necessary connection between the two phenomena. To Mayr the suggestion of a lengthy gap between the origin of a morphological novelty and its ecological or evolutionary success would seem nonsensical, if not bizarre. Mayr focused on the origin of higher taxa as a metric of evolutionary novelty, but novel morphological features often define new clades, and a new clade, whether turtles, grasses, or birds, involves the assembly of a suite of features. The evolutionary history of grasses illustrates that the innovation associated with a clade may be decoupled from the attributes that define the origin of the clade. But evolutionary developmental biologists are in some ways the reflection of advocates of adaptive radiations, for they emphasize the mechanisms that generate novel forms but often neglect the problem of how these new forms succeed and make a living. The origins of novelties, innovations, and clades are distinct issues. They may occur simultaneously, as Simpson and Mayr assumed, but that is a question to be resolved by study of individual cases, not assumed.

Distinguishing the origin of novel forms (novelty) from their success (innovation) is critical to understanding the processes responsible for their formation. I will define these terms more carefully later in this book, but for now we can treat the origin of novelty as the formation of unique features: the carapace of a turtle, the growth habits of grasses, feathers and skeletal changes for birds, language for humans. In contrast, innovations involve transformations of ecological assemblages and long-term evolutionary impact. Innovations are often more complicated than simply the acquisition of a morphological novelty, for they reflect the ecological and evolutionary success of a group, which may require additional adaptations beyond a novel morphology or environmental changes, as was the case with grasses.

The spread of evo-devo has led many biologists to reexamine the driving forces for novelty and innovation, informing the three major issues explored in this book: *How did novel attributes arise, and how did they become successful innovations? Can the same general model explain novelty and innovation across biological, cultural, and technological domains?* Finally, *does novelty simply represent the extremes of adaptive evolutionary change, the sort of thing that happens every day, or are novelty and innovation somehow decoupled from evolutionary adaptation?* As a foundation for what follows it is worth briefly elaborating on these three questions.

No comprehensive account of novelty and innovation in the biological domain can focus exclusively on genetic and developmental changes, on ecological opportunity, or on changes in the environment, any more than a useful account of technological change can focus on inventions to the exclusion of their economic success or failure, or an account of cultural innovation can ignore the broader context in which the changes occur. Exploring the origin of novelties and innovation constitutes chapters 6 through 9, with numerous examples in other chapters as well.

Novelty and innovation have become popular themes in evolutionary biology, business, economics, and culture, among other fields. Biological novelty, evolutionary innovation, cultural transformation, and technological change are usually treated separately by evolutionary biologists, anthropologists, and economists. There is a widely shared (if sometimes a bit inchoate) view that there are great similarities among biological, cultural, and technological innovation. But rarely has the nature of such similarities been investigated in detail. Yet human language, culture, and technology represent some of the most transformative novelties in the history of life. This motivates the second major theme of this book: *Can the same general model explain novelty and innovation across biological, cultural, and technological domains?* If a general approach to novelty and innovation is possible, it must encompass culture and technology.[6]

Charles Darwin famously drew upon the work of Thomas Robert Malthus as inspiration for his theory of natural selection. Modern evolutionary biology and modern economics trace their intellectual history back to the field of political economy in the eighteenth and early nineteenth centuries through Malthus, Adam Smith, Nicolas de Condorcet, and others, with continuing intellectual cross-pollination. Many economists have been deeply influenced by work in evolutionary biology. Economist Joseph Schumpeter has often been credited with distinguishing between invention and innovation in the evolution of technology. Schumpeter was a leading twentieth-century theorist of economic development and entrepreneurship. He emphasized the importance of technological innovation but recognized that inventions by themselves contributed little. It was only when inventions diffused through an economy that they could became innovations and influence economic growth. Just as I view Schumpeter's distinction between invention and innovation as equally applicable to the history of life, I will introduce other approaches from economics and anthropology that may prove valuable to biologists.[7]

Studies in cultural evolution have long made strong claims of analogy between cultural processes and biological evolution. It is hardly surprising that scholars have explored the possibility of a general theory of novelty and innovation spanning biological, cultural, and technological domains, although progress has been stymied by the fact that while there are many metaphorical similarities between novelty and innovation across these domains, we lack a robust theory.

Such a general theory of novelty and innovation could take several different forms:

- A general theory covering biological, cultural, and technological domains is possible, and it is possible to construct a mathematical theory through which we could evaluate the relative importance of different contributions and perhaps even make predictions about the future.

- Commonalities exist across domains, but there are sufficient differences between domains that any mathematical framework (if possible) will be domain specific.
- Commonalities exist across domains, but for various reasons developing a mathematical theory even within domains is unlikely. However, a general conceptual framework covering the three domains can be developed, while acknowledging some degree of domain specificity.
- Despite apparent metaphorical similarities across domains, processes of novelty, invention, and innovation are sufficiently specific within different domains (and may vary so much within domains) that even building a domain-specific framework is a hazardous enterprise.[8]

These alternatives will be evaluated in the final chapter.

A third major issue woven through the book is the seemingly eternal issue of continuity versus disjunction. Does novelty simply represent the extremes of adaptive evolutionary change, the sort of thing that happens every day, or are novelty and innovation somehow decoupled from evolutionary adaptation? Many evolutionary biologists reject any distinction between novelty and the sort of evolutionary adaptation that makes up much of evolutionary change, generally assuming that with sufficient time adaptive change will generate morphological novelty. Recent insights from developmental biology challenge this comfortable assumption, suggesting, for example, that the changes in developmental patterning of the embryo that led to a novelty differ from those that generate adaptive changes. Abrupt transitions, such as the appearance of turtles, seem to lend credence to claims that novel attributes and innovations represent distinct modes of evolutionary change. Defenders of more traditional approaches to evolution dispute the existence of any discontinuities, just as they have disputed the existence of distinct processes of macroevolution since the contributions of Stephen Jay Gould, Niles Eldredge, Steve Stanley, and others in the 1970s and 1980s. From this more traditional, or microevolutionary, view the seemingly abrupt shifts are a description of pattern, not of process. In the case of turtles, one could argue that poor preservation of intermediate forms generated the apparent discontinuity.

In the mid-1990s the evolutionary biologists John Maynard Smith and Eörs Szathmáry developed the idea of the major evolutionary transitions. The eight events identified by Maynard Smith and Szathmáry spanned from the origin of life to the evolution of human culture and language, thus justifying the approach taken here. They believed new evolutionary individuals were constructed during these events by changes in the organization of genetic information and in the means of inheritance, and that these transitions explained the apparent hierarchy in the organization of life. The book by Maynard Smith and Szathmáry was wonderfully stimulating and will be discussed in later chapters. But their analysis fails to adequately incorporate profound changes in the environment from oxygenation of the oceans to increases in functional complexity and the formation of new evolutionary possibilities. Indeed, this book began as a response to the major evolutionary transitions viewpoint. Among the great challenges for research on the history of life is developing an approach to this problem that recognizes the complexity of historical processes and the diversity of information required to understand them. Major transitions encompass a much more diverse array of events than those identified by Maynard Smith and Szathmáry. Their work has been properly influential in articulating a nonuniformitarian view of evolutionary dynamics: that the nature of evolution has itself changed over time.[9]

I have a broad view of the range of evolution, and I find claims of human uniqueness that justify an inviolate separation between humans and the rest of life unsupportable. Social evolution is widespread among many species, even among microbial consortia. There are aspects of human cultural evolution that appear to be unique, and certainly the extent of technology created by humans is unprecedented. But many other species modify their external environment in important ways, suggesting that the differences are of degree rather than absolute. Human evolution, particularly cultural and technological evolution, is interesting exactly because it must be encompassed by any useful approach to evolutionary innovation. And for just this reason Maynard Smith and Szathmáry included the evolution of language as the last of their major evolutionary transitions.

This issue of novelty versus adaptation and continuity and discontinuity has implications for rates of change. So far, I have said nothing about how rapidly novelties arise or how quickly novelties transform ecosystems, beyond noting lags. I would be willing to wager that not a few readers have inferred that novelties must arise quickly. Indeed, as discussed in chapters 2 and 3, many such claims have been made, but in later chapters I will argue that distinctions among adaptation, novelty, and innovation are about mechanism and are independent of the time involved. So, in contrast to the claims of my tribe (paleontologists), estimated rates may be a poor indicator of evolutionary dynamics.

Structure of the Book

Past studies of innovation, whether in biology, culture, or technology, have often relied upon compiling histories of specific cases. The case-study approach allows characteristics of novelty and innovation to emerge from the weight of examples. A more philosophical approach would be to search for definitions of novelty and innovation, emphasizing logical rigor over the messy realities of biology. Indeed, there is a growing literature among philosophers of biology and more philosophically inclined biologists proposing various definitions of novelty. Relying entirely on case studies is often justifiably criticized as anecdotal and ad hoc. The numerous informative examples in this book are in service of a broader conceptual framework. Chapters 2 through 4 examine seven different approaches to novelty and innovation, before I articulate a conceptual framework in chapter 5. Although I began this project with a rigid view of the boundaries of novelty, I have come to see that definitional flexibility is warranted. Chapters 6 through 9 employ this conceptual framework to examine different exemplars of novelty and innovation, but I will endeavor to show that these exemplars represent larger classes of similar patterns and processes.

I explore earlier approaches to novelty and innovation in chapter 2, roughly from the early nineteenth century through about 1990. I have already noted that the common assumption has been that novel morphologies are linked to the diversification of new species. I will argue that while adaptive change may often be a critical component of innovation, adaptive evolution may play less of a role than commonly assumed. Larger-scale or macroevolutionary approaches to novelty and innovation, such as the differential success of species or clades, are also discussed in this chapter. But I conclude that no model of adaptive radiation satisfactorily explains evolutionary novelty or macroevolutionary lags. Indeed, most traditional macroevolutionary approaches have neglected the factors responsible for the generation of morphological novelty.

The latter portion of chapter 2 extends the discussion of novelty and innovation to cultural evolution and technological change. Discussions of cultural and technological aspects are integrated into chapters where appropriate. But one of the challenges is the use of the term *evolution* in different fields and in different ways. Even two biologists may differ over whether the term applies to hereditary changes in gene frequencies mediated by either natural selection or genetic drift, or if selection operates at multiple levels—for example, between populations of a species. The past several decades have seen expansion of the types of inheritance to include cultural, ecological, and epigenetic inheritance. The problems increase with cultural evolution, where the focus shifts to processes distinct from the gene-centric views of many biologists. The term *evolution* becomes more confusing among economists, historians, sociologists, and other anthropologists, who may use it as a synonym for long-term change but reject any parallels with Darwinian or biological processes. While I have sought to identify similarities and differences across the domains of biology, culture, technology, and economics, I make no claim for a reductionist, gene-centric view of evolution, nor an argument for universal Darwinism. Universal Darwinism is an effort to extend a reductionist view of evolution to fields beyond biology, particularly anthropology, economics, and psychology. Instead, the focus is on the nature of novelty and innovation across domains in which the processes of change may be very different. While I use the term *evolution* throughout the book, in chapter 2 on historical approaches, chapters 3 and 9 on the domains of culture and technology, and chapter 4 on cultural spaces I use the term generically as a synonym for change, with no analogy to biological processes. Because human novelties and innovations are found today in a single species, I discuss them together in chapter 9. I make no claims for even a survey of the vast literature on novelty and innovation in economics and business.[10]

Many biologists use *novelty* and *innovation* interchangeably, and writers of popular business books rarely seem to have given any thought to the possibility of a distinction between the two terms. *Novelty* and *innovation* are what cognitive scientist Marvin Minsky termed "suitcase words": words that have a variety of meanings, like *conscience* or *complexity*. Unpacking suitcase words can transform impossible problems ("What is consciousness?") into a series of less difficult problems where progress might be achievable. Definitions of novelty, invention, and innovation differ between different authors and have changed over time. To some, novelty applies to any heritable biological change or is synonymous with adaptation, but such a definition is so broad as to be meaningless. I begin unpacking these words in chapters 3 to 5 as I examine how the concept of evolutionary novelty has been refined over the past few decades. Coincidentally, the beginnings of evo-devo, insights from the fossil record, new approaches to cultural evolution, and new tools to reconstruct evolutionary history led to reassessments of the nature of novelty and innovation, beginning about 1990. These new ideas provide the foundation for chapter 3, which advances the historical presentation over the past three decades.

I highlight how evo-devo studies have provided insights into the genetic and developmental foundations of novelty.[11]

Sewall Wright, one of the founders of the Modern Synthesis, argued that supply of genetic novelties was relatively constant through time, with their success depending on ecological opportunities. Like Mayr and Simpson he saw no useful distinction between novelty and innovation. Wright introduced the concept of an adaptive landscape as a heuristic model to depict the fitness of an organism under different gene combinations. The highs on the landscape represent regions of high fitness or adaptation, while the basins represent low fitness or adaptation. Simpson's adaptive landscapes introduced the concept of a space into evolutionary biology and became one of the most enduring and popular metaphors in evolution. Today spaces of DNA, RNA, and proteins are common, and the idea has been extended to regulatory and metabolic networks. Paleontologists analyze morphology within morphospaces, and other design spaces encompass functional, ecological, or other features of organisms. Not to be outdone, economists have described product spaces and their potential influence on economic development.[12]

We intuitively think of spaces as Euclidean, where the three axes of the space are at 90 degrees to each other and distances can be measured between objects. But many mathematical relationships between objects are not Euclidean spaces or may be only locally Euclidean. The idea of a space is a powerful metaphor that raises important questions: How accessible are different regions of an evolutionary space? Do spaces simply exist, waiting for organisms to fill them, or are they somehow constructed through the history of life? Such questions raise some interesting and poorly appreciated problems associated with the topology of evolutionary spaces, the topic of chapter 4.

Innovation is often described as a search through a space of opportunities, but this is misleading if evolutionary spaces can be constructed. I explore this fundamental difference in some detail in chapter 4 because it has some far-reaching consequences for how we think about evolutionary novelty and innovation. "Search" is the appropriate metaphor in situations where solutions are already present (what are called pre-statable spaces), and movement through the space can be accomplished via simple operations.

The sequence space of 20 nucleotides of DNA or RNA is an example I use in chapter 4. A single nucleotide mutation changes the sequence one position in the sequence space. Ignoring changes in the length of the sequence and considering just single nucleotide changes, then there are 19 possible single nucleotide changes from any starting sequence. And each of these is adjacent to additional single nucleotide changes. Thus, there is a pathway of single nucleotide changes through the sequence space from a starting sequence to a completely different 20-nucleotide sequence. A simple operation (mutations producing the substitution of one nucleotide for another) allows search through the space of alternative sequences. Some biologists argue that similar search processes explain innovation in many other systems. I will return to consider this influential example in some detail in chapter 4. In general, search operates in systems like sequence spaces for proteins and nucleotides, and even potentially for some phenotypes such as logarithmically coiled shells (clams and gastropods). But the metaphor of search becomes problematic where the design space cannot be defined independently.

My view is that most significant evolutionary novelties and innovation involve the construction of new opportunities. Construction involves the formation of a space of opportunities that could not reasonably have been specified in advance. These opportunities could be molecular, developmental, morphological, cultural, or technological, but they often involve more complicated evolutionary changes than in search. Once formed, new design spaces can be exploited by adaptive search. Arthropods, with exoskeletons and jointed appendages, represent the formation of a new space of opportunities, for example. It would have been difficult to define the possibilities for arthropods 600 million years ago, before the morphological novelties appeared. But the origin of arthropods opened a vast range of opportunities, which are continuing to be exploited.

Microbiologist Richard Lenski and his colleagues have followed the evolution of strains of the bacterium *Escherichia coli* for more than 50,000 generations. In 2011 Lenski and Zac Blount announced the discovery of a form of *E. coli* that feeds on citrate. Acquiring this capability required earlier genetic changes that created the potential for the mutation that allowed the bacteria to feed on citrate. These earlier mutations potentiated this change, but they did not themselves establish the novelty. Such potentiating mutations have been associated with other novelties as well and cast doubt on the assumptions of Wright, Mayr, and others: We cannot assume that the supply of novelty is relatively constant through time. Indeed, one of the main arguments of this book is that we need to understand the factors that control the supply of novelty and how this has changed through time. The experimental demonstration of potentiation deeply influenced the conceptual framework of novelty and innovation advanced in chapter 5.[13]

This framework involves four components: potentiation, novelty, adaptive refinement, and innovation. While ecological opportunities are important, in my view they do not play the central role in novelty and innovation proposed by Wright, Simpson, and Mayr. But by distinguishing potentiating factors and adaptive change from the generation of new morphological novelties and the environmental aspects that control whether, and when, the novelties become successful as innovations we may be able to generate a more comprehensive understanding of these aspects of evolution. The conceptual foundations of novelty and innovation are not as well developed in studies of cultural change as in biology or economics, but later chapters extend this approach to culture and technology.

The opportunity-driven view of novelty and innovation assumes that new morphologies depend upon existing genetic and developmental variation. But other evolutionary and developmental biologists have argued that novelty often requires specific genetic and developmental changes. If this latter view is correct, then the origin of novelty may in part reflect the internal dynamics of complex systems, whether they are organisms or cultures, and be largely decoupled from ongoing processes of adaptation. In my view, there is less conflict between these views than may first appear, once the nature of novelty is carefully circumscribed. Chapter 6 will consider specific cases of evolutionary novelty and the role of networks in facilitating change. The primary objective of that chapter is to address the sources of genetic and developmental novelties. The core of the chapter will involve a discussion of how the structure of developmental gene regulatory networks controls patterns of novelty.

The generation of evolutionary innovations in biological, cultural, and technological domains is addressed in chapters 7–9, including the mechanisms responsible for converting evolutionary novelties. Ecological processes are discussed in chapter 7, including those by which organisms construct niches either for their own species or for other species. In addition, that chapter considers several types of diversifications, novelty events, and innovation after mass extinctions or similar biotic crises. Chapter 8 provides case studies of biological innovation that illustrate the importance of a "macroevolutionary triad" of genetic and developmental potential, ecological opportunity, and environmental possibility.

Chapter 9 examines novelty and innovation in culture and technology. Cumulative cultural evolution, language, and cumulative technological evolution are unique human novelties that enable other novelties and innovations. Since these changes all occur among humans and our closely related ancestors, discussing these changes in a single chapter was more effective. Patterns of innovation in cultural and technological evolution exhibit many similarities to processes of biological innovation, but there are also many cases of apparently parallel innovation, as with social complexity such as states, whereas technological innovation more commonly seems to follow patterns of novelty followed by diffusion, more like some evolutionary diversifications in biology. The emphasis is on the formation of new stable structures as a key component of cultural evolution (broadly construed), focusing on macroevolutionary patterns of cultural and technological novelty, new forms of social complexity, the role of institutions, and the extent to which the model developed in chapter 5 may be applicable.

I return to the major themes discussed earlier in this chapter in the concluding chapter, emphasizing the conceptual framework developed in chapter 5, evaluating the prospects for a general theory of innovation and novelty and whether the probability of innovations has changed through the history of life, and why this might be so. One of the central themes of this book is how novelty and innovation reflect the evolution of the evolutionary process itself over the past three and a half billion years or so. The fundamentally historical nature of evolution means that the nature of variation upon which selection can act and the kinds of evolutionary changes that could occur vary across lineages and through time. New mechanisms of regulatory control create new opportunities for evolutionary novelty, while new ecological structures generate new opportunities for evolutionary innovation, just as technological inventions have expanded cultural and economic opportunities. The origin of sophisticated developmental programs in groups with complex cellular differentiation (plants, animals, and fungi) has greatly expanded the complexity of the regulatory genome relative to the possibilities in microbes. In contrast, microbes share huge libraries of genes in ways that are less common among groups with complex development. Thus, the evolutionary process has itself evolved over time, and this is just as true in the cultural domain as it is in the biological domain. I will develop the argument that the opportunity space for both novelty *and* innovation has expanded over time, a view that contrasts sharply with the idea that search through combinatoric possibilities is a sufficient explanation for innovation.

The final chapter also addresses the relationship between innovation and biodiversity. Is novelty primarily responsible for increases in biodiversity? Ecologists frequently invoke the concept of "carrying capacity," an idea derived from population biology, that available resources are limited and constrain total diversity. While there is little doubt that the total resources available to the bacterium *E. coli* in a test tube are limited, the extension of this concept to controls on the number of species is problematic. I am interested in how novelty and innovation influence total biodiversity and changes in the structure and complexity of ecosystems. One possibility is that novelties arise continuously but succeed only if there is sufficient ecological opportunity for successful innovation. Such a situation is analogous to what economists describe as a "demand-pull" innovation, where the need for an innovation drives the process of invention. Alternatively, biological novelty could drive changes in ecological networks, constructing new niches, and thus directly influence innovation through biotically driven positive feedback. The analogue of this in economics is the creation of new markets because of technological innovation. The revolution in personal computers and related personal electronics over the past decades is an example of niche-constructing innovation.

Finally, evolution is fundamentally a deeply historical process, which raises an important philosophical issue. Is the outcome of evolution contingent upon historical events, or would similar outcomes have arisen independently of whatever historical accidents transpired? One view of evolution acknowledges the primacy of natural selection and views long-lived lineages as successful because they were better adapted than competing lineages. An argument for the importance of contingency has been advanced by Jacques Monod in his book *Chance and Necessity*, and by Stephen Jay Gould in *Wonderful Life*, among other evolutionary biologists. A different perspective has been championed by paleontologists Simon Conway Morris, George McGhee, and others. In *Wonderful Life* Gould asked: If one replayed the tape of life again would the result be similar? Is life constrained to be carbon based? To be based on DNA, perhaps even with a similar genetic code? To form animals, even arthropods or highly encephalized primates? Advocates of this alternative describe it as deterministic, because they view the evolutionary options as limited and thus likely to recur, as discussed further in chapter 4. The answers to such questions can reveal much about the opportunity space in which novelty and innovation, and evolution more broadly, operate.[14]

If evolutionary outcomes are limited rather than contingent, the details of novelty may play little role in the outcome of innovation. I will argue that although some aspects of evolution are deterministic, the historical nature of evolution generates a highly contingent dynamic for novelty and innovation. This is particularly relevant for the nature of evolutionary spaces and the distinction between novelty as search versus construction. The topology of an evolutionary space with deterministic outcomes may be very different from one with contingent outcomes. The contingent nature of novelty and innovation is further support for the argument that construction rather than search is the dominant mode of generating novelty. If this argument is correct, it has substantive implications for how we think about evolutionary change through the history of life.

2

Origins

THE VIEW of history accepted by most Westerners from pre-Socratic philosophers (about 500 BCE) until the Enlightenment assumed either endless natural cycles of renewal (with nothing truly novel) or recombination of existing entities. Progressive, cumulative change was rarely obvious on human time scales. For Plato our world was a poor reflection of a more perfect world of forms and ideas. This idea of decline from a more perfect world meshed well with Christianity's idea of a fall from grace. Novelties challenged social stability (parents of teenagers may sympathize), and some societies sought to eliminate or repress anything new. The onset of the Industrial Revolution was retarded for decades by resistance from workers who were afraid of being replaced by technology, and by rulers fearful of insurrection.

The discovery of the New World introduced concepts of change and progress to western Europe. The idea of discovery itself had to be invented, and through the sixteenth to eighteenth centuries scholars wrestled with evolving concepts about facts, experiments, laws, and theories, and even the basis upon which to judge empirical evidence. Realizing change and progress *could* occur, and that new facts and new knowledge awaited discovery, was a necessary precondition to generating theories of change. As David Wooton documented in his incisive history of the Scientific Revolution, from these new views came the recognition that novelty and innovation might be themselves worthy of investigation. For example, artists in the fifteenth century working out perspective contributed the idea of an infinite, abstract space. Only much later were these ideas mathematized. New technology drove advances in the Scientific Revolution, but Wooton makes a convincing case that we may overestimate the contributions of technology. Less obvious are contributions from the new intellectual tools and conceptual scaffolds upon which the Scientific Revolution was built.[1]

This overview of the early history of approaches to novelty and innovation begins with ideas that preceded the publication of Charles Darwin's *The Origin of Species* in 1859, before tracing the work of Darwin's followers through the rediscovery of Gregor Mendel's work on particulate inheritance in peas (the foundation of genetics) in 1900 and up to the onset of what came to be known as the Modern Synthesis around 1930. From the 1860s into the 1920s very different approaches to novelty were championed by three groups: Lamarckians, who argued that organisms could respond to environmental cues in such a way as to cause heritable changes; orthogenesists, who invoked an internal driver of evolutionary change to explain apparent trends found in the fossil record; and a variety of mutationists, including saltationists who argued that evolutionary change could occur via sudden evolutionary transformations. These broad categories mask considerable divergence of perspectives, but all of them stand in stark contrast to those of the transformationists.[2]

The Modern Synthesis marks a transition in understanding novelty and innovation, and thus serves as the fulcrum in this chapter's narrative. The Synthesis arose, in part, through opposition to alternative evolutionary views such as Lamarckianism, mutationism, and the various flavors of orthogenesism. In reconciling modern genetics with Darwin's views of natural selection, the Modern Synthesis effectively killed off the views of Lamarckians and orthogenesists. Key ideas developed during this period include adaptive radiation and key innovation. Despite considerable growth since the 1950s, ideas of the Modern Synthesis remain the guiding conceptual framework for many evolutionary biologists today. Chapter 3 continues the story after 1990. These two chapters demonstrate that current arguments over the nature of novelty can be understood only in their historical context.[3]

Ideas of Change, Novelty, and Innovation to the Nineteenth Century

Many early naturalists identified general architectures that could be found across a range of organisms, often called body plans, or types. Arguments over the nature of types continue to underlie many disagreements over the nature of novelty. When he was not writing *Faust*, penning poetry, or advising the grand duke of Saxony, Johann Wolfgang von Goethe (1749–1832) was a natural philosopher, and for him a type captured the central morphological tendencies of plants, vertebrates, or other animals. Goethe's view of the type has been frequently caricatured by later writers as a rigid, Platonic essence, but recent scholarship has shown that Goethe had a more dynamic view. For some these architectures were tied to empirical examples (in modern systematics the "type specimen" is the exemplar or reference for a described species). Others had a more transcendental view of the concept.[4]

Today the French natural philosopher Jean Baptiste Pierre Antoine de Monet, Chevalier de Lamarck (1744–1829), is remembered for advocating the view that evolutionary change was driven by responses to the environment, but this idea was widespread at the time. Like Erasmus Darwin and others, Lamarck viewed evolution within the framework of the *scala naturae*. The scala naturae was the idea that all life was linked in a Great Chain of Being from minerals through living organisms, from simple to complex, capped with humans, angels, and God. The scala naturae originated with Aristotle in ancient Greece but persisted as a central organizing structure for understanding the diversity of life into the nineteenth century, replacing the Greek gods with the Christian God. The original idea of the scala naturae was that it was fixed and reflected presumed patterns of creation, but by the early nineteenth century Lamarck and others realized that organisms had changed over time, and thus the scala naturae must be dynamic, not static.[5]

But having recognized the transmutation of species, Lamarck faced two problems: First, naturalists scouring the world had found too many species to accommodate a single Chain of Being. But more importantly, if the scala naturae was dynamic, with species moving upward, what force propelled species along their trajectory? In his *Philosophie zoologique* of 1809, Lamarck addressed these problems by employing the idea of the inheritance of acquired characters in response to environmental conditions. The power of the environment was such that organisms would be almost perfectly adapted to their environment, and thus the environment could play an important role in generating evolutionary novelties:

> In the frequent fits of anger to which the males especially are subject, the efforts of their inner feelings cause the fluids to flow more strongly towards that part of their head; in some there is hence deposited a secretion of horny matter, and in others of bony matter mixed with horny matter, which gives rise to solid protuberances: thus, we have the origin of horns and antlers.[6]

But as historian Richard Burkhardt recognized, Lamarck's theory of evolution required both a natural process of organic change and constraining processes to limit the path that change might otherwise take if natural process were allowed to operate unimpeded. The power of this natural process is evident in Lamarck's suggestion that animals would diversify in organization even in a constant environment. His theory is thus one in which form follows function.[7]

The Theory of Form

What is the problem for which an evolutionary novelty is the solution? This question motivated work by generations of evolutionary biologists, from Charles Darwin in *The Origin of Species* to Ernst Mayr and G. G. Simpson. Simpson and Mayr assumed that evolutionary novelties arise in response to an opportunity or need, whether it is ecological, physiological, or cultural. In the *Origin* Darwin argued that natural selection gradually crafted responses to such opportunities, with a new morphology tied to a novel function. Darwin's views were influenced by an 1830 debate between two of France's leading biologists, Georges Cuvier and Étienne Geoffroy Saint-Hilaire, over the primacy of form versus function, and effectively over the nature of evolutionary novelty.

Cuvier and his followers believed that gradual transformations in response to environmental change crafted a close fit between how parts of an organism functioned and their shape, whether the keen eyes of a predator, the tearing teeth of a carnivore, or the shape of a bird's wing. In 1830 Geoffroy proposed an alternative model in which changes in form could be abrupt and less directly influenced by environmental requirements. For Geoffroy, novelties did not necessarily arise in response to an immediate problem. Such novelties might or might not eventually find a problem for which they were a solution. The resulting dispute consumed the attention of many European scientists during the 1830s.

Georges Cuvier (1769–1832) was a French naturalist whose research focused on fossil mammals and fish, through which he was among the founders of both comparative anatomy and vertebrate paleontology. Cuvier must have had a remarkable intellect, for he also helped found the science of stratigraphy, recognized that many of the fossils he studied represented extinct species, and proposed that their disappearances reflected catastrophic events. He emphasized the need for animals to operate as an integrated whole, with function taking priority over form. Cuvier rejected Lamarck's linear classification of animals, informed by the Great Chain of Being. For Cuvier there was not a single scala naturae, a common architectural plan for all life, or at least for all animals (as German transcendentalists argued at about the same time), but through his extensive comparative studies he recognized four major *em-branchements*, or functional groupings, of animals: vertebrates, articulates (comprising arthropods and segmented worms), mollusks, and Radiata (for cnidarians—jellyfish and sea anemones—and echinoderms). To Cuvier, unbridgeable gaps separated these embranchements. He strongly opposed ideas of species transformation and believed that a limited number of functional designs were possible within each embranchement, bounding the range of possible forms.

Cuvier's opponent in the 1830 debate was Étienne Geoffroy Saint-Hilaire (1772–1844), a colleague of both Lamarck and himself. By 1798 Geoffroy was chair of zoology at the Muséum national d'Histoire naturelle in Paris and was invited to join Napoleon's great scientific expedition to Egypt. Geoffroy pursued comparative anatomical studies of invertebrates, of their embryos and development, and studies of various fossil groups. He developed theories about the relationships between different parts of an organism and between different organisms, including his theory of the unity of plan of organic composition, which prioritized form over function. In presentations to the Academy of Sciences during the 1820s, and in subsequent publications, Geoffroy examined similarities in design across different animal groups and emphasized that the same structure could take many different forms, and thus functions, in different animal groups.

In Geoffroy's 1822 monograph *Mémoires sur l'organisation des insectes* (*Memoirs on the Organization of the Insects*) he identified a suite of structural similarities between crustaceans and vertebrates, including the vertebral column and abdominal organs. He provocatively concluded that that arthropods have been inverted dorsoventrally (top to bottom) relative to vertebrates. Plate VII of the monograph uses an inverted view of a lobster from the side to highlight similarities with vertebrates (figure 2.1). In a direct challenge to Cuvier's embranchments, the caption calls attention to similarities in the spinal cord, muscles, intestines, heart, and arteries.[8]

In February 1830 Geoffroy extended his idea of the unity of composition to encompass mollusks, bringing together three of Cuvier's four embranchements. Moreover, Geoffroy did so for a group with which Cuvier was particularly identified (and protective of). The battle was now joined and would play out through meetings of the Academy of Sciences in Paris over the next several

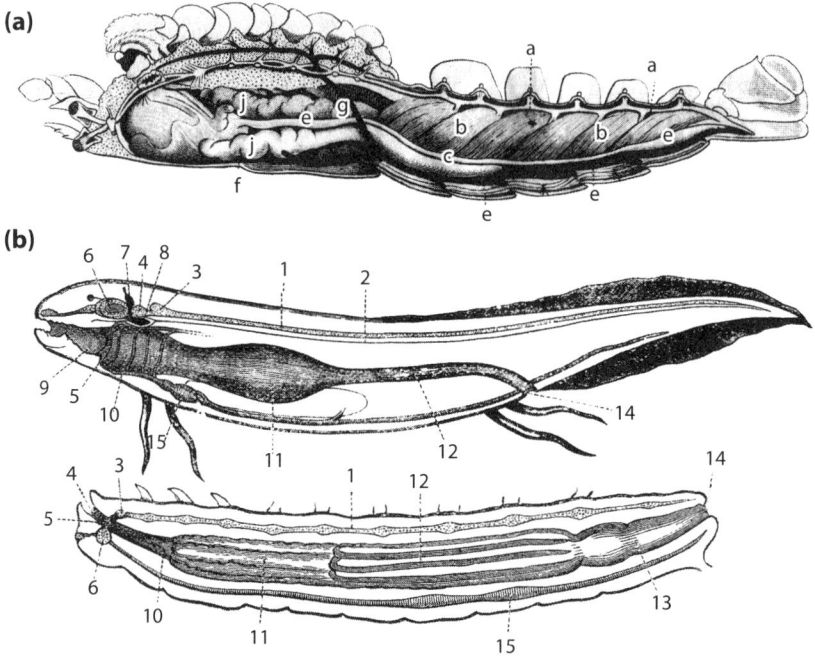

FIGURE 2.1. Geoffroy's depiction of the relationship between the anatomy of crustaceans and that of vertebrates, from plate VII, figure 2, of his 1822 *Mémoires sur l'organisation des insectes*.

months, competing with the events of the Revolution of 1830 in France (in which Charles X was deposed) for the attention of intellectuals across Europe.

Despite being close friends and colleagues, Cuvier and Geoffroy had distinct approaches to comparative anatomy. Cuvier's emphasis on function over form was the opposite to Geoffroy's approach. For Geoffroy the morphology of organisms preserved important information about similarities between species. But because he seems to have largely intuited these similarities, his approach has been described as transcendental. During the public exchange of papers in 1830 Cuvier was fundamentally correct that the evidence Geoffroy advanced was paltry at best, and many historians have viewed Cuvier as the "winner" of the debate at the time.

Yet in the late twentieth century Geoffroy's intuition would be proven prophetic. The discovery of homeobox genes in the late 1980s, followed by comparative studies of the molecular basis of developmental processes, vindicated his 1822 monograph: the expressions of developmental patterning genes in embryos are inverted in vertebrates relative to arthropods. Whereas Cuvier claimed that the only similarity between vertebrates and crustaceans was that both clades were animals, the discovery of highly conserved genes responsible for patterning of eyes, the gut, and limbs exposed deep similarities between vertebrates and arthropods. Geoffroy could not have anticipated the existence of DNA or genes, much less the similarities in regulatory control features, but his deep understanding of animal morphology allowed him to intuit similarities, the basis of his ideas of the unity of composition. The impact of such deep homology is continuing to be absorbed, but helped trigger renewed interest in evolutionary novelty. From our viewpoint, there are substantial errors in how each of these men thought about organismic change: Cuvier rejected any consideration of species transformation, while Geoffroy's ideas came within the framework of the scala naturae. But biologists still recognize the Cuvier-Geoffroy debate as a turning point because of the central role for homology—whether structural resemblance simply reflects similarity of function or whether these resemblances reflect an underlying unity of plan.[9]

For the remainder of this chapter, I will distinguish between transformationist and saltationist traditions. Neither Cuvier nor Geoffroy fits easily into either of these camps, but their debate crystalized differences between form and function as explanations for animal architecture. The ongoing debate over whether novelty and innovation are distinct from adaptive evolutionary change contains echoes of the Geoffroy-Cuvier dispute.[10]

Insights from Embryos

Comparison of embryos played little role in the differing views of Cuvier and Geoffroy. To Geoffroy the embryos from different embranchements were too distinct to provide support for his theory of the unity of plan. But Karl Ernst von Baer's (1792–1876) studies of animal development were leading him to a series of "laws" that provided just the framework needed to incorporate development into studies of broader evolutionary patterns. Baer was an Estonian naturalist who made contributions to many fields of science but is best known today as a founder of embryology. He described the three primary embryonic tissue layers (ectoderm, mesoderm, and endoderm) as part of a program critical of the theory of developmental recapitulation introduced by Johann Friedrich Meckel, a German anatomist. Meckel's theory of recapitulation was an outgrowth of the scala naturae, with successive stages of a developing embryo passing through the adult stages of less complex organisms. To Meckel, development from the egg to adult was a simplified version of the Great Chain of Being.

Meckel belonged to a larger intellectual movement of the late eighteenth and early nineteenth centuries known as *Naturphilosophie*. This thread of German idealism emerged from the ideas of Immanuel Kant, Georg Friedrich Hegel, and Goethe. Naturphilosophie infused the ideas of German Romanticism into a program of scientific research organized around a belief in a unified view of nature reflecting universal laws. One of these universals was a directional, progressive movement from chaos to order, lower to higher, effectively embedding the Great Chain of Being within a metaphysical framework.[11]

Baer criticized Meckel's invocation of the Great Chain of Being and his theory of recapitulation. Baer showed that the diversity of animal embryos could be encompassed by four fundamental animal types corresponding to the four embranchments of Cuvier (although Baer claimed to have independently recognized his types). Moreover, Baer proposed that these types followed branching (rather than linear) programs as they developed into their adult form. From this research he identified a set of general patterns, often known as (von) Baer's laws (although they are really principles rather than laws): the general features, or characters, of a large group appear earlier in development than the more specialized features; within a given form, each embryo does not recapitulate earlier forms but rather becomes increasingly distinct; and the embryos of more complex animals never resemble other adult forms, but rather their embryos. There is no Great Chain of Being, no single lineage of animals.[12]

For Geoffroy, Baer, and many nineteenth-century German naturalists, Naturphilosophie encompassed a transcendental or ideal morphology, with organisms as imperfect realizations of some ideal, transcendental form. The core of Naturphilosophie is environmental modification of an internal drive of the developing embryo to form an individual organism. Goethe believed that underlying the diversity of all vertebrates was a single plan, or archetype, which could be revealed by the comparative study of embryonic and adult anatomy. Transcendental morphology was an expansion of this idealized form of morphology in which just a few ideal forms underlay the great diversity of animal life. Transformative, evolutionary change was accepted by transcendental morphologists not in the Darwinian sense, but rather as an expression of an idealized, internal drive toward perfection. These ideas spread beyond Germany to French scholars, as well as to Richard Owen, Cuvier's equal as a comparative anatomist and one of the most influential nineteenth-century English biologists.[13]

As Darwin drafted the *Origin*, he faced the challenge of confronting these engrained views of idealized archetypes with linear and progressive schemes of development. Darwin cited Baer's "laws" as a useful counter to scenarios of linear transformation in the scala naturae, but Baer could not reconcile natural selection with his views of a creative force driving evolutionary changes and became a strong critic of natural selection. Darwin's genius lay in perceiving a pattern of descent with modification, and in articulating natural selection as the driving force.

Novelty in the *Origin* and the Critical Response

Publication of the *Origin of Species* in 1859 recast the argument over form and function as one over whether similarities represented evolutionary connections or purely functional requirements. Darwin advocated *form follows function*, a mantra adopted by many late nineteenth-century biologists. By the 1930s form follows function became a central tenet of the Modern Synthesis of evolution. From this perspective major novelties solve big problems; small changes solve small problems. There is a close, if probably rather unwitting, analogue among economists and historians of technology, who suggest that a new technology is driven by opportunities generated by consumer demand, often described as a "demand-pull" model. They too are asking how an invention solves a particular problem.

In the *Origin of Species* (1859) Darwin established the reality of descent with modification as the explanation for the patterns of similarity among organisms and the importance of natural selection as a primary mechanism for evolution. Indeed, in the first edition of the *Origin* he was arguably more concerned with challenges from creationists rather than other scientists who accepted transmutation of species. Darwin was cognizant of evolutionary novelties and the challenges they created for his theory. In chapter 6 of the *Origin*, he enumerated "Difficulties on Theory," including the general absence of transitional forms and the problem of explaining the origins of unusual morphologies such as flying squirrels, the complex, image-forming eye, and the transition from the swim bladder of fish into a true lung. In each case he provided a scenario through which small, "insensible" (one of Darwin's favorite words) changes generate new forms over time. For him, gradual changes driven by selective advantage revealed adaptive pathways for the generation of complex morphologies. In the *Origin* Darwin wrote:

> As natural selection acts solely by accumulating slight, successive, favourable variations, it can produce no great or sudden modifications; it can act only by short and slow steps. Hence, the canon of "*Natura non facit saltus*,"

which every fresh addition to our knowledge tends to confirm, is on this theory intelligible. We can see why throughout nature the same general end is gained by an almost infinite diversity of means; for every peculiarity when once acquired is long inherited, and structures already diversified in many ways have to be adapted for the same general purpose. We can, in short, see why nature is prodigal in variety, though niggard in innovation. But why this should be a law of nature if each species has been independently created, no man can explain.[14]

The phrase *natura non facit saltus* (nature does not make leaps) reflects a core principle of Darwin's, and this single paragraph captures his gradualist approach to evolutionary change. He was convinced that novelty could arise only through many small evolutionary changes, guided by natural selection. Darwin argued that if we had a complete record of the history of life, few distinctions between species or genera would be evident, as forms would grade from one into the other. This gradualist approach to evolutionary patterns lies firmly within the tradition of Gottfried Leibniz, describing calculus, and Carl Linnaeus, the father of biological systematics, in his *Philosophia botanica*, from which Darwin probably adopted the adage. What is particularly striking is Darwin's emphasis on the importance of variation, which provides the raw material upon which natural selection operates, coupled with his denial of novelty in nature.[15]

The recognition that species had changed over time predated Darwin and the *Origin*, but with Darwin's work three issues became more pressing: the assumption that the Earth had a relatively brief history, whether species had ever become extinct, and whether biological complexity had increased through time, each of which influenced views of novelty and innovation.

A literal reading of the Bible allowed too little time for evolution to account for the diversity of life. But by the late eighteenth century evidence for a far greater age of the Earth was growing. In *Les époques de la nature* (1778) Georges-Louis Leclerc, Comte de Buffon (1707–1788), recognized that the Earth was once molten and cleverly extrapolated from the cooling of iron balls to estimate the time required for the Earth to reach its current temperature. His estimate of 74,832 years for the age of the Earth provided far greater latitude for biological change than did biblical chronology. Ten years later Scottish geologist James Hutton argued that the Earth was very old, ending his paper with the memorable phrase: "We find no vestige of a beginning,—and no prospect of an end," but Hutton's cyclical view of Earth's history required no estimate for its age. Darwin remained troubled by what he viewed as an uncomfortably short span to evolve the diversity of life. Assumptions about the age of the Earth continued to plague geologists and biologists through the late nineteenth century, with noted physicists such as Lord Kelvin dismissing any suggestion of an Earth older than tens of millions of years. The discovery of radioactivity in 1896 provided an internal source of heat to the Earth as well as a chronometer for determining the age of various rocks. By 1913 Arthur Holmes had published the first geological time scale based on radiometric dating, greatly extending the time available for biological evolution.[16]

The discovery of fossil species with no modern representatives further challenged biblical literalism. Savants explained away such remains by arguing that the species were alive elsewhere on the Earth and had simply shifted their range with changing climate. Thomas Jefferson included searching for living mammoths among the tasks for the Lewis and Clark expedition to western North America (1804–1806). But as explorations of the globe continued such views became untenable, particularly for something as large as a mammoth. In time scientists came to accept the accuracy of Cuvier's claim that most fossils represented species that had become extinct.

Did species change also imply that the complexity of plants and animals had grown over time? Had species originated in a much simpler form and gradually acquired their current complexity? Or were the earliest species of equal complexity to those we find today? Cuvier's periodic revolutions in the history of life could imply either that life had recovered from these extinction episodes or had been episodically thinned from a more complex early history.

Darwin's critics rejected his reliance on the power of small, gradual changes to generate novelties, instead favoring rapid modifications potentially independent of natural selection. In response to these criticisms Darwin adjusted subsequent editions, particularly the sixth edition of 1872. Among the many critiques to the *Origin*, here I want to describe just four: from Darwin's close friend Thomas Huxley; from the Swiss zoologist and paleontologist François Jules Pictet de La Rive, who penned what Darwin himself considered the most perceptive negative review of the book; from St. George Mivart, an English biologist and sometime friend of Huxley; and from Oxford paleontologist John Phillips. The challenges from Huxley, Pictet, Mivart, and Phillips reflected their own studies, and in the cases of Pictet, Mivart, and Phillips, perhaps their religious convictions as well (Huxley was a self-described agnostic—he invented the term). While Phillips remained a creationist of a sort until his death in 1874, Pictet seems to have acknowledged transformation of species even before the publication of the *Origin*. But it is clear from the writings of each man that their primary concern lay with Darwin's failure to acknowledge that the apparent sudden changes in morphology, as documented in the fossil record, might be a critical signal about the tempo of evolutionary change, and therefore about the underlying mechanisms. Thus, the reception of the *Origin* was influenced by the earlier debate between Cuvier and Geoffroy.

Huxley was one of Darwin's most stalwart defenders and immediately accepted that evolution occurred. But he differed from Darwin on the tempo of evolutionary change. In a famous letter to Darwin penned the day before publication of the *Origin*, Huxley wrote: "The only objections that have occurred to me are—1st, That you have loaded yourself with an unnecessary difficulty in adopting *Natura non facit saltum* so unreservedly." In his review of the *Origin* in the London *Times*, Huxley expanded on this: "Nature does make jumps now and then and a recognition of that fact is of no small import in disposing of many minor objections to the doctrine of transmutation."[17]

Huxley's saltationist views remained a source of friction between him and Darwin for the remainder of their lives, and Huxley's belief in discrete types may have been the source of his saltationism. Huxley suggested that spontaneous mutations similar to developmental abnormalities, such as more than five fingers or toes, were the primary driver of significant evolutionary changes. Thus, his position may be due to his research program as a developmental morphologist, uninterested in "the species question" that so consumed Darwin.[18]

Pictet's review of the *Origin* found Darwin's arguments for gradual evolutionary change wanting:

> Mr. Darwin accepts on the one hand the possibility of slight variation and on the other an immense series of centuries. He then multiplies these two factors by each other and concludes that there are powerful and profound variations. . . . Thus he accepts the successive modification of specific characters, next generic characters, then the limits of families, of orders and of classes. . . . Because these bold deductions do not appear to be justified by the facts, it is necessary to have a more powerful argument before accepting them. . . . Nothing which he has said proves to me that slight and superficial variation can in the long run change nature and degenerate in such grave modifications.[19]

While Pictet acknowledged that natural selection may have been responsible for slight changes, he dismissed the possibility that it could explain morphological novelties. He then distinguished between two different forces—natural generation, responsible for changes between species, and a creative force responsible for "giving birth successively to distinct types whose existence is attested to by palaeontology." Although Pictet recognized that the nature and mechanisms of this creative force were unknown, he argued that the history of life required both forces. He dismissed two possible contenders for this "special force," spontaneous generation and creation, and looked instead to a "general law" responsible for the origin of variations (in the best tradition of German ideal morphologists).[20] He closes with a cautious recommendation of the *Origin*:

> Let me say that reading Darwin's book would be truly useful to all zoologists, especially if they exercise constraint in following the author in his seducing argument and stop before they are led from the certainty of slight variations to the dangerous doctrine that over long periods of time these modifications have no limit.[21]

Darwin was particularly concerned by Pictet's review, and in a letter to Harvard botanist Asa Gray, he wrote:

> There has been one prodigy of a review, namely an opposed one . . . which is *perfectly* fair and just, and I agree to every word he says; our only difference being that he attaches less weight to arguments in my favour, and more to arguments opposed, than I do.[22]

The origin of complex structures, including evolutionary novelties, concerned Darwin in the first edition of the *Origin*, and it was this thread that St. George Mivart (1827–1900) exploited so successfully in his book *On the Genesis of Species* (1871). In response to Mivart's critique, Darwin inserted into the sixth edition of the *Origin* a new chapter 7, entitled "Miscellaneous Objections to the Theory of Natural Selection." Mivart initially favored natural selection when he met Huxley in 1859. However, they eventually fell out as Mivart had converted to Catholicism as a teenager, and Huxley was very anti-Catholic. Mivart turned against natural selection, instead advocating a teleological explanation for evolutionary change. I mention Mivart's religious position not to dismiss his critique of Darwin, but because his religious convictions alone are a poor basis to ignore his criticisms. Like Pictet, Mivart identified difficulties in Darwin's explanation of the origins of complex characters, and chapter 2 of his book is entitled "The incompetency of 'natural selection' to account for the incipient stages of useful structures." No subtlety there, as the first paragraph of the chapter makes clear:[23]

> "Natural Selection" simply and by itself, is potent to explain the mainte-nance or further extension and development of favorable variations, which are at once sufficiently considerable to be useful from the first to the indi-vidual possessing them. But Natural Selection utterly fails to account for the conservation and development of the minute and rudimentary begin-nings, the slight and infinitesimal commencements of structures, however useful those structures may afterward become.[24]

For Mivart new structures must be immediately useful. He discusses giraffes, mimicry, the asymmetric eyes of adult flatfish, vertebrate limbs, eyes, and the baleen of whales, among other structures. Several of these examples, such as giraffes, had featured in the arguments of Lamarck and Darwin. The crux of Mivart's argument is that intermediate forms would not have been advanta-geous, and thus they could not have been subject to natural selection. Rather, these structures must have evolved suddenly. On giraffes, he asks, if this was such a wonderful adaptive solution to browsing, why did no other ungulate adopt it? In fairness to Mivart, the power of natural selection did not become fully evident until the rise of quantitative population genetics, beginning in the 1930s.

Darwin's response to Mivart in the final edition of the *Origin* emphasized the power of selection on relatively small differences. Over the past two decades or so, many of Mivart's examples have been restudied. For giraffes, flatfish, and whale baleen, fossil intermediates have been found, and in all cases, evidence suggests the complex structures evolved gradually. Yet not all the questions he raised have been satisfactorily resolved.[25]

Darwin included the chapter "On the Imperfections of the Fossil Record" in the first edition of the *Origin* to address the lack of empirical support in the fossil record for gradual evolutionary change. While he deserves credit for addressing the challenges to his hypothesis (how many speakers at TED con-ferences would do the same?) the chapter did not forestall critiques. By the middle of the nineteenth century geologists had begun to correlate sedimen-tary rocks with similar fossils (and thus presumably of similar ages) across Europe. In drawing such correlations, geologists noticed that new types of fossils often appeared suddenly and were quite distinctive from related fossils. John Phillips (1800–1874) invoked separate episodes of creation for Paleozoic, Mesozoic, and Cenozoic life, but his criticism of the *Origin* was based on his knowledge of the empirical fossil record. From his extensive fieldwork on the fossils and rocks of Yorkshire and Oxfordshire, Phillips knew that most fossil species appeared at the base of a formation (a distinctive unit of rocks) and persisted through the formation and perhaps further, before being replaced by a new species. This pattern was inconsistent with Darwin's theory of descent with modification, for if these fossils were related by evolution, then the rock record, read literally, seemed to suggest pulses of rapid evolution. Darwin suggested global changes in sea level as an alternative explanation that preserved gradual evolution. Intervals when the seas drained from the continents and no fossils were preserved would account for the failure to preserve intermediate forms. Although widely discussed by nineteenth-century geologists, this idea was not fully articulated as the theory of diastrophism until 1909.[26]

Contested Views: Evolutionary Change 1859–1930

Publication of the *Origin* rapidly led to acceptance of the concept of evolution and descent with modification by scientists and indeed by most educated people. But the efficacy of natural selection as an explanation for the diversity of life remained contentious until the development of quantitative methods of population genetics in the 1930s. Between the 1860s and 1930s biologists differed greatly over the "force" of evolution. Did it come from selection, as Darwin had argued, from environmental change, as suggested by neo-Lamarckians, or from something internal to organisms themselves, following the arguments of the German idealistic morphologists? Mivart's criticisms about the origin of complex adaptations focused interest on novelty and innovation.[27]

Complex adaptations such as the transition from the swim bladder of a fish to the air-breathing lung of a land vertebrate were a major focus of George Romanes's (1848–1894) three-volume discourse *Darwin, and after Darwin* (1892–1897). His work is best seen as an expansion of arguments for the importance of natural selection. Examining the origin of many of the same novel structures as Darwin, Romanes emphasized that some structures were well positioned to adapt to new and different circumstances. As fish became amphibious the swim bladder, used by fish to regulate their buoyancy, was already associated with gas exchange, and thus suited for co-option into the lungs of amphibians. Such co-option of structures for a new function is now described as *exaptation*.[28]

Romanes met Darwin while a student at Cambridge and eventually became an important advocate for Darwin's views on evolution. His 1897 book (the second volume of his three-volume treatise) also favorably describes Darwin's observation of the tight integration of different parts of anatomy, and the implausibility that one structure could change significantly without adverse effects upon correlated structures. Indeed, the origin of incipient structures has remained a challenge to transformationist models of evolutionary novelty, with morphological correlations commonly deployed in favor of stepwise adaptive changes.[29]

Evolutionary theories that downplayed natural selection flourished after 1859, invoking a variety of evolutionary mechanisms. Although saltational explanations have generally been derided, lampooned, or ignored for invoking implausible or unscientific mechanisms, these views influenced later views of novelty by emphasizing apparent abrupt discontinuities, by suggesting a decoupling between small-scale adaptive changes and the origin of novel features or new clades, and for proposing that the range of developmentally significant mutations may be greater than is immediately apparent from experimental work, a view that has been reinvigorated since the 1990s by comparative studies of developmental mechanisms. Finally, German Romanticism contributed a school of saltational thought that identified architectural body plans—or Baupläne—as the unfolding of an internally driven process uncoupled from the influence of the environment and thus from selection. Such unscientific mechanisms have long been rejected by evolutionary biologists, but many of those interested in novelty and innovation found utility in the architectural arguments for Baupläne, while rejecting their mystical trappings.

Once Mendel's work was rediscovered in 1900, the significance of particulate inheritance was rapidly appreciated by geneticists such as Hugo de Vries (1848–1935), a Dutch botanist who was one those who rediscovered Mendel's work. De Vries challenged the creative capacity of natural selection and argued that evolutionary innovations were the result of large-scale, nonselective genetic mutations. De Vries's mutational theory of evolution was short-lived, however, as genetic evidence from studies of the fruit fly Drosophila, led by geneticist Thomas Hunt Morgan, revealed that most genetic mutations were of relatively small effect, as Darwin had predicted. But De Vries's work brought theories of saltational evolution into the genetics of the new century by emphasizing the importance of mutations.[30]

Mutationists often underplayed the creativity of natural selection, arguing that its primary impact was to remove less-fit variants. The same claim surfaces today and in part turns on the issue of whether selection can act only on variation initially present in a population, or whether selection impacts the nature of the genetic variation that may arise. If the genetic variation available in a population is replenished by new mutations more slowly than the action of selection, then the efficacy of selection will be limited. In his stimulating dissection of the controversy over the creativity of natural selection, philosopher John Beatty contrasts the views of mutationists and Darwinists:

> Thus, a central difference between the Darwinians and the mutationists was that according to the former, selection brings about directional change all while *shifting and preserving* a wide range of selectable variation. Whereas according to the mutationists, directional evolution takes place at the *expense* of selectable variation: natural selection *reduces* the range of variation that it can act upon. And this has consequences with regard to the creativity of mutation relative to natural selection.[31]

The views of German paleontologists and biologists that rapid evolutionary pulses evident in the fossil record were a primary signal and not, as Darwin argued, a reflection of the inadequacy of the fossil record persisted after publication of the *Origin*, with new explanations for the apparent rapid evolutionary jumps, or saltations. Ernst Haeckel described episodes of rapid increase in morphological diversity (which paleontologists term *disparity*) as "flowering periods," and like many German evolutionary biologists invoked internal driving forces.[32]

Orthogenesists, for example, held that morphological variation was limited, with change driven by a force that predisposed evolution in particular directions. One of the most influential German orthogenesists was Otto Jaekel (1863–1929), a neo-Lamarckian paleontologist who held that an internal force (not simply response to the external environment) characterized each major lineage. Such intrinsic trajectories carried lineages from a youthful stage, characterized by rapid evolutionary diversifications, through a less effervescent adult stage to eventual decline and extinction. As was true of earlier saltationists, Jaekel saw evolution operating on two levels, natural selection for changes within species, and a second level for the origin of distinctive architectures of different animal groups by "jolting" the early and more evolutionarily plastic embryonic stages. Jaekel's conclusions were based, at least in part, on his empirical studies of fossil echinoderms:[33]

> The greatest divergences of types are to be found during the phyletic "youth" of a branch and may well be considered as typical for this phyletic stage. . . . On a given basis, however, only a few structural plans can be realized, and therefore the final stages usually are much fewer in number than the initial tentative types.[34]

Others, such as Othenio Abel (1875–1946), endeavored to purge orthogenesis of its aura of mysticism and ideas of an internal "vital force." For Abel, the directed nature of orthogenesis reflected biotic inertia: the environment influences lineage evolution, and if the influence persists long enough, then the novel (to Abel) features of the organisms will become fixed. This view shared Jaekel's goal directedness in the sense that major lineages each possessed an internal force driving them along a unique evolutionary trajectory.[35]

Orthogenesis found converts in the United States and Great Britain, particularly Henry Fairfield Osborn (1857–1935), a vertebrate paleontologist who dominated the American Museum of Natural History in New York in the early twentieth century. His work on dinosaurs gave us *Tyrannosaurus rex* and *Velociraptor*. Osborn rejected both Lamarckian approaches and natural selection for a flavor of orthogenesis he called *aristogenesis*. To Osborn, evolutionary novelty was generated by "aristogenes," new "biomechanisms" responsible for the gradual, directed generation of new organismal features, independent of either environmental influence or natural selection. Such aristogenes helped establish new classes of animals, which would diversify into a range of different ecological roles through an adaptive radiation (a term Osborn had introduced in 1902).[36]

By 1900 many biologists accepted descent with modification but continued to doubt the creativity and efficacy of natural selection in generating evolutionary novelties. Jaekel's view that evolution occurred on at least two levels was widely shared among evolutionary biologists of the 1920s. But just blocks from Osborn at the American Museum, Thomas Hunt Morgan at Columbia University was in the vanguard of new developments in genetics. Quantitative studies of gene activities showed that genetics and natural selection were entirely compatible. Yet the work of Jaekel and other German paleontologists did not disappear with the Modern Synthesis, but particularly influenced two later evolutionary biologists in the German tradition of saltational evolution: paleontologist Otto Schindewolf and geneticist Richard Goldschmidt. Although each was influenced by Jaekel, Schindewolf and Goldschmidt shared saltational views of evolution and of novelty while rejecting any mystical, internal drive. We will return to Schindewolf and Goldschmidt after examining how the Modern Synthesis explained novelty and innovation.

The Modern Synthesis: 1930–1970

Evolutionary novelty was a primary concern of the Modern Synthesis, particularly during the 1930s, when its founders focused on the power of population genetics, on the primacy of adaptation of organisms to their environment, and

on disposing of "non-Darwinian" accounts of evolution such as those of mutationists and orthogenesists. I will focus here on vertebrate paleontologist
George Gaylord Simpson, ornithologist and evolutionary biologist Ernst
Mayr, botanist G. Ledyard Stebbins, and ornithologist Bernhard Rensch. The
foundational contributions to the Synthesis made by geneticists Sewall Wright
and Theodosius Dobzhansky are taken up in chapter 4.

Julian Huxley's 1942 book *Evolution: A Modern Synthesis* invoked adaptive
radiations as a source of "adaptive trends." A British evolutionary biologist and
grandson of T. H. Huxley, Julian Huxley was less an original thinker than a
skilled synthesizer, and in this book coined the term *Modern Synthesis*. Huxley
observed that any large taxonomic group, such as a Linnean class, will display
"ecological diversification in the grand manner." Although recent work has
tended to emphasize adaptive radiations within a genus or among several
closely related genera, Huxley suggested that "classes and sub-classes provide
the optimum size of group in which the phenomenon may be studied." The
power of adaptation to generate and sculpt novel responses to ecological and
environmental circumstances seems unlimited in Huxley's text, and he illustrates his case with the Cenozoic radiation of placental mammals, the radiation
of marsupial mammals in Australia, as well as smaller-scale events such as the
radiation of cichlid fish in African lakes. In each case evolutionary novelty
emerged from adaptive specialization.[37]

If Huxley did not inquire deeply into the causes underlying adaptive radiations, the paleontologist George Gaylord Simpson more thoroughly investigated them in *Tempo and Mode in Evolution* (1944) and *Major Features of
Evolution* (1953). These books develop his argument that the origin of new
higher taxa (classes, orders) occurs via macroevolutionary rather than microevolutionary modes, a view that relied heavily on his concept of "adaptive
zones" and "adaptive grids," which defined potential physical and biological
environments.[38]

Simpson identified three sources of ecological opportunity underlying
adaptive radiations: formation of a "key" innovation allowing access to a
new adaptive zone, invasion of a new habitat, and extinction of competing
species that limited diversification. These situations reduced natural selection
and increased available resources, thus permitting a greater range of form and
behavior. With time this should foster rapid speciation and increased morphological variation leading to adaptive radiation and novelty:

> On a broad scale we see, even more clearly than Darwin did, that every
> marked expansion of a group, whether it be a genus or a phylum or the whole

animal kingdom, is an adaptive radiation. Each starts with a group of a certain adaptive status in a particular range of environments. . . . Each radiates by a combination of two processes: a parceling-out of a broader ecological range among more specifically adapted separate lines of descent, and the invasion of new ecological niches by modification of the ancestral adaptation.[39]

In *Major Features*, Simpson questioned the applicability of experimental, microevolutionary studies to understanding the origin of higher taxa, an issue that resurfaced in later debates over macroevolution. Simpson's argument for the origin of higher taxa is closely tied to his views of new adaptive zones. Thus, the distinction I draw between novelty and innovation here would have been foreign to Simpson:

> The event that leads, forthwith or later, to development of a higher category is the occupation of a new adaptive zone. As a general rule, the broader the zone the higher the category when fully developed. After initial occupation of the zone, adaptive radiation into its subzones, increased specialization by adaptation to narrower and narrower subzones, weeding out of lineages and in general the whole sequence . . . usually follows. Occasionally it happens that the zone occupied is so narrow that significant radiation does not occur and yet is so distinctive in an adaptive type and correlated morphology of the organisms occupying it that systematists do give it high categorical rank.[40]

Here Simpson ties morphological novelty to the presence of an adaptive zone, with novel morphologies unfolding as the zone is occupied. In this passage Simpson acknowledges that new morphologies could arise unaccompanied by extensive diversification yet have such divergent morphologies that under the Linnean approach to systematics they would be accorded a high rank (orders or classes). This can be read as acknowledging the possibility of macroevolutionary lags with the phrase "forthwith or later," although whether Simpson would agree is unknowable. On the following page he acknowledges that expansion and diversification are not inevitable upon entrance into a new adaptive zone, although while "they may be considerably delayed when they do occur, . . . they do usually happen soon, geologically speaking." What I will describe as innovation seems to roughly correspond to rapid transitions between adaptive zones for Simpson. These transitions involve morphological and ecological discontinuities, generally during the early or "explosive" phases of adaptive radiations. Although today we think about this in different conceptual terms than those used by Simpson, he recognized that the boundaries of an adaptive zone may be established early in a diversification, but that the boundaries of the zone may evolve:

> When radiation does occur in a zone, it usually happens that this entails not merely an occupation but also a considerable expansion of the zone and changes in many of its features. Although the land carnivore zone of today has developed from that of the Paleocene, it is tremendously broader and so different that it does not even include any animals adaptively similar to the earliest carnivores.[41]

For Simpson, morphological novelty and evolutionary innovation were intimately associated with the origin of higher taxa and thus with the occupation and exploitation of new adaptive zones. He emphasized the exceptionally rapid rates of evolution associated with the origin of higher taxa and argued that morphological gaps between higher taxa were not, or at least not exclusively, artifacts of poor preservation but an evolutionary pattern that required a different explanation. Simpson was careful to distinguish his ideas of quantum evolution from the saltational theories of Schindewolf, in which rapid shifts occurred via "key mutations" (Schindewolf's work is described later). Simpson noted that the sudden occurrence of new groups in the fossil record necessarily obscures a more gradual transition, and that there is no fossil evidence supporting Schindewolf's saltational evolution.[42]

Throughout his many papers and books for both scientific and public audiences, Ernst Mayr shaped views of evolutionary biology from the 1940s into the 1980s. Mayr was an ornithologist by training, specializing in the birds of New Guinea. Using his field experience with birds as a foundation, he explored the interactions between genetic and environmental changes and ecological opportunity and was principally responsible for renewed interest in evolutionary novelty through his papers of 1954 and 1960.[43]

Mayr's initial contribution to the Modern Synthesis came in a 1942 book, *Systematics and the Origin of Species*, which concludes with the chapter "The Higher Categories," a discussion of his views on macroevolution and implicitly on innovation:

> In conclusion, we may say that all the available evidence indicates that the origin of the higher categories is a process which is nothing but an extrapolation of speciation. All the processes and phenomena of macroevolution and of the origin of the higher categories can be traced back to intraspecific variation, even though the first steps of such processes are usually very minute.[44]

Like Simpson, Mayr viewed higher Linnean categories such as orders, classes, and phyla as the avenue for studying evolutionary novelty. In his definitive 1963 volume *Animal Species and Evolution*, his discussion of novelty is embedded within a section on the origin of higher categories.[45]

His 1954 paper largely articulates his arguments for the importance of small populations on the margins of the range of a species, which he described as "peripheral isolates" (which later formed the foundation of Stephen Jay Gould and Niles Eldredge's theory of punctuated equilibrium). Mayr viewed peripheral isolates as a key ingredient in the formation of evolutionary novelties, with changes in the "genetic environment" being sufficient to change selection on one gene, triggering changes among other genes, but he rejected arguments for genetic saltations:

> The genetic reorganization of peripherally isolated populations . . . does permit evolutionary changes that are many times more rapid than the changes within populations that are part of a continuous system. Here then is a mechanism which would permit the rapid emergence of macro-evolutionary novelties without any conflict with the observed facts of genetics.[46]

His conclusion is vintage Mayr: "The problem of the origin of higher categories is inseparable from the problem of the origin of new species. Those who have denied this seem to be unfamiliar with the facts."[47]

Botanist G. Ledyard Stebbins's views of evolutionary novelty largely tracked those of Mayr and Simpson, particularly on the importance of adaptive radiations. But two interesting twists to Stebbins's work illustrate the diversity of views at this time. First, he doubted that most adaptive radiations led to increases in evolutionary complexity (what was then termed *evolutionary grade*, which was as poorly defined then as *complexity* is today). Doing a thought experiment using estimates for the number of species, Stebbins argued that "an estimate of 100 major advances that have occurred throughout the evolutionary history of eukaryotes is a reasonable . . . number. If this number is accepted, then we must recognise the probability that only one out of every 32 million adaptive shifts has led to a major evolutionary advance." From this Stebbins concluded that talk of evolutionary progress or evolutionary "laws" was largely vacuous, but perhaps more importantly, that very few transitions opened new evolutionary opportunities. One can quibble with Stebbins's thought experiment, but it supported (as was probably intended) the second twist to his work, the view that evolutionary advances were driven by interactions between new mutations and the existing genome, rather than between organisms and their environment.[48]

Bernhard Rensch's *Evolution above the Species Level* directly addressed issues of macroevolutionary pattern and process. Rensch (1900–1990) was professor of zoology at the University of Münster, where he worked on the formation and distribution of species. Rensch's views mirrored those of Simpson in focusing on rapid diversifications of a variety of higher taxa, principally orders, classes, and phyla, ranging from Ordovician marine snails to Cenozoic mammals. And like Simpson, Rensch argued that increased selection due to changing environmental or ecological conditions was a more satisfactory explanation for novelty than increases in the rate of mutation, or in the size of morphological changes associated with mutations (i.e., the macromutations of Schindewolf, Goldschmidt, and others). Rensch dissected the saltational and typostrophic views of Beurlen, Schindewolf, and Goldschmidt, concluding that the apparent gaps between higher taxa generally reflect inadequacies in the fossil record, rapid evolution following origin of "a new structural type," or small population size. For Rensch, evolutionary novelties reflected change in the environment.[49]

Over the past several decades, critiquing the Synthesis has become a parlor game of evolutionary biologists and historians of biology. But it must be evaluated in the intellectual setting of the time, particularly at its outset in the 1930s. The architects of the Synthesis were trying to build a modern science by integrating insights from the quantitative rigor of population genetics at a time when typostrophism, orthogenesis, and all manner of shoddy theories retained many adherents. In the absence of a mechanistic understanding of large-scale evolutionary processes, attempting to explain as much of evolution as possible via the activity of shifting alleles within populations was sensible and indeed may have been the most plausible course toward reforming evolutionary theory.[50]

Critiques of the Modern Synthesis: 1930–1970

Schindewolf and Goldschmidt: Modern Saltationists

For Otto Schindewolf neither neo-Lamarckianism nor Darwinism could satisfactorily generate the explosive evolutionary changes during the origin of new clades. Schindewolf (1896–1971) was the most important paleontologist of his time in Germany and a significant figure internationally. Schindewolf's unified saltational theory of evolution, which he termed *typogenesis*, incorporated elements of idealistic morphology, orthogenesis, and cyclical evolution. This process was not influenced by environmental or ecological factors but reflected intrinsic dynamics. After an initial evolutionary burst, evolution transitioned to an orthogenetic, driven phase, often involving increased specialization, which ended in the decline of the clade and its extinction. Selection played no role in establishing the major types or architectural body plans within clades, only in microevolutionary changes of the finest details of species.[51]

In *Basic Questions in Paleontology* (published in German in 1950 but not translated into English until 1993), Schindewolf deployed his vast knowledge of the fossil record to illustrate typostrophism. He rejected Darwin's arguments for the insufficiency of the fossil record, instead viewing morphological discontinuities as "natural evolutionary leaps, and not the circumstantial accidents of discovery and gaps in the fossil record." Of the transitional forms between major architectures, he wrote, "In very many instances, for purely morphological and biological reasons, transitional forms between the different structural designs are not even possible or conceivable. In many ways, we have been searching for things that cannot possibly exist." But Schindewolf's saltational views were not based solely on the fossil record, for they were deeply influenced by geneticist Richard Goldschmidt, the happy heretic of the hopeful monsters. In earlier work Schindewolf had written: "The transformation and evolution occur in leaps, through sudden, more or less large mutational steps, which are caused by changes in the gene and genome."[52]

In stark contrast to the emphasis on population genetics among the founders of the Modern Synthesis, Goldschmidt (1878–1958) attempted to integrate genetics and development. Goldschmidt understood Darwin's arguments and those of the Modern Synthesis and largely accepted them. But for him the only route to understanding the origin of new higher taxa was through development. Goldschmidt rejected the transformationist view that cumulative small mutations are sufficient to explain the origin of higher taxa. Instead, he proposed that pervasive genetic rearrangements triggered the rapid appearance of new species or higher taxa.[53]

Key to Goldschmidt's views were homeotic mutations, where one body part forms where another should develop. The classic example is the *bithorax* mutation in flies, which produces a second pair of wings in place of the halteres. Goldschmidt interpreted such mutations as support for the view, which he adopted from Jaekel, Schindewolf, and others, that evolution operated on two levels, with evolutionary novelty driven by systemic mutations involving a genome-wide repatterning of chromosomes and developmental macromutations. The developmental macromutations were changes in developmentally significant genes, which generated large changes in physiology, morphology, or behavior. Goldschmidt did not deny the importance of adaptation and natural selection for change within species, but as saltationists had for decades, he rejected arguments that microevolution explained gaps between species, much less the gaps between higher taxa. Goldschmidt's impact was hampered by the fact that he often reached conclusions beyond the available evidence.

Mayr once wrote: "All saltationists have been typologists, and most typologists have been saltationists of one sort or another," claiming they simply failed to understand modern population genetic thinking. Here Mayr is certainly thinking of Schindewolf and Goldschmidt, and his criticisms rendered the concept of a "type" a charged issue in evolutionary theory. The concept of a type is related to historical kinds and individuation, two important issues for understanding evolutionary novelty, which have been addressed extensively by philosophers. There are three very distinct meanings of a "type," however. The Platonic, or essentialist, view of types that Mayr critiqued was central to idealistic morphology: that groups of similar organisms, such as mammals or mollusks, were united by some idealized concept of "mammalness" or "molluskness." There was no expectation that any living animal would fully represent this ideal, but those who adhered to this idea believed that the perceptive scientist could discern the ideal through comparative study of living species. The second, ontological, alternative rejected Platonic essentialism but recognized historical kinds or individuals as having a birth and death and sharing a suite of architectural features. Species originated at a point in time and persisted until they became extinct. Types were real, not idealistic forms. At the outset of his discussion of the concept of a type, Schindewolf wrote: "The real problem in evolution thus shifts from the 'origin of species,' which until now has been in the forefront, to the understanding and explanation of the far-reaching, unmediated differences between types." For Schindewolf, the types underlying different clades were markers of real evolutionary events, and the distinctions between types were real evolutionary disjunctions. The third alternative is an outgrowth of the ontological view, which recognized that kinds reflect shared networks of interaction, and saw the mechanisms underlying the formation of these networks as a fundamental problem in evolution. Because the different meanings of types were only recognized in the 1990s, earlier work is often confusing.[54]

Mayr's opinion of saltational or macromutational ideas was heavily informed by his view that development required "the harmonious interaction of many, if not all, of the genes of the organism." If so, only the cumulative impact of many small changes could permit the gradual acquisition of new structures. To Mayr the origin of species was the fundamental problem, and the continued insistence by saltationists that evolution operated on at least two levels was not just wrong, but a threat. In contrast, for Goldschmidt, evolution within species was decoupled from the formation of new species *and* the formation of new clades (new types). Schindewolf was less concerned with species than new types.[55]

Explosive Evolution

Fossils have always posed the greatest challenge to transformationist scenarios of evolutionary novelty, as revealed by Huxley's and Pictet's responses to Darwin's interpretation in the *Origin*. Particularly challenging are the numerous episodes of discontinuous, evidently rapid appearances of new clades encompassing evolutionary novelties. But such abrupt appearances could reflect discontinuities in the rock record—gaps where fossils were not deposited. The most extensive development of this approach in the decades after publication of the *Origin* may have been by Charles Doolittle Walcott, the American paleontologist who discovered the Burgess Shale fossils in 1909. Walcott posited a "Lipalian interval" below the first Cambrian fossils to account for the sudden appearance of the first trilobites and other animals. Confident that intermediate forms documenting the gradual evolution of these groups must have existed, Walcott concluded that the most reasonable explanation for such apparently explosive evolution was a long interval of history missing from the geological record. If the seas had drained from the continents, no marine sediments would have been deposited, and there would be no reason to expect a record of the gradual evolution of trilobites and the other Cambrian fossils.[56]

Walcott's Lipalian interval was part of a larger debate among geologists about changes in sea level, the causes of mountain building, and how rock units could be correlated from one region to another. At the time many geologists accepted some version of diastrophism, a theory of periodic and globally synchronous movements of the Earth's crust leading to changes in sea level that in turn controlled evolutionary patterns and the resulting fossil record.[57]

By the middle of the twentieth century, however, diastrophism was being questioned by a new generation of geologists who viewed the fossil record more charitably. Paleontologist Preston Cloud contributed a paper to the journal *Evolution* in 1948 with a lengthy discussion of "eruptive evolution." Based on his reading of the fossil record of marine invertebrates, particularly Devonian brachiopods and the Cambrian appearance of animals, Cloud sketched a three-phase model of eruptive evolution tied to ecological opportunity: (1) rapid appearance of very different forms within a group, (2) the availability of many ecological niches, and (3) increased selective pressure that eliminates much of the variability leading to a less dynamic phase of evolution.[58]

Although the meeting was planned before Cloud's paper appeared, in 1949 the Geological Society of America held a symposium "Distribution of Evolutionary Explosions in Geological Time," a series of critiques by seven notable geologists and paleontologists of claims that diastrophism was sufficient to explain apparent evolutionary bursts. Like Cloud, the participants gave little credence to diastrophism; many of the paleontologists presented evidence for bursts of diversification and innovation early in the history of the major clades they examined. I see this work, as well as Simpson's book, as the beginning of the contested reemergence of novelty and innovation as evolutionary issues.[59]

Novelty Reemerges

The celebration of the centenary of Darwin's publication of the *Origin* provided an opportunity to reconsider evolutionary theory. Mayr published "The Emergence of Evolutionary Novelties" in a centenary volume, which also carried discussions of novelty and innovation by Simpson and by the paleontologist Everett Olson. Mayr swiftly dismissed his twin bogeymen of macromutations and Lamarckianism, downplayed the significance of changes driven by mutation, and eliminated the origin of higher taxa from consideration as being outside the scope of the discussion. His definition of novelty was straightforward: "any newly arisen character that gave rise to [a new function]."[60] He eliminated changes in size or pigmentation as novelties but linked structure and function:

> Tentatively one might restrict the designation "evolutionary novelty" to any newly acquired structure or property which permits the assumption of a new function. This new definition must remain tentative until it is determined how often it is impossible to decide whether or not a given function is truly "new."[61]

Mayr's concerns with rigorously defining "new" were prescient, as that has been a continuing challenge. By 1963 he had modified this definition:

> We may begin by defining evolutionary novelty as any newly acquired structure or property that permits the performance of a new function, which, in turn, will open a new adaptive zone. Many evolutionary novelties, such as new habits and behavior patterns, are not primarily morphological although they may have morphological consequences.[62]

Mayr's expansive view of novelties encompassed not just morphology but also behavioral, metabolic, and physiological changes. He considered cellular and structural (i.e., morphological) novelties to belong to different classes, albeit with overlap, with structural changes involving more mutations and greater changes in developmental pathways. Mayr's challenge in understanding novelty was explaining how small gene mutations can generate a new structure before the structure is sufficiently large for selection to take hold: the problem of incipient structures. He suggests that many novelties may have begun with changes in behavior that changed the selective environment and facilitated subsequent morphological changes. Thus, behavioral changes represent a pre-adaptation to novelty.

If new structures are inevitably modifications of existing structures to serve new functions, Mayr identified three potential mechanisms. First, a new structure might form as a by-product of the pleiotropic interactions between genes. Pleiotropy arises when a single gene has multiple effects. The horn of a rhinoceros was Mayr's example, with the first bump forming as a pleiotropic effect that was subsequently elaborated into a horn. Intensification of function of an existing structure was the second route to a novel structure. As an example, he offered the change in the mammalian foot or hand from the five-toed condition to the odd number of toes on the foot of perissodactyls (horses and rhinos) or the two-toed foot of artiodactyls (cattle, goats, camel, deer, hippos, and others). In this case there are no truly novel structures but sufficient rearrangement to generate a different outcome. Mayr also discussed the evolution of the eye as another example of intensification.[63]

Finally, Mayr viewed the most important source of novel structures as a change of function. If a structure acquires a new secondary function, a gradual transition from ancestral to novel function is possible with no loss of ability. The most obvious cases of duplication and change of function are found in segmented organisms such as arthropods, where individual segments and appendages can adopt new functions. In many arthropods the most anterior appendages have become highly modified feeding structures, while the more posterior appendages have become highly specialized for walking.[64]

Like Simpson, Mayr recognized the importance of environmental and ecological context. In explaining the lack of intermediate forms, Mayr linked novelty to adaptive radiations, writing: "We would say nowadays, that adaptive radiation will not take place until after the evolutionary novelty has reached a certain degree of perfection." He also identified four categories in which "the environmental situation" (as he termed it) would be conducive to the success of novelties. These were changes in physical or biotic surroundings; invasion of a new niche or adaptive zone, which required pre-adaptation to the new setting; behavioral changes leading to phenotypic change; and changes in the structural environment. One point is striking: there was no sense from Mayr's discussion that organisms modify their environment. Rather, the organisms adapt to the environment. Over the past two decades there has been renewed interest in how organisms modify their environment to suit their needs, while also constructing new habitats for other species, a topic we take up again in later chapters. Mayr's linkage between structure and function (essentially a continuation of Darwin's view) was the basis for elaboration of the definition of novelty since 1990.[65]

Debates over Macroevolution: 1972–1990

Simpson's contributions in the 1940s and 1950s provided a conceptual framework for the debates of "explosive evolution" and remained an intellectual influence on American paleontologists through the remainder of the century—an enviable achievement. Since the publication of *Major Features of Evolution* in 1953, the term *macroevolution* had been associated with the large-scale evolutionary patterns documented by the fossil record, and most evolutionary biologists agreed that these patterns were the consequence of small, gradual changes. By the 1970s younger generations of paleontologists challenged this view, and since the 1990s they have been joined by comparative developmental biologists. Here I introduce the debates over punctuated equilibrium because they have often been cited in work on cultural and technological novelty and illustrate the changing discussions over the origin of "higher taxa."

By 1965 the successes of population genetics and modern molecular genetics had eclipsed arguments for macromutations and orthogenesis. Views of macroevolution, including novelty, were captured in a special issue of the journal

Systematic Zoology titled "The Origin of Higher Levels of Organization." Although the papers emphasize higher levels of organization (what is today called increased complexity) rather than higher taxa, the arguments are consistent with Simpson's 1953 view of macroevolution: that evolutionary innovations are generated by the accumulation of small changes that cumulatively generate novel features. The importance of adaptive radiations and key innovations is exemplified by ornithologist and evolutionary biologist Walter Bock, who illustrated the origin of birds using Simpson's adaptive zones. Bock's adaptive trajectory proceeded with pulses of adaptive radiations from quadrupedal, ground-dwelling reptiles to active flight in birds through "transitional adaptive zones." These zones were described as transitional because their ecological circumstances required rapid evolution to the next adaptive zone.[66]

One characteristic of good graduate students is rejecting the assumptions of their mentors. For Niles Eldredge and Stephen Jay Gould, the abrupt changes documented in fossil groups, from snails and trilobites to early mammals, accurately reflected the tempo of evolutionary change. From their fieldwork they drew examples to illustrate patterns of speciation in their famous 1972 paper on punctuated equilibrium. Extending Mayr's view that new species arose via small, isolated populations, Gould and Eldredge proposed that new species would not appear in the fossil record until population size and geographic range increased. Since the early, isolated populations were unlikely to be preserved as fossils, the shifts between species would appear abrupt. Equally importantly, however, they argued that most species showed little net change in morphology during their existence—the equilibrium part of their scenario. Thus, speciation happened in discrete events, and was essentially decoupled from adaptive changes during the lifetime of the species. But if the formation of new species was decoupled from adaptive change within species, then species might become a focus of selection, with selection acting on properties such as geographic range. Species selection was not an alternative to natural selection, but an additional level that might explain patterns in the fossil record.[67]

Several conclusions of this work are relevant for issues of novelty and innovation. First, building off punctuated equilibrium and discussions about species selection, many evolutionary biologists developed a broader view of selection. Today many evolutionary biologists acknowledge that selection may occur on many levels, both above and below the natural selection of individuals. This has shed light on evolutionary conundrums, including the evolution of cooperation, the origin of eukaryotic cells, and even the origin of human

culture. Second, many paleontologists have argued that the decoupling between evolutionary adaptation within species and the success or failure of
species and clades might limit the potential of natural selection to explain
long-term adaptive trends.

The focus on the success or failure of different clades emphasized the importance of mechanisms that changed the diversity and abundance of novel
forms, once they existed, rather than addressing their formation. Paleontologists had access to changing patterns of fossils in time and space, but whether
they could evaluate issues associated with the origins of macroevolutionary
variation was more controversial. The criticism that De Vries leveled against
Darwinists in 1906 might also have been made of macroevolutionary research:
"Natural selection may explain the survival of the fittest, but it cannot explain
the arrival of the fittest."[68.]

This neglect of the "arrival of the fittest" was addressed by studies of evolutionary changes in the timing of developmental events, under the general term
of *heterochrony*. In Gould's first magnum opus, *Ontogeny and Phylogeny*, he
resurrected Baer's view that the development of individuals proceeds from
more general to the more specific, and thus may recapitulate some of the embryonic or juvenile stages of ancestral forms. Gould offered a model in which
progenesis, a form of heterochrony in which sexual maturity is accelerated
relative to other aspects of development, plays a key role in generating morphological novelties. He acknowledged that most such transitions are doomed
to failure, but argued that rare, improbable events may have been significant.
Although this renewed interest in heterochrony led to dozens of studies of
pattern, they provided little insight into the mechanisms through which such
changes arose. A mechanistic understanding of heterochrony would have to
await later insights from molecular studies of development.[69]

Deeper insights into the generation of evolutionary novelties came from
the work of Austrian morphologist and evolutionary biologist Rupert Riedl.
He deeply influenced current views of evolutionary innovation. Although
Riedl was influenced by Schindewolf's typostrophism and ideas of orthogenesis, his work was more firmly grounded in modern evolutionary theory,
rejected Goldschmidt's macromutations, and largely accepted the gradualist,
microevolutionary mechanisms of the Modern Synthesis (unlike Gould, for
example).

Riedl developed his ideas about evolutionary burden to address two issues:
first, the apparent stability of animal body plans over hundreds of millions of
years, and second, the dynamic trends in evolutionary patterns within body

plans, neither of which he felt had been adequately addressed by the Modern Synthesis. Riedel saw development and morphology as hierarchically structured, with architectural features defining the body plan being highly stable, because many functions and characters were dependent upon them. Thus, they had high burden (structure and function were linked). Other characters had fewer interdependencies and therefore less burden, and consequently were more likely to evolve in response to changing conditions. More recently evolved features are less burdened than more ancient characters, and older systems should be more interconnected and thus less evolvable than more recent ones. But as new characters are added they will inevitably depend developmentally or functionally upon already-existing characters. Successive additions of new characters progressively increase the burden on older characters. Thus, the novelties defining a body plan did not arise as body-plan characters but came to be so because they were woven so tightly into developmental systems.[70]

Although Gould and Riedl developed distinctive approaches to incorporating development into macroevolution, neither approach was particularly successful, in part because the experimental techniques needed for a mechanistic understanding of the evolution of developmental processes did not exist in the 1970s and 1980s. More recent insights into the networks of regulatory interactions controlling developmental gene activity have shown that the regulatory genome is indeed structured hierarchically, as Riedl proposed. Most plant and animal clades have mechanisms for ensuring evolvability across the regulatory genome and thus easing the dead hand of burden that Riedl imagined. Later chapters will address the critical role of reconfiguring these gene regulatory networks in generating novelty.

The Origin of Higher Taxa

In *Ontogeny and Phylogeny* Gould summarized earlier views on the origins of new body plans, from Cuvier through Goldschmidt and Simpson:

> Theories of macroevolution have not faltered in explaining sustained trends within adaptive zones and common *Baupläne*. But they have floundered badly in trying to apply the Darwinian, continuationist perspective to transitions between fundamentally different designs in the origin of higher taxa— for how can these transitions be gradual and under continuous selective control[?] The names of Cuvier, Goldschmidt, and Simpson are rarely linked,

but on one thing they all agreed—pure gradualism with conventional control by selection cannot extend across the gaps in basic design. Cuvier, of course, denied that the gaps could be bridged at all: the correlation of parts would not allow it. Goldschmidt postulated a macromutation that would bridge the gaps by sheer good fortune at a single leap: the hopeful monster. Simpson (1944) invoked genetic drift as a contributing agent to quantum evolution across a discontinuity between adaptive zones: the inadaptive phase. It is not likely that such a profound transition can occur by continuous gradation mediated only by direct selection upon morphology.[71]

Saltationist views are less easily summarized because they have been so disparate. Saltationists and transformationists differed in whether the patterns revealed by the fossil record required processes distinct from the gradual, fine-grained changes of microevolution. When Russian geneticist Yuri Filipchenko (1882–1930) introduced the terms *microevolution* and *macroevolution*, he was acknowledging arguments for two levels of evolution that included those of Huxley, Pictet, and Jaekel.[72]

Paleontologist Tom Kemp views the morphological gaps between taxa as evidence for the evolutionary reality of higher taxa. Kemp is a vertebrate paleontologist, and for him much of the disparity between higher taxa reflects extinction of intermediates. In other words, Kemp argues that higher taxa are ex post facto entities, a view consistent with those of Darwin, Mayr, and Simpson.[73]

Kemp's correlated progression model captures an interesting approach to novelty and innovation. He views organisms as developmentally and functionally interacting parts, with linkages of varying strength between different parts. The vertebrate limb and skull are examples of structures where a change in one aspect necessarily entails changes in other dimensions. This modular structure of morphology has been discussed for generations, but Kemp makes three important claims. First, most phenotypic structures are subject to heritable variation, and he suggests that any stability in these structures must reflect stabilizing selection, not absence of genetic variation. He views this assumption as "uncontroversial" (and indeed many other evolutionary biologists make the same assumption), but I see this issue as the crux of the difference between adaptive evolution and the origin of novelty. Second, his correlated progression model allows slight variations in characters without significant harm to the functional linkages. But the limits imposed by the functional correlations across an organism limit the tempo of evolutionary change. Thus, the evolution of major new morphologies occurs via a stepwise progression of small changes, one character at a time, preserving the linkages (although the strengths of specific linkages may vary over time). Third, the sequence of changes is random, in that there is no preferential order in which changes need to occur.[74]

I chose 1990 as the transition between this chapter and the next because it corresponds to this conceptual refocusing on novelties rather than taxa, to the increasingly widespread use of phylogenetic approaches, and because the rise of comparative studies of development generated a new, mechanistic foundation for understanding evolutionary novelties that had been missing from earlier work. While Huxley, De Vries, Goldschmidt, Schindewolf, and the others intuited a problem with a strictly transformationist approach to novelty, little was known of processes of development and genomic change. First, however, we turn to early views of novelty and innovation in cultural evolution, economics, and technology.

Novelty and Innovation in Culture and Technology

Cycladic figures are instantly recognizable: flat, geometric human representations with a nose the only interruption of an otherwise blank face, a trace of arms folded across the body. These representations of a great goddess are striking, but perhaps their most remarkable attribute is the uniformity of design across almost the whole millennium of the Neolithic during which they were carved. There were several different stylistic traditions across the Aegean Sea, but within each tradition one gets the sense that the most highly regarded sculptors were those who deviated the least from some idealized representation. Such stability is characteristic of the Paleolithic and Neolithic. Acheulean hand axes first appeared over a million and a half years ago, but the technology changed little for thousands of years. Perhaps the earliest knappers quickly found the ideal form, but the diversity of later stone tool technologies raises doubts about that claim.

Such widespread conformity is unimaginable today. A visit to Santa Fe Indian Market in late August will find the most creative and skilled artists balancing individual artistic expression with accommodating buyer demand for traditional Navajo weavings, Pueblo pottery, and Hopi carvings. Collectors who appreciate a vivid, impressionistic landscape or a challenging conceptual ceramic are vastly outnumbered by those hunting for Two Grey Hills Navajo weavings based on colors and designs codified around 1911 by the (Anglo) owners of the Toadlena and Two Grey Hills trading posts.[75]

Yet this thirst for novelty and innovation is anomalous in human history. For the hunter-gatherers who dominated most of human history, survival was the preeminent concern. Doubtless innovations arose among Paleolithic and Neolithic bands, but novelties faced two challenges: A bias in favor of what was known and had been successful limited the acceptance of novelties. Moreover, hunter-gatherer bands were small and dispersed, so that when novel behaviors arose, they rarely persisted long enough to diffuse to other bands. Eventually, successive novelties in social interactions and the acquisition of additional resources allowed larger groups to coalesce into more complex entities, a process that eventually led to the formation of the earliest states. Cultural, social, economic, and technological novelties and innovation are ultimately about capturing ideas and knowledge about the world and using that information to construct the human environment. This is true whether the knowledge is of the structural properties of stone to understand how to chip arrowheads or carve sculpture, the properties of clay to build and fire pottery, chemistry in engineering dyes, or political and moral dimensions of human interactions.

When Darwin wrote the *Origin* many viewed humans as endowed with a spark from God (some still do). But as we gain greater insight into the behavior of the great apes, and as details of hominid evolution have emerged from the fossil record, claims for human uniqueness have lost much of their strength. Yet two innovations stand out as fundamental to humans: cumulative cultural change with each generation that builds upon the learning and accomplishments of earlier generations, and language. Learning and communication systems are widespread among animals, and some exhibit elements of culture. But the complexity of human language and culture far exceeds that of other species.

That evolution applied to culture and technology was evident to Darwin when he wrote the *Origin* and *The Descent of Man*. The relationship between cultural and technological evolution, at least metaphorically, was widely discussed by late nineteenth-century scholars as diverse as Karl Marx, Ernst Haeckel, Herbert Spencer, and William James. Since the late nineteenth century, studies of culture, economics, and technology have borrowed heavily from work in biology and evolution as a source of metaphors and models. Each of these fields has also been deeply influenced by analogies to Darwin and subsequent studies of biological evolution, and they have influenced each other as each has developed independent scholarly traditions.

Beyond the evolution of modern humans and the acquisition of language and tools, issues of novelty and innovation turn to the more prosaic concerns of survival. Economists assess well-being by estimating income per person, and by this metric there was little improvement in the human condition between at the latest 1000 BCE and about 1780. In his *Essay on the Principle of Population*, Thomas Robert Malthus showed that the natural rate of population increase was greater than the expansion in food and other necessities. Populations can grow exponentially, while at best the means of subsistence increase arithmetically. This relationship led Charles Darwin to recognize the critical role of natural selection in evolution. But Malthus's point was that whenever the food supply is sufficient, human populations will expand until the poorer segments of society are in sufficient distress, as Malthus put it, that population stabilizes or drops. Eventually, agriculture will expand again, allowing population to increase, and the cycle repeats. The absence of sustained improvement in standards of living through most of human history is known as a Malthusian trap, with most families scrounging a meager existence at close to subsistence levels (the rich, as usual, suffered fewer indignities—at least until pestilence struck). Malthus discounted the possibility that agricultural or technological innovations would be sufficient to resolve the problem, suggesting that only a reduction in the birth rate would allow poorer members of society to escape hunger and disease.[76]

Malthus wrote his *Essay* as this Malthusian trap was beginning to ease. By some measures the Industrial Revolution began in England about 1780 and allowed a sustained and continuing increase in population and in standards of living, beginning in western Europe before spreading to European colonies and eventually much of the rest of the world. The steam engine, the development of factories, canals to speed transportation, mining coal to feed the steam engines that drove the factories, declines in infant death rates, and growth of a scientific community were components of this transformation. Early studies of the role of innovation in the growth of technology focused on the origins of the Industrial Revolution, and on interrelated questions: Why did the Industrial Revolution begin in England? And why was growth in western Europe and many of its colonies so much more rapid than growth in other regions, particularly Asia and Africa?[77]

In 1920 statisticians Raymond Pearl and Lowell Reed applied mathematical rigor to Malthus's qualitative arguments, applying a quantitative description of logistic growth to United States census data from 1790 to 1910. Logistic growth occurs when an exponential increase in population slows as it approaches a plateau, termed a *carrying capacity*. Pearl and Reed imported logistic models from ecology and showed that they fit the census data better than alternatives. Good Malthusians, they assumed that the available land area would cause the population of the United States to plateau at 197 million.[78]

Among early economists, Adam Smith in his book *An Inquiry into the Nature and Causes of the Wealth of Nations* reflected late eighteenth- and early nineteenth-century interest in the forces that generate economic growth, as did the work of Karl Marx. In the *Wealth of Nations* Smith used a pin factory to illustrate the importance of specialization and the division of labor as key to increasing output. Smith recognized that an important component of the division of labor was "the invention of a great number of machines which facilitate and abridge labor and enable one man to do the work of many." How these technological changes were adopted did not greatly concern him.[79]

Inventors and inventions were celebrated during the nineteenth century, from Alexander Graham Bell's telephone and Samuel Morse's telegraph to the Analytical Engine of Charles Babbage and Ada Lovelace. Diffusion of new inventions increased agricultural productivity, powered factories, and drove economic growth. The social context of these inventions and the evolution of technology were of great interest to Karl Marx. He had read widely in German studies of the history of technology, and in *Das Kapital* discussed the relationships between labor and technology (in a broad sense). He is also credited with the argument that the complexity of available technology influenced prevailing social organizations, although the extent to which Marx believed that technology was an independent influence remains contentious.[80]

This section provides a historical framework for understanding early approaches to the roles of novelty and innovation in cultural and social complexity, in economics, and in the evolution of technology to about 1990. This begins the discussion of whether sufficient similarities exist in invention and innovation to plausibly support a conceptual framework across biological, cultural, and technological domains. Moreover, just as ideas from Darwin and later evolutionary biologists have infused ideas about cultural and technological evolution, concepts from economics, anthropology, and sociology have influenced broader discussions of novelty and innovation. Such cross-pollination is evident in approaches to cultural evolution from E. O. Wilson's *Sociobiology* of 1975 to Luca Cavalli-Sforza and Marc Feldman's *Cultural Transmission and Evolution* of 1981 and Robert Boyd and Peter Richerson's *Culture and the Evolutionary Process* from 1985. But this work had its greatest impact after 1990 and will be considered in chapter 9.[81]

Social Complexity

Charles Darwin first used the term *evolution* in the sixth and final edition of the *Origin*. Herbert Spencer (1820–1903), the British polymath who contributed to psychology, philosophy, sociology, biology, and anthropology and is today perhaps best known for the phrase "survival of the fittest," first used *evolution* in its modern sense. In 1857 Spencer used *evolution* for progress or development in the history of the Earth and in human cultures. Shortly after publication of the *Origin*, Spencer expanded his ideas on social evolution, and these had wide influence, as social complexity became a major concern for late nineteenth-century anthropologists. Darwin extended his ideas to human evolution from the *Origin* to *The Descent of Man* in 1871 and *The Expression of Emotions in Man and Animals* the following year. Archaeological explorations were revealing the complexities of earlier societies, as ethnographic studies examined hunter-gatherers, pastoralists, farming societies in villages, chiefdoms, and states. Among the insights were an apparent correlation between levels of social complexity and the extent of technological development, which Marx exploited as an argument for technological determinism.[82]

Visitors to the Darwin family home in Downe are likely to have seen the stones Charles Darwin used to measure how rapidly earthworms turned over the soil, a long-term experiment recounted his final book: *The Formation of Vegetable Mould through the Action of Worms*. Philosopher Trevor Price argues that organism-environment interactions were central to inquiries in biology and philosophy in the late nineteenth century, with Darwin, Spencer, and others understanding that while organisms adapted to the environment, so too did organisms impact the environment. Spencer saw no break between biological and social systems, extending his analysis to problems of social evolution, particularly increases in social complexity. He is often remembered today in caricature, but Spencer's emphasis on organism-environment interactions mirrors more recent concerns with niche construction and ecosystem engineering. To preview the important role these concepts will play in later chapters, these involve the activities of organisms that help construct the environment for the constructing organism as well as some other organisms.[83]

Spencer's view of evolution applied the progressive unfolding of development from embryo to adult to comparisons of human cultures, imposing a unidirectional, progressivist view of the evolution of human societies somewhat akin to orthogenesis. Imputing progress to biological or cultural evolution has always been problematic, and Spencer's views that "advanced" societies had greater moral standing are even more troubling.

How do civilizations arise? This question had been of interest to eighteenth-century scholars such as economist A.R.J. Turgot and philosopher J.-J. Rousseau. Spencer's evolutionary approach to culture tied the establishment of notional "stages" of social complexity to the acquisition of technological inventions, with farmers as more advanced than hunter-gatherers, for example. Although the nature of the stages differed between authors and over time, cultures were envisioned as moving along a trajectory of complexity, organization, and allied technology from primitive to advanced. In principle, these stages differ in population size and in the complexity of social regulation required to govern them, as well as their economic and technological foundations. A related argument concerns whether similar cultural elements arose independently in different areas, or whether cultural evolution was dominated by diffusion from a relatively small number of centers of innovation. Both arguments assumed the superiority of white, western European culture (run by men). (I remain baffled why the polluted, disease-infested London of the late nineteenth century was viewed as the peak of cultural attainment, but there is no accounting for taste.)[84]

The role of technological innovation in cultures was a concern of Lieutenant General Augustus Lane-Fox Pitt-Rivers (1827–1900), a British military officer whose infatuation with firearms and their history eventually led him to the evolution of human technology. To the uninitiated, the Pitt Rivers Museum in Oxford appears as a frightful jumble. But the collections have a logic, with everything arranged by type rather than culture: weapons, masks, musical instruments, and so forth. Pitt-Rivers acquired these vast ethnographic collections, and his intensive study of firearms led him to reject linear sequences of technological improvements. Instead, he argued that most inventions had little real impact on firearms. Innovations were a history of what worked rather than what had been invented.[85]

While Pitt-Rivers focused primarily on sequences of technological inventions, other nineteenth-century archaeologists examined relationships between technological inventions and increased social complexity within a Spencerian framework, as exemplified by the American lawyer turned anthropologist Lewis Henry Morgan (1818–1881). In his *Ancient Societies* of 1871, Morgan compiled cultural information from societies around the world to argue that technological inventions fueled increases in social complexity. He cataloged similarities in technologies as diverse as stone toolmaking and hafting blades to arrows or spears, pottery making, the use of fire, agriculture, and even customs such as marriage and making war. Adopting the then-common division of human cultures into savagery, barbarism, and civilization, Morgan linked specific inventions and changes in political and social institutions to increased social complexity. For example, Morgan saw pottery and attendant changes in agricultural practices as marking the transition from savagery to barbarism.[86]

Otis Mason (1838–1908) broadened the range of phenomena considered as inventions from material objects and institutions to the underlying ideas (what we might term creativity). He also examined the benefits of utilizing and applying inventions, and a more diverse category of "the powers and materials of nature involved." This latter category includes the successive increases in energy captured from fire to wind to electrical power, which figured prominently in his ideas.[87]

Mason's emphasis on energy as a critical component in the development of social complexity was shared by mid-twentieth-century anthropologist Leslie White. White linked social evolution to increased capture of energy per person, or more efficient use of available energy. White, along with Elman Service and Marshal Sahlins, revived evolutionary approaches to culture but worked to strip out the racist and deterministic Spencerian framework while preserving the directional trajectories. By the 1960s focus shifted from economics (food gathering) to social structure, and Sahlins and Service advocated a less obviously pejorative series of social transitions, from bands to tribes, chiefdoms, and eventually states. As with all typologies, intermediary stages multiplied, such as complex chiefdoms and secondary states. Unlike the arguments from biological saltationists for abrupt transitions, however, social transformations were seen as occurring gradually.[88]

These cultural hierarchies shared some common assumptions and problems: First, that a single historical trajectory could encompass our remarkable human cultural diversity. Later evolutionary anthropologists developed multilinear evolutionary schemes but retained the typologies of bands, tribes, and so forth, assuming that stages represent steps along a historical trajectory. These trajectories initially assumed that cultures described as "barbarians" in different parts of the world shared a suite of characteristics, or that chiefdoms had common qualities. This has led, for example, to ongoing debates about whether the kings and queens of Hawaii represented a complex chiefdom, or a state, as anthropologist Pat Kirch has recently argued. These views are an extension of the "logic" of the Great Chain of Being to human cultures, with cultural complexity analogous to an organism's development from embryo to adult, and civilization evolving from barbarism and states from complex chiefdoms. The failure of this approach led many anthropologists to reject any evolutionary approach to culture and more generally to borrowing ideas and concepts from biology.[89]

This antipathy to trajectories is reflected in disputes over the origin of major cultural and technological innovations, from the adoption of languages, religious practices, and institutions to packages of technology. If archaeologists identify similar technology in different areas, does this indicate independent inventions or the transfer of knowledge from one group to another (possibly via migration of some members of the first group to the second)? Do these similarities represent independent discoveries, or did most inventions, whether of technology or cultural practices, occur relatively infrequently and diffuse to adjacent societies?

Arguments over diffusion occupied anthropologists through the later part of the nineteenth century and into the twentieth, highlighting distinct expectations about people and their responses to environmental and social challenges and opportunities. These expectations in turn informed subsequent debates about innovation. Those favoring independent inventions in different groups assumed that humans were basically the same everywhere (itself a novel thought for many at the time) and found it hardly surprising that different cultures discovered similar solutions to similar problems. But diffusionists' arguments were based in a rejection of evolutionary approaches to understanding culture. I will abbreviate discussion of these contrasting scenarios, in part because many who contributed to these discussions were replete with racist stereotypes.

Franz Boas (1858–1942) was a pioneer in American anthropology. Raised in Germany, he received his PhD in physics, but an avid interest in geography and cultures led him to fieldwork in Canada and the Pacific Northwest. Moving to the United States in 1887, Boas became professor of anthropology at Columbia, where he mentored leaders of the next generation of anthropologists, including Margaret Mead and Zora Neale Hurston. Boas opposed claims for a biological basis for racial and cultural differences, emphasizing the importance of social learning and culture. He spurned proposals that societies progress through hierarchical stages, inevitability leading to the cultures of western Europe. He also rejected arguments for frequent independent innovations in favor of models emphasizing diffusion of ideas and technology between societies through learning. In a way, diffusionists such as Boas share similarities with those wedded to stories of great inventors today. Invention is viewed as relatively rare. Most people are not particularly creative, and diffusion, either through simple imitation or more creative adaptation, is the most likely route to the acquisition of new cultural traits. Many anthropologists revere Boas for his arguments against the endemic racism in the anthropology of his youth and for his championing of cultural relativism (rejecting any hierarchical ranking of societies), but there is a curious aspect to his assumption that invention was rare. This can be interpreted as viewing most people or cultures as not clever enough to discover new things on their own.[90]

In 1953 anthropologist Homer Barnett (1906–1985) explored the social and cultural contexts of innovations and the factors that determine whether they succeed in *Innovation: The Basis of Cultural Change*. For Barnett, innovation was: "any thought, behavior or thing that is new because it is qualitatively different from existing forms." Some of these will remain ideas, while others have "tangible expression," but in either case they arise from the reorganization of existing ideas. Creativity, whether social or individual, was an essential component of Barnett's view, reflecting his focus on the processes that generate innovations, rather than the resulting entity itself. Thus, while he acknowledged that some inventions may have greater impact than others (comparing the wheel with the hairpin), he rejected the possibility of distinguishing between "radical" and "minor" innovations.[91]

Today cultural evolution is a vibrant and exciting field, but it is conceptually very different from that of the late nineteenth century. Evolutionary approaches went into a long period of decline, despite the efforts of a few, notably anthropologist Leslie White, to revive them. Ideas of trajectories persisted, as interest in the transition from the foraging societies of hunter-gatherers to the onset of agriculture. But recent fieldwork with modern hunter-gatherers has complicated earlier proposals. A recent book by the late David Graeber and David Wengrow summarized this research and concluded that claims for an "Agricultural Revolution" have been exaggerated. They argue that many hunter-gatherer groups had considerable social complexity, that domestication was often adopted when societies were faced with starvation, and that widespread agriculture was a consequence of the growth of urban centers, not its cause. Much of Graeber and Wengrow's argument addresses issues beyond the remit of this book, but they rebut linear evolutionary trajectories in social complexity in favor of greater diversity in social organization.[92]

Anthropologists were not the only researchers concerned with the dynamics of social change. Historical sociologist William Ogburn (1886–1959) objected to the evolutionary stages of anthropologists and to grand theories in general but emphasized the central role of inventions and their diffusion in social change. He studied the sequential process of invention from idea to adoption, and while he rejected the deterministic stages of anthropologists, he favored temporal stages in the success of an invention. The nature of these sequences changed over the years, but essentially involved invention, accumulation (essentially the idea that useful inventions will accumulate over time), diffusion, and finally adjustment of society to the change generated by the innovation. The debate over diffusion versus independent invention had an interesting if perhaps quixotic result: Ogburn coauthored work documenting 148 examples of multiple, independent discoveries of ideas or technologies, from solutions to the three-body problem in mathematics (no. 1) and the nebular hypothesis (no. 9) to the typewriter (no. 144) and use of gasoline engines in automobiles (no. 148). This work, *Are Inventions Inevitable?*, has become a classic for addressing the issue of how readily novelties or inventions can be discovered and whether such discoveries are predictable.[93]

Discussions of technological change often make little distinction among change, innovation, and progress. Although some authors viewed inventions as discontinuous, paralleling the distinction between transformationist and saltationist views in biology, discontinuity does not seem to have been a significant issue. For inventions the patent system provided a set of exemplars of invention, even if the US Congress never defined the term. The objects or processes covered by a patent must be novel, useful, and nonobvious, but ongoing battles over definitions of these terms illustrate that even when discontinuity is part of the definition, it can be challenging to recognize in practice.

Economics

The economist Joseph Schumpeter was one of the twentieth century's foremost authorities on technological innovation and entrepreneurship, and his ideas continue to inform our thinking (the weekly business column of the *Economist* is titled "Schumpeter"). Schumpeter (1883–1950) is perhaps best known for his 1942 book *Capitalism, Socialism and Democracy*, where he argued against what he regarded as specious theories of economic equilibrium. The book provided voluminous examples relating economic growth to the destruction of earlier enterprises, hence the widely known phrase "gales of creative destruction" (which was perhaps not original to Schumpeter). He briefly served as the Austrian finance minister after World War I and, even less successfully, as a banker. But as an economist he had deep insights into business cycles, the nature of entrepreneurship, and the role of technological innovation in driving economic growth. He left Austria in 1932, settling on the faculty at Harvard University. A dedicated teacher, he numbered among his Harvard students Alan Greenspan, former chairman of the US Federal Reserve System, and MIT economist Paul Samuelson.[94]

Schumpeter's work on technological innovation is important for three reasons: First, he distinguished between invention and innovation, with innovation a key driver of growth, and economic change

> due to the unremitting efforts of people to improve according to their lights upon their productive and commercial methods, i.e., to the changes in technique of production, the conquest of new markets, the insertion of new commodities, and so on. . . . This historic and irreversible change in the way of doing things we call "innovation" and we define: innovations are changes in production functions which cannot be decomposed into infinitesimal steps. Add as many mail-coaches as you please, you will never get a railroad by so doing.[95]

Inventions, on the other hand, he dismissed as almost irrelevant: "'Invention' or experimentation . . . are quite another matter and do not in themselves exert any influence on business life at all." Schumpeter recognized that technological inventions did not necessarily lead to economic innovations, that innovations often lagged behind inventions by many years, and that the economic and social impact of an innovation could vary widely. This quote reveals his view that innovations, if not inventions, represent discontinuities. The ongoing (if sometimes puzzling) flow of papers employing patent records of inventions as metrics of innovation ignores Schumpeter's distinction between invention and innovation and his position that records of inventions provide little information about their role in economic activity.[96]

Second, in championing the idea of "gales of creative destruction," Schumpeter argued that economic innovation often required the removal of incumbents, whether products or firms, and the redeployment of those resources into new areas. The math of an equilibrial economy might be easier, but it provides little insight into the dynamic nature of economic development that concerned Schumpeter. Finally, while Schumpeter provided little beyond assertions for his view that technological innovation was the primary driver of economic development, he influenced others to develop quantitative models and to collect supporting empirical data.

Schumpeter's connection between technological innovation and economic growth was a departure from the views of Adam Smith and led to a distinction between Schumpeterian and Smithian growth. The former is driven by technological change and innovation. In contrast, Smithian growth is due to increased specialization of labor and increased trade. In *Culture of Growth*, economic historian Joel Mokyr described Smithian growth as largely about relationships among people, via either trust or legal institutions. Schumpeterian growth is a game with the laws and practices of nature. But Mokyr acknowledges that successful technological innovation is dependent upon social interactions, and that institutional frameworks may impact innovation and thus Schumpeterian growth.[97]

Today there is a broad range of approaches to understanding technological innovation. Economists have produced sophisticated quantitative models exploring the role of technological change in economic growth. Historical economists have sought a more robust empirical foundation for some of Schumpeter's intuitions about technology and evidence of connections between fundamental (or basic) scientific research and the generation of inventions and innovations.

Technological Novelty and Innovation

Although Schumpeter downplayed the importance of technological inventions and largely ignored the impact of scientific research, others interrogated these issues more thoroughly. In the wake of the Manhattan Project, as well as work on radar, cryptography, and other technology projects during World War II, in the United States harnessing successes to benefit the civilian economy became an important national goal. The 1945 report *Science: The Endless Frontier* is recognized as the founding document of postwar federal support for scientific research. This report strongly argues that fundamental scientific research is the basis of economic growth, with funding such research a core responsibility of the federal government.

One of those who worked on the report was MIT economist W. Rupert Maclaurin (1907–1959). Maclaurin was deeply involved with issues of science policy and the question of government support for fundamental research as a driver of economic innovation. He took a field that had largely been the province of historians and historical sociologists (such as Ogburn) and began applying the perspective of an economist. From his studies of individual industries, Maclaurin proposed that scientific research influenced invention through applied research, followed by engineering development, leading to innovation and creation of a new product, followed by diffusion. Through studies of the development of radio, as well as the glass, paper, and electric-lamp industries, Maclaurin and his colleagues concluded that fundamental scientific research was as critical as financing to technological innovation. His ideas evolved as his research progressed, and he came to appreciate disconti-nuities in technological innovation, much as Schumpeter had previously.[98]

Ideas of evolutionary trajectories in cultural evolution influenced early dis-cussions of the relationships among scientific research, invention, and innova-tion. Maclaurin's work and similar work by other economic historians coalesced into what became known as "the linear model" of technological change:

$$\text{Fundamental research} \rightarrow \text{Applied research} \rightarrow \text{Development} \rightarrow \text{Commercialization}$$

Invention "increases technological possibilities," while innovation puts a new production process into use, and diffusion is the adoption of the new approach and replacement of older approaches (where appropriate). This formalized the link between basic scientific research and consequent inventions that had impacts as innovations, leading to economic growth. Many scientists continue to believe this model, as it provides a rationale for the support of fundamental scientific research through organizations such as the US National Science Foundation and the National Institutes of Health.[99]

By the 1960s reaction against the linear model had developed, and Benoit Godin, a scholar of the history of studies of innovation, has argued that sup-port for the linear model was short-lived. Critiques of linear models were mo-tivated by a desire to maximize the efficiency of research organizations and the effectiveness of funding. The most immediate replacement for linear models was the inverse: the demand-pull model. Demand reveals a need or opportu-nity in the market, which can be met by appropriate inventions followed by innovation and diffusion. The conclusion that needs rather than basic research often drove technological innovation in weapons development eventually crystalized into the "need-pull model":

$$\text{Market need} \rightarrow \text{Development} \rightarrow \text{Manufacturing} \rightarrow \text{Sales}$$

The contrasts with the linear model are obvious: needs push technology, and the role of fundamental research is equivocal.[100]

This need-pull model was supported by economist Jacob Schmookler in his book *Invention and Economic Growth*, where he marshalled patent records of inventions in railroading, agriculture, petroleum refining, and the paper industry. Previewing arguments that reemerged later, Schmookler emphasized the importance of knowledge for innovation (an argument like that of Mason in 1907). It was only later, after economists identified knowledge as a public good, that its critical role in invention, innovation, and economic growth would become apparent. He also analyzed whether major inventions stimulated further invention, concluding that scientific discoveries and major inventions were largely irrelevant, which supported his important claim that economic opportunity was the primary driver of invention. Schmookler's work evidently had little impact during the 1960s but was later resurrected as a counterpoint to Schumpeter's "technology-push" arguments. Yet they were asking different questions: Schumpeter about innovation and Schmookler about invention.[101]

It is easy to caricature these models of technological innovation, but those involved in this research understood that both technology-push and need-pull processes occurred, likely varying in their importance depending on the industry. But these models suffered from two significant defects: each invokes a single factor as a trigger, either fundamental research or need, and each assumes linear causality. Moreover, as Godin discusses in some detail, there was considerable confusion about the nature of "demand": did this mean a general societal need or a narrower market opportunity? Feedback, multiple causes, and the broader social and economic framework are missing from these scenarios, with one critique titled "Innovation Is Not a Linear Process."[102]

I have touched on only a fraction of the enormous literature on technological innovation, highlighting major conceptual approaches and how they have evolved over the past century. One lesson from this story is that linear models (in the general sense) capture too little of the dynamics of technological innovation, whether they are driven by scientific discoveries or by consumer demand. Critical to the success of innovations is their diffusion after they are introduced into an economy.[103]

TYPES OF INNOVATIONS

The English philosopher Francis Bacon (1561–1626) identified three inventions as transformative for the English economy: the magnetic compass, the printing press, and gunpowder, each of which originated in China. Schumpeter distinguished innovations in products or goods, new methods of production, new sources of supply, the discovery and exploitation of new markets, and new ways to organize a business or firm. Although hardly identical, this list has several similarities to Simpson's list of factors involved in generating adaptive radiations, particularly key innovations/new products, and new areas/new markets. Schumpeter was primarily interested in the first two types of innovations, and they have received the greatest subsequent attention. His concerns focused on the overall product rather than specific attribute, and on the processes generating an invention. Each concern was less common among biologists.

Of the many subsequent typologies of technological innovations, one of the broader distinguished four categories of innovation, acknowledging that the boundaries could be fluid, and that one type of innovation could trigger changes in other categories. These categories are *product or service innovations* provided by a company; *production-process innovations* from introducing new or refined elements to a production line or changing technology; *organizational innovations* in the formal structure and procedures of an organization; and *people innovations*, including changing personnel, education, and training. Although these were initially focused on innovations within a firm, the ideas can be generalized.[104]

A somewhat orthogonal categorization of innovations comes from a recent book by Vaclav Smil, in which he distinguished simple tools; machines (or complex tools); new materials, whether bronze, plastics, or nanotubes; and new methods of production. Although he did not include process, organizational, or people innovations, Smil did identify the importance of incorporating new materials for the growth of technology. I could enumerate other efforts, but these suffice to illustrate the range of innovations that might be considered.[105]

Francis Bacon was making a much different claim than the classifications described in the last two paragraphs. He identified the compass, printing press, and gunpowder as transformative innovations. Like Bacon, Schumpeter was explicitly interested in radical innovation or technological revolutions, as he viewed these as more likely to be responsible for changes in business cycles. Just as paleontologists use morphological disparity as a metric for differences in form between organisms, any innovation differs to a greater or lesser degree from existing alternatives. These differences allow estimates of what Knight described as *performance-radicalness* and *structural radicalness*, which are roughly equivalent to differences in form and function in biology. This issue of radicality will be central to chapter 5.

Biological metaphors have inspired work in evolutionary economics, with sequential models of technological innovation explicitly based on analogy to the life cycle of organisms from birth and growth through maturity and decay. Nonetheless, there are compelling differences between studies of technological innovation and studies in the biological realm. First, those interested in technology placed far greater emphasis on *processes* rather than the product. Second, although discontinuity or apparent discontinuity has been a major feature of biological discussions, it has played a lesser role in evaluating technological innovations. Knight, for example, discusses the cumulative effect of many small changes in generating apparent discontinuities in a way that is strikingly reminiscent of Mayr and others. Third, the importance of feedback and the limitations of linear models for understanding innovation became apparent to those studying technology much sooner than was the case in biology.[106]

Novelty, Innovation, and Growth of Diversity

Are novelty, invention, and innovation required for growth, whether in biodiversity, cultural diversity and complexity, or economic well-being? The inventions of the Industrial Revolution, the transformation of the Agricultural Revolution or the rise of states, the spread of flowering plants, and the origins of human language each would have attracted attention even if they were not linked to transformative periods of growth. But that link to growth has enhanced the prominence of these episodes. Just as Morgan identified social, institutional, and technological innovations as responsible for the transformation of human societies, and Schumpeter credited technological innovation as the primary driver of economic growth, much of the interest in novelty and innovation reflects the view that such transformative changes underlie increases in diversity in biological, cultural, and technological domains. The Industrial Revolution has served as an exemplar for studies of economic growth and development, in much the same way that the Ediacaran–Cambrian Explosion of animal life or the early Cenozoic diversification of birds and mammals has for evolutionary studies. The long-term stability of the Cycladic culture in the Aegean Sea, of hunter-gatherers until the spread of domestication, and of single-celled eukaryotic clades through the long "boring billion" of Earth's history from 1800 to 800 million years ago have each been interpreted as indicating not just a lack of innovation but a lack of growth.

But growth is defined differently in these different systems, limiting the utility of comparisons across domains. Some biologists view growth as an

increase in number of species or clades, while ecologists often consider increased primary productivity (the rate at which energy—usually from the
sun—is converted into biomass) as informative. Anthropologists may measure
growth as increased settlement size, while economists tally the number of jobs
or gross domestic product (GDP). One common question, however, is whether
increased complexity requires novelty, invention, and innovation.[107]

MIT economist Robert Solow examined the roles of labor and capital accumulation in generating growth. In early studies, after growth due to increased
capital and labor was accounted for, the remaining growth was ascribed to technological innovation, although quite how this worked was poorly understood.
Later economists expanded Solow's models by incorporating the creation of
new knowledge. French economist Philippe Agihon and his colleagues expanded the paradigm of creative destruction with economic growth arising from
general-purpose technologies that eliminate previous innovations. Another approach to growth has been explicitly historical, examining the Industrial Revolution and its diffusion outward from Britain and western Europe, and comparing
the rapid industrialization in Europe and its colonies to growth in China and
elsewhere. Indeed, there are shelves of books exploring the "Great Divergence,"
as historian Ken Pomeranz described it, between Europe and Asia. Some of the
models building on Solow's work will be relevant later, but the drivers of economic growth and the relationship between growth and technological novelty
remain a great challenge for economists.[108]

This chapter has traced four fundamental approaches to the debate over form
and function through the views of transformationists and saltationists: the
views of the German idealistic morphologists that form was a response to
some transcendental ideal; a comparative approach to morphology championed by both Cuvier and Geoffroy, which serves as a basis for analysis even
today; evidence from fossils, including both morphology and the abrupt appearance of new clades; and information from development trajectories from
the embryo to the adult. Although transformationists and saltationists used
similar techniques and even the same examples, their strikingly different theoretical frameworks yielded distinct models for the generation of novelty.

Transformationist and saltationist perspectives have changed markedly
over the past two centuries, and each encompasses diverse and sometimes
conflicting views over the sources of evolutionary novelty. The range of saltationist theories is perhaps the most obvious, from neo-Lamarckians who emphasize the role of the environment in mediating evolutionary changes to

orthogenesists advancing various internal forces driving evolutionary trends to challenges to the Modern Synthesis. But transformationists were never as unified as some historical accounts have tried to paint them. Many disagreements between transformationists and saltationists were conceptual. Morphological novelties were easily conflated with the origin of new higher taxa, apparently sudden evolutionary change, or the ecological success of clades. I have suggested that we must distinguish among morphological novelties, the foundation of new clades, and ecological success. In some cases, they may coincide. Indeed, separating novelty and innovation can be challenging for critical events in human evolution. But cases where they are decoupled may be more revealing about the tempo and mode of evolutionary novelty and innovation. Critical tasks for the next several chapters will be rigorously defining novelty and innovation and distinguishing them from the origin of higher taxa.

The eclipse of novelty as a discrete issue for evolutionary biologists was largely associated with the Modern Synthesis. The focus on the origins of adaptation generated more restricted scenarios of evolutionary dynamics. One might have thought that the continuing emphasis on macroevolution following Simpson's work in the 1940s and 1950s might have heightened interest in novelty, but the issue of the origins of novel structures was subsumed within discussions of speciation and the origins of higher taxa. Evolutionary biologists never entirely lost their interest in evolutionary novelty, despite Mayr's claims in 1960. Indeed, concerns about the tempo, drivers, and impact of evolutionary novelties and discontinuities between adaptive change and novelty have been almost constant for the past two centuries.

Economists and anthropologists faced similar tensions. The growth of social complexity and the controversy over the relative importance of the generation of new ideas and new technology versus the diffusion of inventions have been persistent challenges since the nineteenth century. While Smith and Marx each considered the role of technological innovation in economic growth, Schumpeter's influence significantly increased the visibility of problems of technological innovation. In turn, Schumpeter's emphasis on innovation facilitated the growth of linear models of technological innovation, while the reaction against them led to "demand-push" scenarios. There is an echo here of the differences between biological transformationists and saltationists. By the 1950s and 1960s the field was dominated by intensive studies of specific examples, which might be described as "the natural history of technology." In economics, Robert Solow's transformative theories of the 1950s incorporated technological change as a factor, along with labor and capital.

The next chapter will carry this story forward by considering evidence for the deep roots of some developmental processes. These discoveries led many evolutionary biologists and philosophers of biology to inquire more deeply into the nature of novelty. Is there just one sort of novelty? If there are different types of novelty, do they have different evolutionary consequences? One of the most challenging and contentious issues is whether the generation of novelty involves processes distinct from the generation of adaptive change.

3

Novelty and Innovation

GRASSLANDS COVER great swaths of Africa, Central Asia, the pampas of South America, and the Great Plains of North America, providing food and shelter to horses in Asia, bison on the Great Plains, and zebra, giraffes, and antelope in Africa. Absent turfgrass, baseball, soccer, and cricket would be more exciting if less predictable. Domesticated grasses include wheat, rice, and barley, and beer, vodka, and whisky derive from fermented grasses. And until recently, bamboo was the primary source of scaffolding in China. I even have a bamboo safety hat that I picked up while doing fieldwork in rural China in the late 1990s. All in all, grasses seem to have been one of evolution's good ideas.

If grasses are such a good idea, then one might expect them to serve as a paradigmatic example of an adaptive radiation, exploding in diversity and abundance soon after they first evolved. Yet grasses originated well before grasslands became widespread. Fossil and molecular evidence reveals that several of the major subclades of grasses originated soon after grasses evolved during the Cretaceous, yet the spread of grasslands began only after about 30 million years ago. This lag of 20 to 30 million years between the origin and initial diversification of the different groups of grasses and their ecological success poses a problem.[1]

The lag between the acquisition of the morphological characters that compose grasses (the clade Poaceae) and the ecological success of grasslands illustrates decoupling between the formation of evolutionary novelties and the ecological or evolutionary success of the clade that possesses the novelty. Grasses adapted to open habitats had appeared by about 55 million years ago, but only about 30 million years ago did early grasslands appear, dominated by the standard form of photosynthesis known as C_3 photosynthesis. The spread of grasslands varied among continents. The invasion of the dryland habitats where they are so common today required a new form of photosynthesis better adapted to high temperature and lower levels of atmospheric carbon dioxide. This required changes to the standard photosynthetic machinery to increase the concentration of carbon dioxide and thus improve the efficiency of photosynthesis. These changes produced what is known as C_4 photosynthesis. C_4 plants make up only 3 percent of all vascular plant species, but they account for 25 percent of terrestrial photosynthesis. More importantly, the C_4 pathway evolved independently between 45 and 60 times in many different clades (in both grasses and other plant groups) because it was an accessible adaptation from many different evolutionary histories. Yet despite the evident success of C_4 grasses, there was also a lag between their origins and their ecological diffusion. While the rise of C_4 photosynthesis was linked to declining carbon dioxide levels and rising temperatures, grasslands themselves affect climate because of their high albedo (reflecting much of the incoming solar radiation back into the atmosphere) and their low storage of carbon.[2]

Since the 1990s seven distinct (although partly overlapping) approaches to novelty and innovation have been mooted in fields as diverse as evolutionary and developmental biology, genomics, behavior, culture, and economics. Discussion of these approaches in this and the following chapter establishes the foundation for the conceptual model of novelty and innovation presented in chapter 5. Simpson's approach to novelty centered on the concept of an adaptive zone and was widely adopted from the 1950s onward. Since then, conceptualizing evolution as occurring within a space has become common and underpins several approaches to studying novelty, which are the focus of chapter 4.

I choose 1990 as the transition between the more historical preceding chapter and this more current chapter because of four advances that collectively forced a reformulation of the arguments described in chapter 2. First, rapid advances in sequencing DNA began a revolution in understanding the mechanisms underlying development. Comparison of the mechanisms patterning developing embryos in the fly *Drosophila* (a favorite of geneticists since the early 1900s) with those of vertebrates, principally chickens and mice, revealed unexpected deep similarities. Related genes were found to drive developmental processes in the formation of the eyes, brain, limbs, and gut. Gradually cnidarians, sponges, and other animals were examined. The wealth of this data has impacted the arguments of advocates of both saltational and transformational perspectives and has forced most biologists to adjust their views of novelty and innovation, with insights leaking into other disciplines as well.

The second advance was a revolution in how biologists examine and define evolutionary relationships and generate evolutionary trees. Within the systematic paradigm that began with Carl Linnaeus in the late eighteenth century, taxa were groups in an inclusive hierarchy of species within genera, genera within families, up through orders, classes, phyla, and kingdoms, with a bewildering panoply of subfamilies, infraorders, and superclasses interleaved as seemed appropriate. Both Mayr and Simpson began their careers as systematists, evaluating the evolutionary history of birds and fossils mammals, respectively. Like most systematists of their generation, they inferred evolutionary relatedness from overall morphological similarity, which came to be known as evolutionary systematics. Thus, if the overall shape of two bird species, including the curve and robustness of the bill, the form of the body, and coloration, were sufficiently similar, the two species might be considered closely related. Cacti (from the western hemisphere) and euphorbias, or spurges (from Africa), have species that appear almost indistinguishable, yet the two clades are only distantly related. These species have converged in form as they adapted to hot, dry conditions.

If such convergence is common, similarity provides a very misleading view of evolutionary relationships. In fact, convergence has been ubiquitous. Some examples of convergence are coarse: insects, birds, pterosaurs, and bats all have wings and fly, but no one has any difficulty distinguishing them. I am a dreadful botanist, but I have little difficulty in distinguishing between the cacti and euphorbs in my garden. Other cases of possible convergence on similar form require closer anatomical comparison: there are remarkable similarities between mammalian eyes (including of humans) and those of squid, but these similarities obviously arose independently. (Incidentally, the architecture of squid eyes is far superior to that of mammalian eyes, a fact that, surprisingly, remains unacknowledged by creationists.)

Overall similarity has given way to more reliable inferences made by assessing specific characters, including molecular data. Shared characters between species are a more reliable marker of ancestry and thus evolutionary relationships. Morphological similarity and dissimilarity often played a more important role in decisions on the assignment of a group of species to a higher clade than did strict evolutionary descent. The spread of phylogenetic approaches led to more careful definitions of evolutionary novelty.[3]

Birds had long been separated from dinosaurs (class Aves versus class Reptilia) because they have feathers, most can fly, and they share other skeletal adaptations. But birds are descended from therapod dinosaurs, so recognizing birds as a separate class from the remaining dinosaurs is artificial. Modern practice emphasizes the importance of defining groups that share a common ancestor and encompass all that ancestor's descendants, termed *monophyletic clades*. Birds form such a clade, but removing birds from the larger entity of dinosaurs so that each has the equivalent Linnean rank of class distorts evolutionary history, since dinosaurs are no longer a monophyletic clade.

Unique and bizarre specimens in the history of life illustrate the importance of character-based approaches to phylogeny, and such specimens pose challenges for understanding evolutionary novelty. Consider *Tullimonstrum*. Found by amateur fossil collector Francis Tully near Chicago in a Carboniferous-aged deposit (~310 million years ago), it has long defied paleontologists. Tully's common monster has a sluglike body with a long proboscis at the front ending in two claws, fins to the rear, and a transverse bar amidships, evidently with an eye at each end. The eyes on the transverse bar and the proboscis represent novelties, and the overall morphology of the beast is like nothing else in the history of animals (which is why all invertebrate paleontologists know of *Tullimonstrum*). Fully understanding *Tullimonstrum* requires carefully parsing the characters found in the fossils and deciding which may be preservational artifacts and which are phylogenetically significant. Since it was first described in 1955, *Tullimonstrum* has been interpreted as a worm, a mollusk, a basal vertebrate, and an arthropod, as well as several other groups of animals. Each of these claims depended upon identifying features joining *Tullimonstrum* to these other clades. For example, one group of researchers claimed evidence for a notochord (a flexible rod through the body that defines the chordates), which aligned the fossil with early vertebrates. A second group of researchers supported this scenario with fine-scale details of the eyes also suggestive of a vertebrate affinity. Other paleontologists argued that comparing *Tullimonstrum* to a broader array of vertebrates raises questions about these purported shared characters, and they concluded the fossils might be related to mollusks.[4]

Even if *Tullimonstrum*'s place on the animal tree remains uncertain, we accept that resolving the question depends upon establishing homologies. And the nature of homology was central to the third major development: a renewed interest in more precisely defining evolutionary novelty. *Novelty* and *innovation* are used loosely and often interchangeably in everyday language and in fields as disparate as genomics, developmental biology, behavioral studies, cultural and linguistic evolution, economics, business, and even psychology. Each discipline focuses on its own questions, encompasses different research traditions, and is characterized by distinct cultures. So, the variability in definitions is hardly surprising. Within biology, focusing on the origin of new phylogenetically informative features provides a less squishy definition of evolutionary novelty. Much of this chapter focuses on the progressive refinements to this view of novelty over the past several decades among both biologists and philosophers of biology. But this character-based definition of novelty has interesting consequences. One of these is that changes in size have been largely ignored as a source of novelty. This seems curious. Whales accessed their adaptive zones through substantial increases in size, as did dinosaurs. The origin of birds involves a remarkable trend over tens of millions of years of the reduction in size of theropod dinosaurs. Simpson would have easily captured these size changes as allowing access to new adaptive zones. Essential new homologous characters are associated with whales, dinosaurs, and birds, but ignoring the role of changes in size leaves out critical dimensions of evolutionary change. Focusing novelty on new characters also puts aside questions about the ecological and evolutionary success of clades containing these new features, which I describe as innovation. Later chapters will provide many examples of lags between novelty and innovation. That novelty and innovation *may* be decoupled does not mean that they will always be decoupled. Moreover, as I use the terms in this book, novelty need not entail innovation, nor does innovation require novelty. Investigation of specific events may reveal novelty directly connected to innovation, but such a happy fact is something to be demonstrated, not assumed.

Finally, conceptual developments in economics and studies of cultural evolution around 1990 ushered in new ideas as well. One example is a seminal paper published that year by economist Paul Romer, which reformulated how economists view technological innovation and highlighted the critical role of public goods (which are *not* the same thing as "the public good").

Expanding Views

Here I present seven different approaches to novelty and innovation: the structure-function distinction of Darwin and Mayr; Dobzhansky's postulation that novelties reflect the structure of ecological opportunity; Simpson's view of the cumulative effect of adaptive radiations in generating evolutionary novelty; the extension of Simpson's ideas so the origin of "higher taxa" became a metric for evolutionary novelty; novelty as the origin of new homologous characters, such as the origin of bird feathers, the carapace (shell) of turtles, or eyes; novelty as new combinations; and an approach to novelty that arose from studies of the variability of behavior and proposes that new behaviors may lead to new functions, which may be accommodated by the success of new structures, which in a sense inverts the structure/function views of Darwin and Mayr. Finally, I introduce major evolutionary transitions, or evolutionary transitions in individuality. Although the classic major evolutionary transitions are few through the history of life, this represents an end-member to evolutionary novelties.

These seven different classes represent the most common ways of looking at these problems, but they are *not* mutually exclusive, although some of the proponents of these ideas tend to downplay connections to related ideas. Even this abbreviated list makes clear the diversity of evolutionary phenomena that have been categorized as evolutionary novelties (or innovations). One of the principal differences among these classes is that some focus on the novelty itself, while others are more concerned with process. Three of these seven different approaches will be considered only briefly here: the ideas of Dobzhansky and Simpson about adaptive spaces and exploring the view of novelty as a combinatoric search through possibility spaces. Chapter 4 addresses novelty and innovation as a space of possibilities, and considers these three approaches in greater depth. Later chapters in this book will explore some of the relationships between these different aspects of novelty.

Structure and Function

Novelty as a new structure with a new function, following the traditional views of Darwin and Mayr, remains widespread among evolutionary biologists today. An oft-cited example is the evolution of birds, with particular focus on the evolution of wings from reptilian forearms as illustrated in the preceding chapter. This scenario supports a gradual, progressive origin for novelties, addressing the challenge raised by St. George Mivart to Darwin about the origin of complex adaptations. Small changes can facilitate adaptation of a structure to a new function. If these changes are advantageous, selection will continue as the function of the structure improves. There has been a long-standing debate over whether wings initially evolved for gliding or as small dinosaurs pursued insects through the Cretaceous underbrush. An alternative model of this process has been to invoke pre-adaptations, where changes in a structure for one function fortuitously improve the suitability of the structure for another function. Other proposed pre-adaptations leading to novelties include bird feathers, which probably originated as insulation; the origin of tetrapod lungs from the swim bladder of fish, a buoyancy control device; and drawing as a pre-adaptation for writing.

Despite the frequent conflation of novelty and innovation, the structure-function distinction has led some evolutionary biologists to see novelty and innovation as distinct phenomena, if not necessarily distinct evolutionary processes. As we saw in chapter 2, some discussions of novelty have emphasized new structures, others new functions. This has led several evolutionary biologists and philosophers to distinguish between novelty as a new form and innovation representing the corresponding function. For example, in a 2006 paper, philosopher Alan Love employed the classic distinction between form and function to distinguish between novelty and innovation. For Love, innovations involve the origin of new functions, while novelties involve the origin of new forms, following the criteria of nonhomology as described in the next section.[5]

Wrasses are a diverse group of small fish, particularly common on tropical reefs, where they dart around snapping up invertebrates. Many are quite colorful, and males often defend territories even against much larger fish or the hand of a diver. Wrasses have a unique structure of the four jawbones, which can be easily modified for different functions. For functional morphologist Peter Wainwright:

> Innovations are evolutionary novelties, or new traits or combinations of traits that result in enhanced performance of the organism in some tasks. While we see a continuous range in nature between minor adaptations and major innovations, as functional morphologists we regard innovations as adaptations that provide a novel functional mechanism that underlies increased performance.[6]

Wainwright has shown that functional innovations may lead to increased ecological specialization, but that this may not generate increased numbers of species or ecological impact. In other words, these morphological novelties (by my terms) are largely evolutionary dead ends.[7]

Pigliucci expanded Mayr and Simpson's definition: "Evolutionary novelties are new traits or behaviors, or novel combinations of previously existing traits or behaviors, arising during the evolution of a lineage, and that perform a new function within the ecology of that lineage." Pigliucci makes several important moves here. First, he rejects the linkage between novelty and adaptive radiations, but retains the view that novelty is necessarily linked to ecological function. Second, he explicitly rejects the idea that novelties are continuities with prior adaptations, yet he distinguishes novelty from adaptation. Curiously perhaps, while I would place this view of novelty as largely within the tradition of the Modern Synthesis, it is one of several papers Pigliucci has written arguing for the insufficiency of the Modern Synthesis, and the need for an enlarged view of evolution.[8]

A relationship between new structures and new functions is also common in discussions of the origin of new genes via gene duplication. Many domesticated crops are associated with whole-genome duplications, including asparagus, corn, and wheat. Gene duplications can have several possible outcomes. Most duplications are quickly eliminated, but genes that have multiple functions have long been seen as a significant input for novelty because they have the potential of preserving the ancestral function of genes while simultaneously providing new material for novel expression.[9]

This structure-function approach to novelty has lacked a rigorous definition of *new*, assuming that new functions will be linked with new structures, and has rarely tested whether new structures might arise without a new function. Often the term seems to be applied when an author wished, hoped, or expected to find novelty. Despite the long history of this approach, the absence of any causal structure and the spread of more mechanistic approaches has made it increasingly passé.[10]

A Question of Character(s)

A new structure with a new function. A new "higher" taxon. But what did *new* mean? Rigorously defining *new* seemed unattainable, and few evolutionary biologists explored the problem in much depth, content to accept arguments of respectable systematists that A or B was a novelty. Such arguments from authority are not science (or at least, not good science), which became increasingly clear as a younger generation of biologists challenged traditional systematic practices during the late 1970s and 1980s.

The resolution to the problem of *new* began with the spread of phylogenetic approaches to systematics during the 1970s and 1980s. Over a couple of decades fraught with controversy, systematists developed a rigorous system of analyzing evolutionary relatedness based on shared characters. All mammals share a suite of features, such as being warm-blooded, having hair covering the body, and providing milk to the young. These features help to define all mammals, but because all mammals share them (or their ancestors did) they are of little use in establishing the relationships among horses, or among whales. These shared, derived characters diagnostic of individual clades are described as *homologous*.

Homologous characters were probably recognized in a general way by Aristotle. Darwin and his successors realized that homology could provide evidence for evolutionary descent, but this required similarity due to a shared history. If a particular protein structure is highly advantageous, separate evolutionary lineages might evolve convergent structures. A classic example comes from the antifreeze proteins (glycoproteins) in northern cod and Antarctic notothenioid fish. Glycoproteins are a large family found across the tree of life with many different roles. But two evolutionarily distant groups of fish inhabiting opposite poles have independently repurposed specific glycoproteins to generate nearly identical structures that act as antifreeze proteins.[11]

The eyes of a human and a chicken are homologous because each descended from the last common ancestor of birds and mammals (likely a Triassic archosaur), which also had eyes. Indeed, we can trace the history of the vertebrate eye deeper in phylogeny (or further back in time) to Devonian fish and Cambrian cephalochordates such as *Amphioxus*, all of which share a similar eye. In brief, two structures are homologous if they derive from a common ancestor that also possessed the structure. In the case of vertebrate eyes, this presumably first appeared in the last common ancestor of vertebrates and cephalochordates. Vertebrate eyes are quite easy to distinguish from those of insects, or scallops. The morphological structures we describe as eyes evolved multiple times in many different lineages (several times in scallops alone, for example). But the underlying developmental pathways to produce eyes share regulatory controls, even across distantly related animals such as insects and vertebrates. So, while the eyes are not homologous, there do appear to be homologies at a genomic level. The arm bones (radius, ulna, and humerus) are homologous across vertebrates, so birds and bats have homologous arm bones, but bats and birds independently evolved wings, which are not homologous.[12]

Assessing homology in more distantly related species can be challenging. Distinguishing the eyes of flies and chickens is easy. Fortunately, a raft of sophisticated computer programs parse the data and identify likely evolutionary relationships. To confuse the matter further, some morphological characters that are clearly homologous can often have strikingly different developmental patterning mechanisms. Resolving issues of homology is critical to the work of reconstructing the phylogenetic history of organisms.

In 1990 Austrian evolutionary biologist Gerd Müller defined novelty as a quantitatively new structure with a discontinuous origin, severing the issue of function. Müller's perspective placed novel phenotypes outside the range of expected variability within a species or a larger group. But as with the traditional new-structure–new-function definition, it was unclear how different a structure must be to qualify as novel. The following year Müller and Günter Wagner "reformulated" novelty as follows: "A morphological novelty is a structure that is neither homologous to any structure in the ancestral species nor homonomous to any other structure of the same organism." This homology-based definition as the individuation of new characters has since been refined but has remained central to many discussions of evolutionary novelty.[13]

This definition helps position the first appearance of a novelty on an evolutionary tree. If some morphological feature, such as a feather on a bird, is homologous to a similar structure in an ancestral form, then the novelty first appeared in an ancestral bird. *Homonomous* is a technical term for repeated instances of the same feature, such as vertebrae or the legs of a centipede (also known as serial homology). Müller and G. Wagner use the example of pectoral fins in fish.* If a species of fish adds two more pectoral fins than were present in the ancestor, that is something new, but arose by duplication of a feature already present. New homologous structures are quasi-independent, individualized parts of a body that have an independent history. Examples of novelties include feathers of birds; the various appendages of arthropods, from the massive claws of crabs to the individualized mouthparts of trilobites; and petals of flowering plants. Müller and G. Wagner also noted that the sequence of novelties is often important and, in some cases, can predict morphologies that may be observable in the fossil record.

Adopting this definition has some interesting consequences. First, novelty had been so broadly defined as to include virtually everything, or so restrictively that the problem of novelty was overly circumscribed. Müller and G. Wagner's definition makes no assumptions about the mechanisms generating a novelty. Second, they emphasized that new homologous characters form despite developmental constraints. Development is biased toward generating like from like. Thus, the factors generating a novelty must overcome whatever has limited generating new attributes of an organism, and this can provide some insights into the nature of developmental or phylogenetic constraints.

---

*Andreas Wagner, a former student of Günter's but otherwise unrelated, will soon appear in this and later chapters, so Günter will appear as G. Wagner and Andreas as A. Wagner.

Third, this character-based definition of novelty emphasizes the origin of new forms or structures rather than their ecological or evolutionary success. In a sense these authors neatly decoupled the problem of form and function that had so bedeviled biologists since the early nineteenth century: focus on the new form and ignore the issue of function. This decoupling of form and function let some define novelty as the origin of new characters, and innovation as the origin of the associated new function. Finally, homologies are historically unique: they arose in a population at a particular time and have been preserved (unless they are subsequently lost) in descendant lineages. This homology definition rapidly became the de facto standard, leading one biologist to ask, "Does novelty begin where homology ends?"[14]

Feathers also serve as a useful example of the difference between the origin of a novelty and subsequent evolutionary changes. Feathers represent a novelty, a new structure derived from reptilian scales, but feathers come in an incredible variety of forms, from the downy under feathers of geese to the rigid flight feathers of hawks and eagles and the showy plumes of male peacocks. These different types of feathers are not novelties by G. Wagner's definition, since they are not semi-independent, individuated structures. Rather, they represent alternative states of a feather. Similarly, the forewing and hindwing of insects have different character identities. In dragonflies and butterflies the fore- and hindwings are obvious and adapted for flying, although in butterflies the wings have an important signaling role as well. In flies the forewing is the flying wing, and the hindwing is reduced to a small balancing structure called a haltere. The forewing of beetles is now a hard structure that covers the hindwings, which are the ones used for flying. I think of novelty as constructing a new space of potential character identities (types of feathers, or different types of insect fore- and hindwings, in these examples).[15]

Some may object to the consequences of such a definition. For example, since the wing of a bird is a derived version of the forelimb of a theropod dinosaur, and not a new structure, it is not a novelty. Limbs are already highly individuated structures, and the formation of a bird wing continues this evolutionary trajectory. Moreover, since evolutionary novelties are discrete features, changes in body size do not represent a novelty. The origin of birds is associated with a 50-million-year-long reduction in the size of a particular lineage of dinosaurs. To accommodate such challenges, G. Wagner distinguished between type I and type II novelties. The first encompass new character identities, and Wagner offers the vertebrate head and insect wings as paradigmatic examples. But feathers are transformations of reptilian scales, and tetrapod limbs represent transformation of the paired fins of fish. Wagner describes these as type II novelties, encompassing deep transformations of the original character identity.[16]

NOVELTY AND DEVELOPMENT

Focusing on novelties as new characters coincided with the rapid expansion of tools to study Hox genes and dozens of other developmentally significant genes. G. Wagner's 2014 book on innovation was an important and substantial achievement, representing the culmination of a long period of thought about homology and novelty. He also tied individuation and novelty to a particular mechanism of change in how genes are expressed during development. Providing such a mechanistic underpinning to novelty was possible only because of the expansion in understanding of animal development beyond reliance upon a few "model" organisms, particularly the nematode, fly, chicken, frog, and a few others.

In the previous chapter I acknowledged the significance of Gould's *Ontogeny and Phylogeny* in 1977 in recognizing the importance of development in evolution. Others made similar discoveries about the same time, including Rudy Raff and Tom Kaufman (developmental biologists with a hobby of collecting fossils), Jim Valentine, and vertebrate biologists such as David Wake and Brian Hall. While studies of heterochrony demonstrated that it could contribute to the origins of novelties—for example, in larval sea urchins or in the size of the human brain (the focus of the final chapter of Gould's book)— they lacked sufficient specificity about the nature of evolutionary novelty, or an understanding of the mechanisms generating changes in development.[17]

Rapid DNA sequencing, the ability to visualize gene expression within developing embryos, and compare such patterns among different organisms, as well as increasingly sophisticated techniques for tracing the interactions among genes, were just a few of the tools driving the expansion of comparative studies of development. Evolutionarily inclined developmental biologists had speculated about changing patterns among corals, scallops, or birds but could interrogate these patterns experimentally. By embedding this developmental information within an increasingly robust phylogenetic framework, it was finally possible to explore the processes that generated evolutionary novelties at a molecular level.

In elucidating his view of novelties, G. Wagner connected them to subcircuits of genes, wired together in such a way as to ensure the formation of the morphological novelty. Beyond providing a rigorous definition, Wagner emphasized that his distinction between characters and character states paralleled differences in the mechanisms of change in gene regulation responsible for the origin of novelties versus their later modification into a variety of different states.[18]

Curiously, these subcircuits, which Wagner described as character homology identity networks (ChINs), are essentially identical to structures that the late Eric Davidson and I described based on the work of Eric's group at Caltech on sea urchin development (both papers were published in 2006). The attraction of these network subcircuits is that they are recursive, so they once they become active, they help lock a cell into a particular regulatory state. This helps explain how a novelty can become stable, whether it is a new cell type, or a more complicated pattern such as a feather. Our understanding of the evolution of gene regulatory networks has greatly expanded since then, and I am unwilling to claim that specific subcircuits of regulatory interactions are the only genetic basis for generating phenotypic novelty. Wagner and I agree on two points, however: First, that understanding evolutionary novelties requires an understanding of evolutionary changes in how genes are turned on and off through development. Second, that the processes of regulatory change underlying the generation of novelty will, in most cases, be different from those responsible for adaptive evolution. One of the claims I will advance is that understanding cultural and to some extent technological novelties similarly requires an understanding of the patterns of regulation that facilitate such advances. Recently, Wagner and colleagues have updated and expanded his approach to suggest that character identity networks exist at multiple levels, with ChINs defining cell types, as discussed in chapter 6.[19]

In his 2014 book, G. Wagner provides several case studies supporting his individuation argument, including the origin of new cell types; of characters derived from vertebrate skin, including hairs, scales, and feathers; of hands and digits; and of various features of flowers. In each example a new character type emerged from an ancestral form, and enough is known about the developmental architecture of these examples to illuminate the critical role of changes in gene and protein interactions in facilitating the success of each novelty. There are other obvious examples of novelty via character individuation that he does not discuss, particularly arthropod limbs, as well as some more complicated examples where similar novelties develop in different clades, including nervous systems and eyes.

Feathers are an interesting example because they arise from local interactions between the skin (epithelia) and the underlying tissue called mesenchyme (epithelial-mesenchymal interactions are critical to many aspects of development). Reptilian scales have the same origin, so scales and feathers share many developmental signals. Thus, feathers represent a type II novelty, a deep transformation of reptilian scales. But once feathers evolved, opportunities for further novelties arose, including the barbs, or the individual filaments of a feather, and the central rachis from which the barbs emerge (on most feathers).[20]

But is there only one definition of novelty? Indeed, this has been a central issue—how different is different enough to qualify as a novelty? In later work Müller distinguished among novelties involving a new body plan, such as the new animal body plans that appear in the fossil record during the Cambrian Explosion (~535 million years ago); those introducing a new structural element with no homology to an element in ancestral forms, including the horns emerging from the heads of dung beetles, vertebrate teeth, feathers of birds, and the various parts of a flower; and individualization of a character, epitomized by the tusk of a narwhal (derived from a canine tooth), or the false thumb in pandas (derived from repurposing the sesamoid bone in the wrist). The latter two types of novelty are both the most numerous and quantitatively discontinuous from the ancestral condition.

INSIGHTS FROM THE GENOME

Before genome sequencing became so widespread and so cheap, the duplication of genes, gene segments, or entire genomes had been identified as the major emphasis for genomic novelty. Since many genes have multiple functions during the life of an organism, duplication can allow for splitting of these functions. This process of specialization (known as subfunctionalization) appears to be a common means for the origination of new types of cells. But more frequently one of the two genes fails to acquire a distinctive role and is eliminated, presumably because it is redundant and unnecessary.[21]

With expanded sequencing the variety of changes associated with novelties has grown. Some novelties represent genes that have no detectable sequence similarity with any other gene in related species. In most cases this will be because the sequence has evolved so rapidly that it is difficult to identify related genes. Another type of novel gene shares similar sequences with genes in related species but contains a novel domain (a domain in a gene is a gene segment, which often codes for a particular structure in the resulting protein; many complex genes contain multiple domains). Homology thus plays a central role in the definition of each of these types of novelties. But genes may also arise through recombinations, in which the same domains are present but with a novel architecture. These illustrate the variety of pathways to individuation of new genes and show that although Wagner developed his view for new morphology, a definition of novelty focusing on individuation of new characters can be expanded to genomic novelties. Later I will argue that we can also encompass some cultural and technological novelties as well.[22]

CHALLENGES

Although individuation of new characters has become widely accepted over the past decade, there are several consequences. First, this addresses only discrete characters, rather than more qualitative changes such as increases or decreases in size, changes in shape, or other adjustments that are challenging to parse into discrete characters. Second, this definition is restrictive and eliminates many evolutionary changes that have long been described as evolutionary novelties. Changes in the color patterns of butterfly wings, changes in the jaw structure of vertebrates, and adaptive radiations from Mesozoic mammals to Hawaiian flies have all been described as evolutionary novelties. Whether such a change should truly be considered a novelty will depend upon one's perspective. For many, a more rigorous definition provides a useful bound on what might comprise novelty and facilitates comparisons between different episodes of novelty in the history of life.

Adaptive Spaces

Dobzhansky's claim that novelties reflected the structure of adaptive spaces was introduced in the previous chapter, and this view has adherents today. If the distribution of adaptive opportunities is clumpy rather than smooth, this will be reflected as apparent discontinuities between ancestral and descendant clades, a view accepted by Mayr, Simpson, and others. In this scenario, novelties arise as species move to new adaptive opportunities, but these transitions are sufficiently rapid and involve such small populations that they would be essentially invisible to the fossil record. Increased emphasis on a phylogenetic perspective and on development has led to reformulation of Dobzhansky's argument, so that novelties arise from the interaction of phylogenetic and developmental constraints with ecological opportunity. Under appropriate conditions, this can allow species to cross valleys in an adaptive landscape. This

approach shifts the discussion from the definition of novelty itself to the processes generating novelty (like those for genomic novelties), and will be covered in detail in the next chapter.[23]

Adaptive Radiations and Key Innovations

Key innovations provide new characters and new character combinations,
allowing the occupation of new niches. Moreover, key innovations have been
linked to increases in the number of species. As I described in chapter 2, adaptive radiations have been seen as a generator of evolutionary novelty through
their association with the opening of a new adaptive zone. Novelties might
arise through progressive adaptive divergence to the new opportunities presented by this adaptive zone, or through key innovations. As Mayr noted, behavioral shifts might precede morphological adjustments to the new adaptive
zone. If the adaptive zone were sufficiently broad and the resources sufficiently
rich, then the adaptive zone might generate new higher taxa.

The shift to a phylogenetic perspective led to greater emphasis on characters and concerns over homology, eventually leading to the work on character-
based definitions of novelty by Müller, G. Wagner, and Moczek as described
above. But key innovations have been identified at levels ranging from species
within a genus to families of mammals or flowering plants, with the flower as
the key innovation, and in most reasonably diverse clades from tropical orchids to tortoises in southern Africa to continental-scale radiations of lizards
and eucalyptus trees across southern Africa. Key innovations figure in only a
minority of these adaptive radiations, but well-studied examples include the
fourth molar cusp, or hypocone, of mammals, which has evolved some 20
times in different clades to allow these groups to become herbivorous; nectar
spurs on columbine flowers, which increase the effectiveness of insect pollinators; and the previously mentioned case of C_4 photosynthesis. In each case,
key innovations are viewed as a disproportionate impact on the evolutionary
trajectory of the clades that possess them, an argument that seems particularly
persuasive in cases such as the mammalian hypocone and C_4 photosynthesis,
when the new features have arisen many different times.[24]

Such scenarios are amenable to testing within a phylogenetic framework,
and more rigorous tests of key innovation scenarios seek to establish which
characters are responsible for increased diversity, and for access to new resources or to new habitats. Many purported key innovations are unique, however, and all too often, they turn out to be some feature of particular interest

to the investigator (pure coincidence, of course). As with any ecological or evolutionary change, the success of a key innovation will generally depend on the ecological and environmental context. The same change in different clades may play out differently for reasons having nothing to do with any intrinsic advantage. Key innovations can also be tested by performance in the new clade. This is easiest (at least in principle, if not always in practice) with living species, but there are several cases where proposed key innovations have been tested in fossils—for example, with the bite of saber-toothed cats.[25]

Just as key innovations are involved in only a portion of adaptive radiations, adaptive radiations cover only part of a greater variety of evolutionary diversifications. I take the same delight in Darwin's finches and in the silverswords of the Hawaiian Islands as other evolutionary biologists. But the term *key innovation* has been applied so broadly, and invoked almost ritualistically, that the great diversity of styles of evolutionary diversification has been obscured. Since the early 1990s many evolutionary biologists have critiqued the adaptive radiation framework and proposed a variety of other classifications of diversifications. As with Darwin's finches and the silverswords, many classic adaptive radiations have been identified on islands or archipelagos, yet a recent paper found that the morphological diversity of birds associated with island adaptive radiations was not higher than in comparably aged radiations on continents or among widely distributed birds. In fact, the authors concluded that the pattern of disparity provided no support for claims that key innovations, ecological opportunity, or other factors were necessary to explain island diversifications. More generally, relatively few adaptive radiations generate evolutionary novelty if one relies upon a homology-based definition.[26]

Another proposal was to define an innovation as "an apomorphy [that] possesses some specific characteristic or downstream consequence of interest." Donoghue and Sanderson purposely left this definition vague, but like many claims for "key innovations," this makes an innovation whatever an investigator is interested in, precluding any general statements about innovation.[27]

Higher Taxa and Macroevolution

The Cambrian Explosion of animals and the diversification of Mesozoic and Cenozoic mammals, and of flowering plants, have been described as adaptive radiations, just on a vaster scale and over longer intervals of time than the adaptive radiations of Galapagos finches. In the last chapter I used the work of Kemp and of Valentine to illustrate the interest among paleontologists in the origin of higher taxa as a metric of evolutionary novelty. Their work was part of a set of robust studies of large-scale, macroevolutionary patterns in the fossil record, where adaptive radiations and key innovations have been only part of a larger suite of questions about the origin of higher taxa. The success of an innovation depends upon whether ecological opportunities were available for exploitation. Geneticist Sewall Wright suggested in 1982 that the rate of novelty might be essentially constant through time, but that the success of these novelties (as innovations) varied because of ecological opportunity.[28]

The fossil record allows such hypotheses to be tested. An early example was a project during graduate school to better understand the sudden appearances of so many new clades and unusual morphologies during the Cambrian and Ordovician radiations. All well-skeletonized marine phyla appeared during this interval, as well as (at that time) 54 of 56 classes and 152 of about 235 marine orders. While the extent of early Paleozoic innovation was quite clear, so was the absence of new phyla and the appearance of only a few classes during the Mesozoic after the end-Permian mass extinction—the most extensive of the Phanerozoic, removing upwards of 85 percent of all marine species. One possible explanation was that developmental patterning systems had "hardened" to such a point that extensive morphological changes were no longer possible by Triassic time. But the other possibility was that ecological constraints limited success.[29]

We assumed if a new family of marine invertebrates appeared but just as quickly went extinct, this could serve as a marker for an attempted novelty. I compared their number during the Cambrian and the Ordovician diversifications to an equivalent length of time in the Mesozoic after the end-Permian mass extinction. To translate this idea into the framework used in this book, I was testing the idea that pervasive novelties were still possible in the Triassic and would show up in the fossil record but fail to become innovations. It is worth emphasizing that we were employing short-lived families as a useful proxy for morphological novelty.

The results showed roughly similar numbers of short-lived families during the early Paleozoic and Mesozoic, despite the significant differences in the production of phyla, classes, and orders. We found no evidence for a burst of novelty in the aftermath of the end-Permian mass extinction. This pattern suggested that the generation of the sorts of phenotypic novelties characterizing higher taxa might be decoupled from the generation of new families.

No one would be daft enough do anything like this today, and not just because Linnean categories are not necessarily monophyletic clades. Today's databases of the fossil record would allow this study to be done in an afternoon. Beyond this, however, our conception of both the developmental and the ecological hypotheses was incredibly primitive. Few of the groups we studied were monophyletic clades. Rather, they were a phylogenetic hodgepodge. While later analyses of diversity have suggested that monophyletic taxa might not always be a requirement for such studies, they are certainly preferable. But more importantly, the test we set up had no unique predictions to distinguish between the ecological and developmental hypotheses.

Similar studies during the 1980s and 1990s examined whether specific environments were more likely to generate new orders of marine organisms, again using higher taxa as proxies for evolutionary novelty. Some work suggested that new orders were more likely to have arisen close to shore, rather than in deep waters, before they diffused across continental shelves. During the post-Paleozoic, orders of arthropods, crinoids, sea urchins (echinoids), various mollusks, and even bryozoans commonly first appeared in shallow waters, although whether these new morphologies originated in shallow waters or simply first became sufficiently common there to be visible as fossils is difficult to establish. Moreover, some of these studies identified decoupling between novelties among orders and novelties among smaller groups such as families and genera. This decoupling rendered improbable claims that patterns at lower levels, such as adaptive radiations within a genus, can be smoothly extrapolated to larger scales.[30]

By 1990 four discrete views of novelty and higher taxa had emerged: Many evolutionary biologists followed Darwin and others in viewing the apparent distinctiveness of higher taxa as an artifact of the extinction of intermediate forms. Under this scenario a more complete fossil record would reveal gradual evolutionary transitions between apparently distinct forms, so no special mechanisms were required to explain their origin. As described in the previous chapter, beginning in the 1970s rebuttals challenged the efficacy of this claim by more thoroughly documenting the fossil record and the relationship between clades of Cambrian trilobites, Devonian early amphibians, and birds, among others.[31]

Others accepted that phyla, classes, and orders represented biologically distinctive architectures and viewed their origin and persistence as legitimate evolutionary questions. This group encompassed divergent views about the implications of higher taxa for novelty. Adherents to microevolutionary per-spectives did not view novelty as distinct from adaptation. Some have argued that extinction of intermediate forms exacerbated the apparent distinctiveness of some clades and adopted a microevolutionary approach as a sort of fallback. In contrast, many paleontologists, and later, members of the evo-devo com-munity, recognized higher taxa as indicators of an interesting evolutionary problem. The third view accepted the origin of higher taxa as a legitimate evo-lutionary problem reflecting the structure of ecological and environmental opportunity at the time, essentially expanding Dobzhansky's argument from the 1930s, although greater in scope. Finally, the long history of seeing the unique architectures characterized as body plans as reflecting developmental mechanisms has grown since 1990. Riedl, Gould, Valentine, and others men-tioned in chapter 2 viewed the novelties associated with new body plans as fundamentally different from later novelties in animals. Many novelties can be identified within each of the many different animal body plans, but for Riedl only those that become fully integrated with other characters are components of the body plan. Müller and Wagner emphasize that this collective nature of body plan architectures means that it is irrelevant whether the integration is functional or developmental.[32]

Two key points emerge from this work: First, higher taxa as Linnean taxo-nomic units are too coarse. With the spread of phylogenetic analyses, the ritual invocation of groups defined by Linnean taxonomic ranks (phyla, classes, orders) has largely disappeared, as focus shifted to the dynamics of monophyletic clades. Where phyla or classes are employed, it is as a marker for a monophy-letic clade. Large clades do reflect successful innovation. Most large clades contain novelties, often many of them, but as will become evident in chapter 5, we must focus on the novelties themselves, while the clades to which they contribute are the focus of studies of innovation. Second, despite the apparent controversy over architectural body plans, few biologists dispute their existence. As developmental morphologist Brian Hall succinctly pointed out: "The need is not to regard the *Baupläne* as the idealized, unchangeable abstrac-tion of Geoffroy, but to treat it as a fundamental, structural, phylogenetic organization that is constantly being maintained and preserved because of how ontogeny is structured." Discussions should focus on the origins, persis-tence, and evolutionary significance of such distinctive architectures.[33]

The new phylogenetic methods encouraged biologists to examine evolu-tionary diversifications with new eyes. Key innovation models often assume a single change is responsible for subsequent events, whether an increasing number of species, a broader range of morphology, or ecological dominance. Plotting characters on trees increased concerns about the validity of "key innovation" hypotheses. Suites of novel characters and accompanying adaptations may be spread across multiple nodes, and novel characters often create opportunities for other changes. Phylogenetic trees are necessarily calibrated against time, and this requires a means to assess the ages of branching points. But such calibration allows greater interrogation of the environmental context in which diversifications have happened.

Contextual information such as climate change or other environmental effects, or ecological changes such as the introduction of a predator or movement into a new region, can be incorporated into phylogenetic studies to identify confluences, or points where environmental changes interact with new characters to impact diversification rates, and synnovations, a neologism of synergy and innovation, for interacting combinations of traits that may drive diversification. This approach helps identify specific loci at which important events occur, whether the solidification of a novelty or changes in species diversification rates. I will use a derivative of this approach in chapter 5 and later in the book.[34]

Do Behavioral Changes Drive Novelty?

If the Darwin-Mayr view of novelty is a new structure with a new function, which comes first? The structure or the function? Many plants and animals will grow in different ways depending on the conditions. Saplings near established trees bend away into the sun, and the size and shape of leaves may vary depending upon the intensity of sunlight. One tropical African butterfly produces very different color morphs in wet and dry seasons, while in social insects, the caste of a developing embryo will depend on the nutrition with which it is supplied. These are examples of phenotypic plasticity: the ability to generate different phenotypes depending upon environmental conditions, and a ready source of variation upon which evolution can act. Changing environmental conditions may favor one phenotype over another, leading to subsequent genetic changes. Particularly in animals, some changes may begin with shifts in behavior, at least to start, rather than being encoded in the form of an organism. For example, when home delivery of milk was common, great tits in the United Kingdom discovered a fine source of liquid protein available in the early morning by pecking through shiny pieces of paper atop glass jars. Similarly, in 1953, a Japanese macaque named Imo began washing sweet potatoes before eating them, a behavior that soon spread through her troop of monkeys. Three years later Imo started throwing mixtures of wheat and sand on the water, then scooping up the floating wheat. These cultural innovations persisted in Imo's band long after she died.[35]

Plasticity allows species to develop new behaviors or new morphologies, induced by particular environmental conditions, without needing to abandon already-successful phenotypes. It has been examined in hundreds of species, fueling questions by animal behaviorists about the circumstances generating new behaviors, and whether they are more common in larger-brained animals than among, say, earthworms. I am a bit suspicious of such studies, if only because birds and macaques are seen as individuals by animal behaviorists, while invertebrates such as earthworms and snails are seen as collectives, so a particularly creative earthworm is unlikely to be recognized.[36]

Ecologist Mary Jane West-Eberhard has done extensive work on the evolutionary impact of interactions between organisms and their environment. Her seminal 2003 book exhaustively documented how plasticity in development allows organisms to respond to shifting environmental conditions, and how this may facilitate evolutionary novelty. West-Eberhard focused on organisms sensitive to an environmental change or a mutation that induces a phenotypic (but not yet genetically based) response. If the new variant is favored, a process of genetic accommodation may change expression of the environmentally induced trait. In this way behavioral and other environmentally induced changes gradually come under genetic and developmental control, leading to evolutionarily stable novelties.[37]

West-Eberhard defines novelty as "a discrete phenotypic trait that is new in composition or context of expression relative to established ancestral traits." But as with other similar definitions, it is difficult to distinguish novelty from adaptation. Indeed, she has no interest in making such a distinction, for she argues that novelty is a result of environmentally induced changes in an organism, producing "retrospective gradualism": the apparent rapid appearance of a morphological novelty that masks an underlying more-continuous transformation.[38]

How might a variable response to a new environment translate into a stable evolutionary novelty? If plastic responses are more efficacious than the standard responses, selection may favor mutations that incorporate them into the genome, so that they occur even in the absence of an external environmental stimulus. Subsequent adaptive changes can integrate the new phenotype into development. Two features of this scenario are worth emphasizing. First, the environment triggers the expression of potentially useful variants. Second, developmental processes essentially mask the phenotypic variability until the environment changes.[39]

The involvement of plasticity in the origin of tetrapods from fish was suggested by a study of the living bony fish *Polypterus*. The shape of *Polypterus* is reminiscent of Devonian fish that were ancestral to the earliest tetrapods, and like them it can use its fins to move on land. Specimens of *Polypterus* raised on land show modifications to the pectoral girdle, the bones supporting the pectoral fins (which become forelimbs in tetrapods), which are like those in Devonian early tetrapods such as *Eusthenopteron*. Thus, it is plausible that the deep transformation from fins to limbs at the base of tetrapods began with initial plasticity in behavior, allowing primitive locomotion on land, which induced changes to bones in the pelvic girdle. Genetic accommodation stabilized these changes. A similar scenario has been advanced for the evolution of the shoulder girdle in early birds.[40]

Plasticity-first evolution has been investigated experimentally with the North American spadefoot toad (*Spea*). *Spea* has both a normal "omnivore" phenotype and a second, "carnivore," morph with different behavior and mouthparts, adapted to feeding on fairy shrimp and other tadpoles in rapidly drying ponds. The carnivore morph represents a novelty among toads, and the facultative switch in behavior leading to this morph has been taken as emblematic of more fundamental novelties. We return to the possible role of plasticity-first evolution in the origins of novelty in chapter 8.[41]

New environments can be stressful to organisms, and these stresses can even be reflected in plasticity-induced novelty in types of cells. Cell types are a fundamental novelty and have been essential for many innovations in the history of life. A characteristic element of eutherian, or placental, mammals (the dominant clade since the end of the dinosaurs) is the implantation of a developing embryo into the wall of the uterus, leading to formation of the placenta to provide nutrition. But implantation provokes a stress response among the cells of the uterus. The decidual cell appears to have formed to manage the inflammation and ensure that pregnancy continues to term. In contrast, the same signal in an opossum triggers birth of the young. This led Günter Wagner and his colleagues to develop a model of stress-induced evolutionary novelty, where activation of a stress response is a form of phenotypic plasticity that can lead to the origin of new characters, in this case the decidual stomal cell type, and not simply adaptive changes in existing cell types. Support for this hypothesis has come from studies of ricefish, a small fish of Southeast Asia with some species that inhabit rice paddies. The female of these species broods its eggs externally, and while it does not have a placenta, external brooding involves a plug of material with connections into the body. The generation of this plug triggers a stress response involving some of the same genes activated in mammalian pregnancy. In the evolution of eyes, light may have been the stressor triggering expression of opsin genes, which are related to those for melatonin production. The proteins that form the lens, known as lens crystallins, evolved to protect mitochondria from damage due to oxygen. Taken together, these studies support the involvement of stress responses as a source of phenotypic plasticity to generate evolutionary novelty. In later chapters I will provide further evidence supporting behavioral changes and morphological adjustments as a driving force for new novelties.[42]

Combinatorics Opens Possibilities

When Andrew Balducci moved his family fruit and vegetable shop from Brooklyn to Greenwich Village in lower Manhattan after World War II, he wanted to stock as many different types of produce as would sell. Rather than simply seeing what sold, if Balducci had known the number of fruits and vegetables required by the most frequently used recipes of his customers, he would have been able to predict not just what would sell but roughly in what abundance. This illustrates the power of combinatorics. The number of recipes you can cook in your kitchen depends in part on your skill as a cook and on the equipment in your kitchen, but primarily on the number of available ingredients. As more ingredients become available, the number of combinations of ingredients, the repertoire of possible outcomes, increases even more rapidly.

Imagine two kitchens. The first kitchen has 381 ingredients, from almonds to zucchini, while the second kitchen is a bit more modest with only 127 ingredients, just one-third of the larger kitchen (one might think of these as pre-Balducci and post-Balducci). Obviously, the larger larder allows more recipes, but how many more? The difference is surprising. Comparing ingredients against databases of recipes yields two clear results: First, the number of recipes that can be created in the large kitchen is 56,498, but only 597 (1 percent) are possible in the smaller kitchen (a chef would not be constrained by recipes and could create more in both kitchens, but the relative proportions still hold). Second, the complexity of the recipes as measured by the number of ingredients per recipe is also much greater in the larger kitchen—only 8 in the small kitchen but up to 21 in the large one (which I imagine to be a nice Oaxacan mole). The menus of the Middle Ages were limited not by culinary ignorance, but by the absence of whole classes of foods. This is an example of what is known as a combinatorial explosion—as the number of available components increases, the combinations of those components, what might be thought of as the space of possibilities, explodes even more rapidly. So, as Balducci added new items to his market, the cooks of Manhattan could explore an increasingly wide range of culinary space. This is the power of combinatorics, and the principles apply to biology as well as culture and technology.[43]

Here is a more straightforward example. With 10 different building blocks, construction of 1013 combinations is possible; with 20, 1,048,576; 30, 1,073,741,793; and 40, 1,099,511,627,775. This applies to language (the number of letters in a language, the number of phonemes, and the number of words), sudoku and similar games, poker (the number of different hands of five-card stud), technology (the number of components), the growth of economies (a "product space" is discussed in the next chapter), ecological communities (the number of functionally different species), the complexity of plants and animals (the number of different cell types available), and even the scope of mathematics.[44]

Combinatorics raises an important issue. If we envision evolution as exploring a space of possibilities, a legitimate response to questions about novelty and innovation is that a steady, gradual addition of components, whether cilantro and peppers to your kitchen, words in a language, or genes to a genome, will inevitably lead to novelty. Arthropods seem to represent a combinatoric explosion, even before insects are evolved. They have endlessly specialized appendages into mouthparts or structures for swimming, walking, and flying, added gills for respiration and sprouted horns. This supports the intuition of some participants in the Modern Synthesis that the gradual accretion of new genes and new adaptations would inevitably generate novelties. In some cases, an increase in the number of components and thus the number of combinations may be all that is necessary.

But combinatoric approaches to novelty also encompasses scenarios in which new entities are constructed by combinations of existing entities rather than simple accretion. This is one of the most significant sources for novelties, ranging from the origin of new genes via the fusion of preexisting genes to the co-option of genes or structures for one function to serve a different function, and horizontal transfer of genes between often distantly related species. The incorporation of an α-proteobacterium to form mitochondria in early eukaryotic cells and cyanobacteria to form photosynthetic chloroplasts represent two of the more consequential combinations in the history of life. Human cultural evolution may be dominantly driven by combinatorics, from incorporating new foods into a cuisine to borrowing words from another language (*algebra* from the Arabic *al-jabr*, for example) and borrowing technology and ideas from different cultures. Evolutionary biologist Todd Oakley has suggested the general term *fusions* to capture combinatoric means to form evolutionary novelties.[45]

Short sequences of RNA have long served as a powerful model to understand the power of combinatorics for generating evolutionary novelty. The next chapter examines the combinatoric space produced by an RNA sequence of 20 nucleotides, which is adjacent to 20 other sequences. These short sequences of RNA form a vast combinatoric space of sequences, each differing by just a single change. Here novelties are RNA configurations that are difficult to access. But the idea can be extended to DNA, RNA, and protein sequences in general.[46]

Plant and animal development combines different patterning processes, which generate stripes, tubes, cavities, invaginations, and related structures. Combining stripes with invagination can produce a tube, for example, such as the notochord of chordates. Various combinations of these components would yield the panoply of development, with some new combinations generating small changes from an ancestral form, as with changes in butterfly wing patterns, while others might yield greater novelties, such as the eye bar or the proboscis of *Tullimonstrum* described in the introduction to this chapter. The fullest exploration of combinatoric novelty has been with studies of the evolution of new cusps on mammalian molars, which result from the interaction of developmental fields. These cusps lack the genetic individualization required by the definitions of Wagner. Combinatorics can play an important role in novelty at the level of the DNA sequence. Cryptic or hidden variation can allow a lineage to build up considerable unexpressed variation, which can be exploited when confronting new challenges or opportunities. This sort of combinatorics is just how economist Brian Arthur envisions the evolution of technology, via combinations of what he calls "primitives" to yield new technologies. Google (or Alphabet, whichever) follows a similar process in allowing employees to devote part of their time to their own projects, thus continually generating variation hidden from the world beyond the Googleplex.[47]

Major Evolutionary Transitions

If Günter Wagner's conception of evolutionary novelty centers on individuation of new characters, then the ultimate extension of this approach is the origin of new evolutionary individuals. At the simplest level, an evolutionary individual is an entity subject to natural selection. In law and economics, corporations are discrete entities, with an economic and legal existence beyond that of a single proprietor or a limited partnership of individuals. This idea is not new. The Catholic Church and colleges and universities have been recognized as legal individuals since the Early Modern period, and both profit-making and non-profit corporations are recognized as legal and economic entities.

In 1995 evolutionary biologists John Maynard Smith and Eörs Szathmáry proposed eight major evolutionary transitions in the history of life, each of which changed the packaging of information and incorporated previously separate entities into a larger whole, generating a new reproductive individual with a distinct evolutionary identity. The book was recommended to me by a colleague (admittedly somewhat given to hyperbole), who described it as the most important twentieth-century book on evolution.

The first transitions identified by Maynard Smith and Szathmáry were three associated with the origin of life: (1) the incorporation of independently rep-licating molecules (DNA, RNA, protein) into compartments to form the first cells; (2) packaging of DNA into chromosomes; and (3) the transition from RNA as both an enzyme and a transmitter of genetic information to DNA, and the origin of the genetic code (which transcribes DNA into RNA, and then translates RNA into proteins). The next four events each significantly increase complexity: (4) the origin of the eukaryotic cell by incorporation of bacterial and archaeal cells as symbionts; (5) a transition from asexual clones (as in bacteria) to sexual reproduction; (6) complex cellular differentiation in plants, fungi, and animals; and (7) combining separate individuals into colonies, with some previously independent forms yielding their ability to reproduce. Ants, bees, and other social insects are the exemplar of this transition, but it has oc-curred in other clades as well. Finally, the eighth transition in their scheme is from primate societies to human societies with the evolution of language.[48]

Each transition involves the generation of a new evolutionary entity upon which selection can operate. With the origin of the eukaryotic cell, for exam-ple, selection can act upon the cell as well as formerly free-living symbiotic constituents: the chloroplast and mitochondrion. The emphasis on individu-ation and the formation of new entities capable of reproduction and evolution is consistent with the homology- and character-based definitions of novelty. Each of these evolutionary transitions also involves an increase in the division of labor among the components of the new entity, and thus the prospect for conflict—for example, between symbionts and their host eukaryotic cell. After each transition the new evolutionary individuals must develop means of regulating and moderating such conflicts. Moreover, the transitions reflect a trend of increasing complexity and centralized control (if, for the moment, one puts aside the very real challenges of defining complexity). Maynard Smith and Szathmáry focus on changes in information flow as a primary feature of these transitions, to the exclusion of any consideration of their ecological impacts (they were not unaware of the impacts, but those simply were not central to their concerns). In the framework developed here, Maynard Smith and Szathmáry focused on the novelties and neglected the innovations. Finally, these transitions involve what Maynard Smith and Szathmáry described as "contingent irreversibility," meaning that once the transitions have solidified, it is difficult or impossible to retreat. They are an evolutionary ratchet, a feature of particularly significant episodes of novelty and innovation.[49]

The idea of major evolutionary transitions sparked great controversy among evolutionary biologists and philosophers of science. One virtue of the proposal, however, was the focused attention on a discrete suite of events that have had profound impacts on the course of life on Earth. By any measure, they represent evolutionary novelties associated with transformative innovations. The question taken up in later chapters is whether these events can be encompassed within a more general framework for novelty and innovation, or whether they represent singular events in the history of life.[50]

The information-centric approach to major transitions of Maynard Smith and Szathmáry spurred two alternative views from paleontologists, one focused on ecology, the other on metabolic power, as well as a more comprehensive integration of views of evolutionary transitions. A significant lacuna from Maynard Smith and Szathmáry's argument was ecology, how organisms make a living and interact with other organisms. The ecological scope of organisms has expanded through the past 3.5 billion years, and six transformative expansions of ecological complexity, or megatrajectories, were proposed in 2000 by paleontologists Andy Knoll and Richard Bambach: (1) prebiotic metabolism; (2) prokaryotic diversification; (3) the diversification of unicellular eukaryotes; (4) aquatic multicellularity of animals and red, brown, and green algae; (5) the invasion of land by plants, animals, and fungi; and (6) the evolution of intelligence. These six events identify ecological expansions in the possibility space for evolutionary innovations.[51]

For example, the fourth transition, to aquatic multicellularity, generated ecological networks with longer pathways and additional levels; larger "packages" of biomass; an increase in the maximum size of individuals to tens of meters or even over a hundred meters; and expansion of the physical, chemical, and other interactions between life and environment. In each case an increase in ecospace was followed by an adaptive expansion, as species explored the new opportunities. But ultimately there is a limit to the ecospace available with each transition that limits further expansion. Subsequent transitions are not inevitable, they are a contingent historical effect with the property of "contingent irreversibility" that Maynard Smith and Szathmáry described.

For Gary Vermeij, a paleontologist at the University of California at Davis, the struggle to acquire nutrients and energy, what he describes as metabolic power, has driven the evolutionary transformations. He sees species with greater capacity to acquire and utilize energy replacing those with lower capacity. His argument is worth quoting at some length:

> From time to time in Earth history, physiologically favorable circumstances together with biochemical and organizational innovations, triggered larger-scale increases in the control that living things wielded over each other and over the supply of raw material on which life depends. . . . Those species whose individuals, either alone or in coalitions, transformed the greatest quantities came to dominate and therefore to exercise effective control of other, less energy-intensive life forms. Such expansions, including episodes centered on the Cambrian explosion some 540 Ma and the Middle Cretaceous renewal about 120 Ma, involved cascades of feedbacks which collectively brought about profound changes in the metabolism and complexity of the natural economy. Growth eventually stopped, or at least slowed down, as potential directions of improvement became ever more constrained by functionally conflicting requirements. . . . Conservatism became the hallmark of economically dominant life forms just as it had already done among the subordinates on the biosphere's economic margin.[52]

Vermeij's argument echoes the earlier work of anthropologist Leslie White, who tied increased human social complexity to technology that expanded energy capture or efficiency. For Vermeij, the success of innovations involves positive feedbacks to both metabolic activity and ecological complexity. By "control that living things wielded over each other," Vermeij means the energy and power available to species of ecological assemblages; a valuable, if somewhat unique perspective. Use of a common currency allows comparisons of groups across the evolutionary tree and through time.

The information-centric approach of Maynard Smith and Szathmáry, the macroecological transformations posited by Knoll and Bambach, and the bioenergetic focus of Vermeij each propose punctuations to the expansion of evolutionary opportunities. They differ in focus, however, with Maynard Smith and Szathmáry primarily concerned with novelties, while the megatrajectories and Vermeij's argument address how novelties play out as evolutionary innovations. Vermeij explicitly ties the success or failure of novelties to ecological and environmental conditions: "Novelties still arose, but the prevalence of negative feedbacks often prevented their establishment until rare and unpredictable events, such as mass extinctions temporarily disturbed the economic status quo."[53]

The chemistry of Earth's oceans and atmosphere has experienced two major transitions, each setting ecological and evolutionary boundaries on what life might accomplish. I think of these as three very different Earths, with each transition between Earths mediated, at least in part, by biological innovations. The first began with the origin of the Earth, followed by the oceans, an atmosphere, and eventually life. Just when life evolved is uncertain, but the presence of crystals of the mineral zircon indicates that oceans may have been present by 4.4 billion years ago, because zircons are formed by subduction of crustal materials during plate tectonics. The oceans and atmosphere lacked oxygen until the Great Oxidation Event (GOE) about 2.4 billion years ago, which generated the second Earth. Photosynthesis that generated oxygen as a byproduct had evolved earlier, but oxygen began to gradually build up in the oceans and atmosphere only during the GOE, which marks the transition to the second Earth. Oxygen was present in the oceans and atmosphere between 2.4 billion years ago and 600 million years ago, but in very low amounts, varying temporally and geographically. The oceans were largely low oxygen and sulfur rich, a noxious brew inimical to most forms of life. Some 800 to 600 million years ago a second oxygen revolution occurred, with oxygen slowly and fitfully building up in the oceans and atmosphere. Oxygen likely did not reach something approaching modern levels (21 percent oxygen in the atmosphere) until about 400 million years ago and the third and final Earth. Even after this time, episodic anoxic events occurred, sometimes associated with biodiversity crises.[54]

The available energy was far more limited in the first and second Earths than it is today. To a first approximation life runs on energy from the sun, energy that is captured by photosynthesis through bacteria, which in turn provide food to the rest of the atmosphere (recall that plants and various algae captured photosynthetic bacteria as chloroplasts). Early forms of photosynthesis did not produce oxygen, nor did they generate as much energy. Thus, the first Earth was driven largely by chemical interactions and anoxygenic photosynthesis, with just a fraction of the energy available to life today, by one estimate perhaps 100 times less. Each subsequent transition stepped up the energy available.[55]

A continuing supply of energy from photosynthesis presupposes a recycling of nutrients and other critical resources (otherwise limiting nutrients would wind up sequestered on the seafloor). Thus "recycling revolutions" must have occurred through which geochemical cycles and microbial processes progressively increased the efficiency of recycling. From more efficient capture of energy from the sun by photosynthesis to worms breaking down leaf litter in forests, a complex web now ensures the resupply of critical resources. Four such "revolutions" have been proposed since the origin of life: an initial reliance upon energy from chemical gradients in simple, prokaryotic ecosystems. Perhaps 2.7 billion years ago this was succeeded by novelties associated with oxygen-producing photosynthesis and the increases in available oxygen during the GOE. The third revolution encompassed greater biological complexity, including the origin of the eukaryotic cell and the diversification of eukaryotes, eventually leading to multicellular eukaryotes and the origin and diversification of animals in the Cambrian Explosion. The advent of human culture is the most recent revolution.[56]

There are obvious overlap and connections among these seven approaches to novelty and innovation. Behavioral shifts and phenotypic plasticity can generate new functions, facilitating selection for new forms, and connecting with both the structure-function approach and the individuation of new characters. Many of the discussions of higher taxa are an outgrowth of Simpsonian adaptive radiations but can be reframed in ways that are congruent with, for example, newly individuated characters contributing to the definition of new clades. Some combinatoric approaches to evolutionary novelty generate considerable hidden or latent variation before it is expressed in a rapid shift in the phenotype. So these scenarios are not silos, although I think they accurately reflect differing epistemic commitments.

Novelty in Human Behavior, Culture, and Technology

No discussion of evolutionary novelty and innovation can be complete without addressing the origins of human culture, from language to the generation of technologies and social complexity. These events have also driven innovations extending to other species through domestication of plants and animals, hunting and fishing, and changes in land use, collectively impacting almost every species on the planet. Human cultural and technological developments also provide insights into other cases of novelty and innovation as well as evolution in general.

Big brains, our capacity for extensive cooperation (prosociality), the cognitive abilities encoded in our brains, and tool-making ability each have had their advocates as the key attribute behind the success of our species. But untangling the causes of our success requires greater subtlety. We are not the only cultural species. Many species exhibit some degree of culture (depending on how culture is defined), but no other species can rely on accumulated learning, institutions, tools, and artifacts as extensively as we have. Our success has exploited two evolutionary systems, genetic and cultural, to generate complex cultural novelties, inventions, innovations, and adaptation.

If paleontologists have much of the fossil record to themselves, human novelties and innovations are addressed by scholars across a vast range of disciplines, from psychology, cognitive science, and anthropology to economics, the history of technology, and even mathematics, through the debate over whether mathematics was discovered or invented. As we saw in the previous chapter, archaeologists are interested in the early human tool kit, and anthropologists have employed this work to explore increases in social complexity. But inventions have been of just as much interest to economic historians and historians of technology. Physical anthropologists, cognitive scientists, philosophers, and others have explored the origin of the brain, while the origin of language has been pursued by linguists, cognitive scientists, philosophers, and others. Some historians have recently extended their remit into "deep history," enveloping these topics. Cultural evolution and its connections to biological evolution have attracted their own fractious coterie of specialists.

This section identifies some major themes of human cultural and technological change that have emerged over recent decades, which address novelties associated with the origins of modern humans, including cumulative culture and language, and the subsequent cultural, social, economic, and technological novelties and innovations. These will bear on the conceptual framework developed in chapter 5. Chapter 9 applies this framework to specific examples of cultural and technological novelty and innovation. One challenge is the disparate range of perspectives. Some scholars approach these issues from an evolutionary perspective, generally some variant of the Modern Synthesis (although often ignorant of the latest work in evolutionary theory). But the link between evolutionary theories of culture and racism as described previously led many to reject evolutionary frameworks in favor of a historical or humanistic framing.

CULTURE

Multiple conceptual frameworks for understanding human behavioral evolution have appeared over the past few decades. The broadest approach is sometimes described as cultural evolution, applying Darwinian selection to cultural phenomena much as Darwin intuited, but shorn of the progressivist views of Spencer, Morgan, and others, and updated to reflect more modern approaches to evolution. Studies of cultural evolution draw parallels to biological evolution, applying selection across a wide range of cultural and social problems, often with the same theoretical techniques, to examine how socially transmitted beliefs, values, and knowledge have influenced behavior.

Memes, sociobiology, genetic determinism, human behavioral ecology, evolutionary psychology, and gene-culture coevolution represent distinct intellectual traditions, each with its own advocates. There is a dizzying diversity, with the affinities of the protagonists often challenging to pin down. In *Sociobiology* Wilson laid out a broad theory of the genetic basis of social behavior from protists to humans. Building on the work of many others, he showed how cooperation could evolve and how selection could favor activities that might benefit related individuals. The real challenge of the book came in the now infamous final chapter, provocatively entitled "Man: From Sociobiology to Sociology," Wilson's effort to synthesize the biological and cultural sciences. Although much of the book was laudable, it was heavily criticized for suggesting that human behavior could be reduced to genetic effects. This controversy has been covered in detail elsewhere, but in the aftermath of the sociobiology debates three distinct research programs emerged to analyze the evolutionary dynamics of human behavior: human behavioral ecology, evolutionary psychology, and gene-culture coevolution, in addition to cultural evolution. These approaches share a common motivation in seeking to understand why humans cooperate with one another. Cooperation is always puzzling, and biologists have devoted careers to scrutinizing it in other species. We do not cooperate with just our kin, but with people to whom we are unrelated (except in the most distant sense). This prosocial behavior is one of our novelties that allows others to propagate, from cooperative hunting to the existence of nation-states.[57]

Evolutionary psychology developed among psychologists during the 1980s in reaction to perceived excesses of sociobiology, principally by those who felt that innate behavioral attributes had been ignored. They emphasized that humans had evolved in very different physical and social environments than we live in today and claimed that many human behaviors are rooted these ancestral conditions. Understanding human behavior and human minds required understanding the problems that confronted Pleistocene hunter-gatherers. Gene-culture coevolution examines the interactions between genetic and cultural evolution and is the most mathematically rigorous of the approaches described here. Digesting milk and milk products as adults required the lactase gene to continue functioning into adulthood. The coevolution of this genetic change with dairy farming is one example of this approach, which has generated numerous models of behavior, learning, and cooperation. Finally, human behavioral ecology grew out of the new ecological models of the 1960s and 1970s, which offered anthropologists new tools to investigate relationships between different environments and human behavior. There was often the assumption that humans would follow optimal strategies, a very different view to that of evolutionary psychologists, for example. Even as this approach waned among ecologists, it persisted in anthropology, but has had little influence on work on novelties or innovation.

This background helps interpret differing perspectives on novelties and innovations. For evolutionary psychologists, the critical novelties are domain-specific cognitive modules that evolved among early humans. The gene-culture coevolution approach closely parallels the transformationist approach of adherents to the Modern Synthesis, as described earlier, aided by the quantitative population genetic models that provide the foundation for the Modern Synthesis. One of the most useful insights from gene-culture coevolution is the idea of niche construction, through which organisms influence their own environment, and thus the future course of their own evolution. The idea is not new. Charles Darwin's last book investigated the impact earthworms have on the soil, and George Gaylord Simpson discusses the process at length. Niche construction will figure prominently in later discussions of both biological and cultural innovation.[58]

How different types of cultural novelties interacted to lead us to where we are is a difficult problem, but an interesting approach was provided by a computer simulation. "Lucky-leap" innovations generated new cultural tools and further opportunities for exploration. The authors of this simulation imagined that the lucky leaps are the result of some sudden insight, hence the name.

Exploring lucky-leap innovations leads to a variety of different tool-kit novelties that could arise. If fishing is the lucky leap, then nets, weirs, and fishhooks might all be different tool-kit innovations, but the generation of tool-kit innovations is dependent upon the initial formation of the main, lucky-leap innovation. The cultural context allows people to recognize that an innovation in one area might be applied to a very different context: a direct analogy innovation. When the magnetron (a high-energy vacuum tube that emits microwaves) was invented in the 1920s, it was not immediately obvious that the technology would lead to a more compact variant, the cavity magnetron, that would be a crucial component to the invention of radar in 1939, nor that the magnetron would lead to the invention of microwave ovens in the 1940s (much less the attendant demise of cooking among a later generation). Finally, recombinations of tools may also yield an innovative product. This scheme of cultural innovation is focused on function, rather than form, so there is no clear distinction between novelty and innovation. The recognition that some innovations may generate the opportunities for other innovations echoes suggestions in biology that some innovations may be particularly generative.[59]

LANGUAGE

Among the communication channels discovered by animals are scents, visual displays by butterflies or peacocks, birdsong, the infrasonic rumblings of elephants, and the songs of humpback whales. Many animal communication systems are innate, rather than learned socially, and whether any of these forms of communication should be regarded as language remains controversial. Certainly, some apes have been taught to communicate, but human language encompasses a suite of features unexploited elsewhere in the animal kingdom. Among the attributes of human language that distinguish it from animal communication systems are the ability to embed thoughts within thoughts, as in "I think, therefore I am", or recursion; talking about things that may not be real, or events in the past or future; and the open-ended, combinatoric aspects of language, which enable the manipulation of ideas about the world and symbolic thought.

When human language arose is controversial. Some trace early language to at earliest the origin of anatomically modern humans some 200,000 years ago, while others have argued that a protolanguage had evolved by 500,000 years ago, with elements of language deriving from our sister apes. Such a deep history permits the possibility that language abilities may have coevolved, to a greater or lesser degree, with changes in the human brain. Other scholars place the origins of language much more recently in human evolution, perhaps less than 50,000 years ago, with language as a largely cultural construct.[60]

This debate over the timing of the origin of language and the extent to which it is genetically determined or socially constructed divides two distinct research traditions. The first is defined by the work of Noam Chomsky, and his computational view of language as underpinned by purportedly universal features, including recursion, hierarchical structuring of words, and unbounded combinations. Chomsky believes that the core features of language evolved for internal, cognitive functions, with the use of language for social communication being secondary and derived. Chomsky and his allies also argue that a computational procedure they term "Merge" is the "key innovation" (in biological terms) that provides the functionality of human language. A Merge operation between two words or between a word and a phrase generates the hierarchical structure of language. But because Merge is either present or not, this scenario rejects an incremental evolution of human language from the communications of apes and our human ancestors. Thus, language is a rapid, essentially saltational novelty. Some adherents to this framework acknowledge that the adoption of language as a form of interpersonal communication may have been more incremental (which could be interpreted as distinguishing novelty from innovation).[61]

Alternatively, language may be a property of populations, to facilitate communication. Language is an extension of earlier communication systems, crafted by natural selection, albeit with enhanced properties enabled by social intelligence, cognition, and intergenerational learning. In this case, while language may have a genetic component associated with the origin of modern humans, it could have greater time depth than under the Chomsky scenario.[62]

There is an enormous variety of approaches to the origins and evolution of human language. But if a comprehensive framework for understanding novelty and innovation is possible, it would need to include, at least in a general sense, language, and the cumulative human culture that it permits. In chapter 9 I will propose that distinguishing the capacity for language from the acquisition of language and the spread of the novelty could be helpful. One of the great challenges in understanding these processes, however, is that much like the origin of life, of eukaryotes, and of animals, we lack an intermediate record of the steps associated with their origins. We may interpret them as more sudden than was the case, and the paucity of evidence certainly encourages speculation. But the evolution of language is linked to the evolution of modern humans, human culture, and to our capacity for symbolic thought.

Underpinning novelty in culture and technology is collective and individual creativity. Whether through individual human inventions or the generation of cumulative culture, from mathematics to music, creativity is essential. There is a rich body of work on human creativity, but building models of human novelty and innovation requires some understanding of the sources of creativity. Psychologist Dean Keith Simonton defines creativity as generating ideas that are both novel and useful. Leonardo da Vinci famously generated many novel ideas, but they remained on pieces of paper until the nineteenth century. Simonton, like Schumpeter, is interested in the *introduction* of something new, which does not require that the innovator has created the innovation. Creator and innovator might be the same (Mozart, Einstein), but often the innovator is realizing someone else's creative notion. While Thomas Edison is often described as America's greatest inventor and is credited with many inventions, perhaps his greatest was the industrial research laboratory, where he integrated the creativity of many inventors.[63]

Simonton follows earlier work by Donald Campbell in advocating an explicitly Darwinian view of creativity, involving blind variation and selective retention. Humans vary greatly in the extent of their creative ability, but one of the more striking results from Simonton's work is that some of the most creative individuals also have the greatest number of failures. The greatest hitters in the Baseball Hall of Fame also have a vast number of strikeouts: Ted Williams hit .344 lifetime, but he struck out 709 times. Inventors almost always have a long history of trials behind them before things come together in the invention for which they are known.

Histories of human creativity often elevate individual inventors as driving forces behind economic progress and human achievement. Such stories will be absent here. Accounts of "heroic inventors" often downplay both the long histories of prior work that provided the foundation for "breakthrough" inventions and the necessary changes to most inventions before they could become widely adopted. Moreover, invention is a frequently collaborative enterprise, with the nature of this collaborative enterprise the most interesting aspect. The Scientific Revolution beginning in the sixteenth century and the Industrial

Revolution of the nineteenth century generated self-sustaining systems for the generation of novelty and innovation, which economist Edmund Phelps described as endogenous or indigenous innovation.[64]

ECONOMICS, FINANCE, AND BUSINESS

Earlier I noted the critical role of technological innovation in economic growth. But technological innovation was not fully incorporated into these models. In other words, new technology was exogenous and unspecified, in contrast to capital accumulation and changes in labor or population size, which were the major focus. Technological progress was essentially a black box. A seminal paper in 1990 recast the role of technological innovation in a way useful for understanding both biological and cultural evolution.

Incorporating technological change within a model of economic growth, or making it endogenous to the model, was the theme of work by economist Paul Romer in his paper "Endogenous Technological Change" in 1990. The only surprise in 2018 when he shared the Nobel Memorial Prize in Economics was that it had taken 28 years for the Royal Swedish Academy of Sciences to recognize Romer's achievement. Ideas were central to his work. Not more capital; not population growth. Ideas. Romer suggested that ideas need to be understood as an economic good with properties that enhance technological progress.[65]

Economic goods are normally thought of as something that one can buy and sell. I can buy a pound of carrots, some butter, maple syrup, and a frying pan to turn them into glazed carrots. Once I have bought the carrots, butter, and syrup they belong to me, and I'm the only one who can use them—they represent private goods. But anyone can use a recipe for glazed carrots. That recipe, like ideas in general, has the property of being *non-rivalrous*. Use of a non-rivalrous good has no impact on someone else using the same good. Economic goods also have a related property reflecting the ease with which the owner of a good can limit someone from using a good, known as *excludability*. Computer code can encapsulate ideas, but copyright, encryption, and other systems can limit use of the code and make it excludable. Excludability is thus a property of both technology and institutions, in this case a legal system for enforcing copyright protections.

The different types of economic goods are illustrated by a two-by-two matrix of excludability and rivalrousness (figure 3.1). Private goods are both rivalrous and excludable and represent what we normally think of as an economic

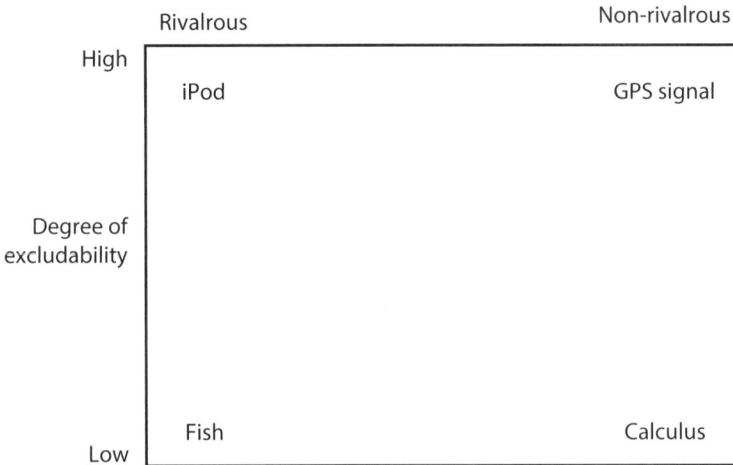

FIGURE 3.1. Classification of economic goods based on excludability and rivalrousness.

good: something easily bought and sold, from food and clothing to surfboards and garbage cans. Public goods, like calculus and math more generally, are both non-rivalrous and non-excludable (lower right-hand corner of the figure). Once Newton and/or Leibniz (partisans take your pick) discovered calculus, anyone interested can solve problems requiring calculus. My solving calculus problems has no impact on you solving calculus problems. Public goods encompass ideas, scientific research, technical know-how, and national defense. As we will see, this endows public goods with interesting properties. Non-rivalrous but excludable goods (upper right corner) are those where members of a club can control who can access them. Computer software is such a club good, as are toll roads, movies on streaming services, and theme parks. Access to club goods can be controlled, which means that money can be made from providing access. Finally, non-excludable but rivalrous goods (lower left corner) include fish in the oceans, clean water, and similar common goods. Garret Hardin's "Tragedy of the Commons" chronicling despoiling of the environment refers to this quadrant.[66]

Ideas, including cultural and technological inventions and innovations, underpin economic growth. But once the idea, blueprint, recipe, or what have you is in hand, the cost of reproduction is essentially nil, so it is difficult to make money from non-rivalrous and non-excludable goods. This explains the existence of copyright, patent law, and other legal machinery that tries to move goods from public to private (and is a principal reason why basic scientific research has been largely funded by governments or foundations since World War II).

Romer's work firmly established the importance of technological growth, with ideas as central components, although he later came to appreciate that rules and social norms can play equally important roles in economic growth. The critical issue here is how different types of economic goods contribute to novelty, innovation, and growth. I have argued that evolutionary biologists need to consider biological analogues to economic goods, and particularly the equivalents to public goods. One of the best examples of a biological public good involves the pangenome. For many microbial species, a few individuals are not a sufficient sample of the genome. In fact, as one samples more individuals new genes are discovered, even though the total size of the genome remains about the same. Such species are described as have a pangenome divided into core genes, which are present in all members of the species, and a large array of accessory genes, which are swapped in and out of individual genomes depending on environmental and ecological requirements. These accessory genes are public goods, available to all members of the species. Cumulative human culture, including language and other fundamental human novelties and innovations, is also driven by ideas, and thus by public goods. Romer's insight about the significance of public goods will play a prominent role in later chapters.[67]

The idea of different economic goods is most fully developed in a capitalist society, but economic approaches to novelty and innovation extend more broadly, and not just through the idea of economic goods. Significant transitions in social and economic activity occurred from the Neolithic agricultural revolution onward through agriculture, the rise of states, and other innovations in economic institutions.[68]

Over the past few decades, one could not walk through an airport bookshop without seeing the latest innovation title promising riches and renown. One of the most prominent of these authors was the late Clayton Christensen of the Harvard Business School. His *Innovator's Dilemma* (1997) articulated his model of disruptive innovation. So, when you hear claims that the newest widget is going to "disrupt" books, newspapers, and vacuum cleaners, you have Christensen to thank.[69]

Christensen asked how is it that smaller, upstart companies come to displace older, more established, often vastly larger, and certainly better financed companies. The telephone disrupted the telegraph, and email the telephone.

Personal computers replaced mainframe and mini-computers (and thus Apple and Dell Computers disrupted IBM and DEC (Who? Exactly.)) only to have laptops and then tablets replace desktop personal computers. Christensen identified similar patterns in the steel industry, with mini mills replacing integrated steel mills, and discount retailers replacing Macy's, Gimbel's, and other large department stores (and then Walmart replacing Kmart—one of Christensen's examples). Most new firms fail, of course, but Christensen was more interested in the failures of large firms than the disappearance of start-ups. Similar questions have been asked in biology, where big, established clades are described as having "incumbent advantage." How do new clades (bivalves, sea urchins, or mammals) replace older, more established clades (trilobites or dinosaurs)?

Mass extinctions and other biotic crises have often been the decisive factor ending the inherent advantage incumbents have over challengers, but Christensen developed a more nuanced (or at least, less catastrophic) model. Most firms continue to innovate, generating incremental or sustaining innovations (what biologists call adaptations), but this drives up costs and often adds features that many customers do not need (see the latest smartphone). The largest profits come from the high end of the market, and these are the customers pursued by companies as they get larger. But pursuing the high end creates an opportunity for low-cost products or services. These disruptive innovations initially take market share that the incumbents had essentially abandoned. As the new companies gain converts, often opening a market for goods or services that had been too expensive for many to afford, their quality improves. By the time incumbents notice the new firms, the latter have already captured a considerable share of the market, and it is difficult for the incumbents to respond. And then the cycle repeats. One of Christensen's insights is that the managers of the incumbents are not stupid, they are making rational decisions throughout the cycle, and in fact may have competitive advantage over the disruptors for much of the cycle. The world simply looks different to managers of incumbents than it does to the upstart firms.

Christensen had a huge following in the business community as a lecturer and consultant (I once gave a talk just before Christensen, and to say the audience was eager for me to depart somewhat understates the situation). Everyone wanted to be a disruptor—that was the route to success (and the path to inevitable destruction in this deterministic scenario, but that would be a problem for someone else). The challenge for a big firm was to somehow incorporate disruption as part of a business plan. Venture capitalists loved

Christensen's ideas because it gave them a way of assessing the likelihood that new firms might be successful: Who they were going to disrupt? Schumpeter and his "gales of creative destruction" were not mentioned in *The Innovator's Dilemma*, but Christensen later acknowledged the relationship between disruptive innovations and Schumpeter's work. Christensen was aware of this, in fact, and in his lectures and writing he never claimed to have provided a general theory of innovation. He was always clear that he was focused on a particular sort of innovation and how it impacted business. Some of his enthusiasts, however, were often more expansive in their claims.

The challenge to any work that relies upon case studies (including this book) lies in the reliability and generality of the cases. As readers of Christensen's work will know, Harvard historian Jill Lepore published a pointed critique of disruptive innovation in the *New Yorker*, challenging many of the analyses for selective use of data and a failure of proper historical analysis. Christensen extended his approach beyond business to universities and other enterprises, which Lepore criticized for failing to appreciate the differing goals of such institutions. A valid part of Lepore's critique was as much about innovation as a popular religion as it was about Christensen's work, for she questioned whether the concept has any predictive power, either for business or for society more broadly.[70]

Bounding Novelty and Innovation

Novelty has been variously described in ways indistinguishable from adaptive change to a more restrictive class of evolutionary changes. Underpinning such attempts is the assumption that a single definition can serve to encompass the phenomena of novelty. Among the challenges in sculpting a general, broad, and inclusive but operationally useful definition of novelty is that many are situational: tied to the focus of individual research programs. Philosophers Ingo Brigandt and Alan Love accepted "definitional diversity" in their efforts to understand how biologists have used the term. This section sketches some criteria for bounding definitions of novelty and innovation.[71]

One qualitative but well-thought-out approach to identifying novelty and innovation articulates four criteria: First, a novelty trait is a qualitative departure from the ancestral trait *in complexity or information* (more on this below). Second, the novelty is associated with increased performance in the current niche or utilization of a new niche; third, the new trait has positive fitness and so is likely to spread. Finally, the novelty should significantly impact population growth and eventual evolutionary diversification of the clade containing the novelty. In this scheme there is a close relationship between a new structure and a new function, and it distinguish three phases: inventions, novelty, and innovations. Inventions encompass biological mutations and recombinations, and new ideas, discoveries, and devices in culture. These may combine to generate qualitatively new phenotypic traits within individuals. Finally, innovations have an impact on populations of a species, ecological communities, and eventually on evolution, roughly equivalent to my use of the term *innovation*.[72]

Advocates of this approach realize that most adaptive changes do not lead to innovation, and some significant changes may not lead immediately to innovation. Consequently, they recognize a fourfold classification of changes. This distinguishes innovations that impact populations or ecosystems from those with no immediate effect, which they term *latent* innovations. If they are not actively selected against, latent innovations may persist, possibly becoming innovations following later inventions or favorable changes in the environment. Novel traits may generate innovations by enhancing the performance of the group within the same way of life, or by generating a new niche for the group (a niche innovation). The latter are essentially equivalent to "key innovations," allowing access to new ecological opportunities. One challenge to this scheme is the difficulty of establishing what constitutes a "qualitative phenotypic change." Applying the term *invention* in a biological context could provide aid and comfort to creationists, and in any case such events are already commonly described as evolutionary novelties, so in biological and cultural contexts I will use the term *novelty* and reserve the term *invention* for discussions of technology.

There is no necessary reason why there should be a single, universal definition of novelty, even within biology. New genes are not necessarily expressed by morphological or behavioral changes, and novel genomic architectures or developmental processes might have different origins than novel behaviors. More importantly, the formation of a developmental novelty, for example, does not necessarily entail a morphological or behavioral novelty. The reverse is true as well, that behavioral novelties need not generate morphological or genomic novelties. Some novelties might be more consequential, opening new evolutionary opportunities, while others may primarily exploit these opportunities.

To some, *any* evolutionary change might be considered novel or innovative. Yet few evolutionary biologists would advocate that every nucleotide change in a DNA sequence is an evolutionary novelty, nor any shift in the color patterns of a butterfly. Any of these changes might serve as the basis for selection and adaptive change, and some might drift via random processes to greater frequency.

The need to bound novelty is captured by Günter Wagner's emphasis on novelty as the origin of a new character, rather than any change in character state. But distinguishing characters from character states can be fraught. Among the eighteen whisks currently in my kitchen, four are traditional balloon whisks of different sizes, but these are just different character states of a basic whisk and cannot be considered novelties.* I also have two French whisks, which might be confused with a balloon whisk but have thicker wires and a more compressed profile. My flat whisks, for preparing roux and sauces, represent an adaptation of the basic whisk. But what of my coil, dough, and twirl whisks (the latter for frothing hot chocolate)? Here things become more complicated, with the character analysis dependent upon the scope of the study. If I were studying tools in general, I might lump all of these together as whisks, since they are easily distinguished from hammers, socket wrenches, or a Pulaski axe. But if kitchen tools were the focus, a more intensive study of whisks might be helpful.

Even if agreement could be forged on what we mean by *novelty* and *innovation*, the boundaries of the concepts may remain vague or fuzzy. This was the conclusion US Supreme Court Justice Potter Stewart famously reached in 1964 when attempting to define obscenity. If asked to decide how bald is bald (10 hairs? 100 hairs? 500?), how old is old or how many wheat grains are needed for a pile of wheat, no answer is possible. If one grain of wheat is not a heap, then neither are two, nor three . . . Philosophers know this as the sorites paradox, credited to the fourth-century BCE philosopher Eubulides. If there is a gradation of incrementally different values, among which, taken separately, no discontinuity appears, how can we say when we have a heap of wheat? Just because one cannot discriminate whether adding an additional grain to 225 grains of wheat makes a heap does not mean that a pile of 1 million grains of wheat is not a heap.[73]

Such ontological vagueness is meat to philosophers. The arguments of saltationists were based largely on empirical data rather than theoretical arguments, but their emphasis on discontinuities as fundamental to evolutionary novelty is the easiest way to avoid the sorites paradox. If gaps are real and a

---

* Except when justifying a purchase to my wife.

product of the generation of novelty, the difficulties in defining novelty disappear. This is the approach taken by those who define novelty as the individuation of new characters. Others rely on qualitative and subjective descriptions of novelty and innovation as sufficient, if fuzzy. Pigliucci opts for a fuzzier definition, with an argument that the only practical approach to understanding novelty is for biologists to agree on a limited number of "model systems," much like the fruit fly *Drosophila melanogaster* and the frog *Xenopus laevis*, to serve as model organisms for molecular and developmental biology. Concentrated effort on a few systems of novelty could lead to greater understanding, and he offers eight potential model systems: the turtle shell, bird wings and feathers, the vertebrate eye, the mandible–ear transition, heart, the ability of some great tits in England to open milk bottles, flowers, and Hox genes. The list is focused on vertebrates, ignoring the incredible novelties among invertebrates (arthropod limbs and cephalopod eyes, for example). In-depth investigation of individual cases is undoubtedly illuminating, and many of the systems included in Pigliucci's list will appear in later chapters. But as I demonstrate in this book, an accumulation of case studies alone does not allow us to resolve the broader questions about novelty and innovation posed in chapter 1.[74]

Novelty versus Innovation

In opening this chapter, I used the evolutionary history of grasses to illustrate the problem that macroevolutionary lags posed for the Simpson-Mayr adaptive radiation model of novelty. If novelties capitalize on ecological opportunities, why do lags arise between establishment of a novelty and its ecological or evolutionary success? This conundrum reflects the curious decoupling between evolutionary and developmental biologists' approaches to novelty and innovation recounted in the past two chapters. For Mayr, Dobzhansky, Simpson, and other transformationists discussed in chapter 2, the origin of novel morphologies was a far less important problem than the ecological opportunities that controlled their success. Ecological opportunity could come from many different sources: key innovations, migration to new areas, or extinction of other species. Whatever their origins, ecological opportunities were the driving force behind evolutionary novelty. Schumpeter had a similar view of economic growth, emphasizing innovation rather than the supply of inventions. Earlier I quoted from Schumpeter: "Add as many mailcoaches as you please, you will never get a railroad by so doing." Schumpeter, like Mayr and Simpson, had little interest in what generated inventions. For Schumpeter the factors that allowed innovations to thrive underpinned economic growth. Indeed, his emphasis on entrepreneurship rather than inventors reflects his understanding that it was not necessarily the inventor, but rather the entrepreneur, who was the more significant. In both biology and economics, the supply of new variants was assumed to be sufficiently regular that it could be safely disregarded.[75]

It may already be apparent that by defining novelty as the origin of new homologous characters, developmental biologists such as Müller and G. Wagner adopted the opposite view from the evolutionary biologists of the 1950s and 1960s. Müller and Wagner were not unique, for the central thrust of evolutionary developmental biology has been a far greater concern with the mechanisms responsible for variation in general and specifically for the sources of the variation leading to novelty. At the same time, developmental biologists have cared little for how or whether such novelties found homes in ecological communities. I suppose developmental biologists would argue, with some justification, that success is not really their problem—ecologists and evolutionary biologists must sort that out.

The discussions of novelty and innovation in the past two chapters have differed in what would constitute an adequate account of the problem. In other words, what are we trying to accomplish by defining novelty and innovation, and what are the criteria by which someone could determine whether we had succeeded? In mathematics the math must work, and chemical equations must balance. The philosophical equivalent for conceptual problems is criteria of explanatory adequacy.[76]

I can distill the following seven criteria for a unified conceptual framework, which are based on criteria initially articulated by philosopher Alan Love. First, an adequate explanatory scheme must encompass both the generation of new forms or structures (e.g., the vertebrate eye, or a new gene) and the origin of new functions. Second, any adequate scheme should provide insights into ecological and evolutionary success without conflating them with the supply of new variants. Third, as novelties arise at all levels of biological organization, from components of genes and the structure of a genome to morphology and behavior, a sufficient conceptual framework should address the generation of new variation across both spatial and temporal hierarchies. Fourth, useful definitions of novelty and innovation must distinguish one from the other, and each from regular processes of evolutionary adaptation. Fifth, evolutionary novelties and innovations are often recognized retrospectively. But a more useful definition would allow the prospective recognition of a novelty, and, ideally, the circumstances under which the novelty could contribute to an innovation. The converse of this is the ability to recognize innovations that did not occur, and why. As biologist Jonathan Losos observed, any adequate account of adaptive radiations would need to recognize the adaptive radiations that did *not* occur, as well as those that did. Sixth, can studies of a novelty or innovation be generalized to other cases, or do they apply only to unique or particular circumstances? Are investigations of examples informative about deeper nodes on a phylogenetic tree? In other words, can research on developmental mechanisms of living arthropods, mollusks, and vertebrates be informative about events associated with the early evolution of animals (much deeper on an evolutionary tree), or can studies of modern languages reveal something of the processes associated with the initial evolution of languages? Finally, for the purposes of this book, definitions of novelty and innovation need to apply across the domains of biology, culture, and technology. This is a test of the possibilities specified in chapter 1. This criterion assumes a level of generality in the definitions that would apply to episodes across each of the domains, and thus that there are some conceptual similarities to these episodes greater than the metaphorical claims that have been made previously.[77]

There is one last and possibly unattainable criterion. An ideal explanatory scheme would provide appropriate metrics, allowing novelties to be distinguished from adaptations. Such quantification would resolve endless debates over aspects of novelty, but I fear that issues of novelty may lie within the remit of the sorites paradox, rendering hopeless any thought of such quantification.

Articulating these criteria makes evident that my goal is not to reduce all novelty and innovation to a single underlying theory (what philosophers describe as "theory reduction"), but rather to acknowledge the rich variety of patterns and processes, while still articulating a common conceptual framework.[78]

4

Evolutionary Spaces

THE TOPOLOGY OF INNOVATION

IN "THE LIBRARY of Babel," Argentine author Jorge Luis Borges (1899–1986) imagined the universe as a vast array of hexagonal rooms, doors on two walls connecting adjacent rooms, while the other four walls were each laddered with five shelves, each shelf with 32 books. Each book has 410 pages with 40 lines to a page and 80 characters per line. The texts are composed of the 25 symbols of Spanish: 22 letters, period, comma, and space, in all languages that can be constructed with those symbols, including languages that have never existed. Although the books are in random order and no two books are exactly alike, those who live in the Library believe that it contains all possible variants of books of the appropriate dimensions, from segments of Tolstoy's *War and Peace* to books of nothing but the letter *A*, or *m*. But the vast Library also contains all possible versions of *War and Peace* with one misspelling, two misspellings, and presumably a condensed, 410-page version.[1]

Borges's Library is a vast space of opportunities. It contains biographies of everyone who has ever lived, millions of biographies of people who have never lived, the histories and false histories of everything that has ever happened or might happen sometime in the future, as well as novels written and as yet unwritten. Fake news indeed. But most of the books in the Library are pure gibberish in any language. Somewhere in Borges's Library is a copy of this book and a copy of this book in the style of Cormac McCarthy (fairly easy to distinguish, I should think).[2]

The ideas behind the "Library" have attracted philosophers, mathematicians, and even movie makers. In the 2014 film *Interstellar*, Matthew McConaughey, as Cooper, is trapped in a universe based on Borges's Library. The Benedictine monastery in Umberto Eco's novel *The Name of the Rose* has a labyrinthine library and a blind monk: Jorge of Burgos. Although Borges could not have been thinking of the multiverse view of reality when he conceived of the Library of Babel, it incorporates the multiverse in one entity.[3]

Borges's Library is a vast array of combinations of letters, books, and rooms. Since the size of each book is fixed, and each employs only 25 symbols, the size of the Library is finite (if a bit on the vast side). Umberto Eco once wrote that the size of the Library was irrelevant to Borges's purpose. Eco was doubtless correct, but the combinatoric possibilities of the Library are too great for mathematicians to abjure, and the extent of the Library has been determined rather exactly: $25^{1,312,000}$ books, or in the more familiar notation $10^{1,834,097}$ (10 followed by 1,834,097 zeros—more zeros than the number of characters in a book in Borges's Library). Some $10^{1,834,013}$ universes the size of ours would be required just to contain the books of Borges's Library, and vastly more universes would be required if the books were arrayed in Borges's hexagonal rooms. So, the Library is vastly large.[4]

Genes, proteins, genomes, and even the shapes of plants and animals involve just the sort of enormous possibility spaces envisioned by Borges. Instead of letters, books, and rooms, these evolutionary spaces are combinations of nucleotides to make DNA and RNA, strings of amino acids folded into proteins, genes to make genomes, individuals to form populations, and species assembled into ecological communities. Different cell types combine to form tissues, and tissues form organs, which contribute to the modules and sub-modules that form the bodies of most animals. Combinatoric modularity also applies to DNA and to the regulatory structures governing the activities of DNA. Combinations underlie large spaces of cultural and technological possibilities. Cross-cultural analyses have been based on a wealth of ethnographic and archaeological data from cultures around the world—for example, in the Human Relations Area Files at Yale University. The variables in the files, from types of houses, whether the wife moves in with the husband's family (patrilocal) or the reverse (matrilocal), and styles of hunting or agriculture, help define cultural spaces. Technologies are combinations or recombinations of libraries of subcomponents.

Gottfried Leibniz's doctoral dissertation *Dissertatio de arte combinatoria* is a foundational study of combinatorics, building on work of René Descartes to form a universal representation of ideas, with more complex ideas constructed from simpler components. Philosopher Daniel Dennett drew on Leibniz and Borges's Library as the inspiration for his Library of Mendel, composed of all possible DNA genomes. Although the structure and organization of the

Library of Mendel is unclear, later we examine how Dennett employs it to illustrate the idea of the vast space of biological possibilities in which we find embedded the sparse distribution of actual genomes, an idea that has been explored by many evolutionary biologists. Since the 1960s paleontologists have investigated the shape spaces (morphospaces) of fossil groups as disparate as clams, snails, and early plants. Using various measurements of cats, dogs, thylacines, Jurassic beavers, and other mammals we could define a shape space. To track evolution across such a space we can project an evolutionary tree into a morphospace, connecting different species based on their evolutionary relationships, or phylogeny. As with Borges's Library, where the space of possible books is vastly greater than the space of books that have been written, so the spaces of possible trilobites, genomes, or proteins are vastly greater than the spaces of those that have ever evolved.[5]

Simpson envisioned evolutionary novelty as the exploitation of new adaptive spaces. While his approach to adaptive spaces was largely metaphorical, he used the fossil record of horses and other vertebrates to illustrate his ideas. More recent work has extended Simpson's ideas to the combinatorics of many other evolutionary spaces and explored the roles they may play in novelty and innovation. Here we explore the diversity of evolutionary spaces, from fitness and adaptive spaces to others defined by sequences of DNA, RNA, or proteins, and morphology, and some of the questions they raise: How do novelties, or different sorts of novelties, arise within spaces? Do novelties generate new spaces or change how spaces are explored by evolution? Do cultural and technological spaces have different properties than biological spaces, or can we identify general properties that span spaces in all three domains? Evolutionary spaces can differ greatly in their properties, which influences our views of whether a novelty involves a discontinuity. Dobzhansky's and Simpson's discussions of the adaptive possibilities provided by the structure of ecological communities within adaptive spaces spawned related approaches to defining novelty, as discussed in chapter 3. Adaptive landscapes incorporate the standard Modern Synthesis views that discontinuities, if real, represent the structure of the adaptive landscape. This chapter discusses visualization of adaptive landscapes beyond the attractive imagery of Simpson's "adaptive zones."

Two combinatoric approaches for recognizing novelties and innovations have emerged from ideas about spaces: Stuart Kauffman's influential idea of the "adjacent possible," and a view that embedded within a sequence space of DNA, RNA, or proteins are large networks where most changes have no effect on the resulting phenotype, just as most small changes make little difference to the overall impact of *Pride and Prejudice*. Particularly significant has been analysis of short strings of RNA to illustrate movement through a space driven by neutral changes, rather than driven by selective advantage as in adaptive landscapes. In contrast to the ideas explored in chapter 3, in these RNA spaces novelties are defined not as a new structure or new homologous characters but rather as possibilities that are hard to access and are "generative" of further evolutionary change. This approach combines Simpson's insights about new adaptive zones with a neutral, microevolutionary approach, while also revealing that although we live in a flat, Euclidian world, few evolutionary spaces are Euclidean.

But not all possibilities are real, and not all paths are possible. Arthropods may have many appendages—3 pairs of legs in insects, 4 pairs in spiders, and over 100 pairs in millipedes—but vertebrates are limited to a paltry two pairs (raising a question Gould might have asked regarding *It's a Wonderful Life*: Just how was Clarence going to get his wings?). Limitations to evolution in general, and novelty more specifically, arise because much of any space may not be accessible. Part of a space may represent proteins that could never form, or inside-out snails. Such issues of accessibility and evolutionary constraints suggest that the variety of possible novelties may be limited.

Other questions have been posed about evolutionary spaces: How does evolution explore these various evolutionary spaces? For example, what is the distribution of actual genomes within the spaces of all possible genomes? Is there just one space of all genomes, or all proteins, or are the spaces somehow disconnected from each other? Are all possible regions of the space biologically realistic, or are there limitations? How do these spaces grow and change? Given the vast sizes of the biological spaces, how much of protein space or genomic space has been explored during the four billion years of life on Earth?

From Mount Fuji to Kansas: Evolutionary Landscapes

There is a glitch in Borges's Library as a metaphor for evolution, and this glitch involves the central point of this chapter. Searching Borges's Library is fruitless. Imagine you chance upon a book beginning: "It is a truth universally acknowledged, that a single kumquat in possession of good knife, must be in want of a grape." You might think "Hmm, well that's interesting, if a bit off," unless you knew that somewhere in Borges's Library was a copy of Jane Austen's *Pride and Prejudice* beginning "It is a truth universally acknowledged, that a single man in possession of good fortune, must be in want of a wife," with the rest of the book just as Austen wrote it. But starting with an almost perfect copy of *Pride and Prejudice* in Borges's Library, there is no way to find a more accurate copy, much less the perfect copy. In fact, it is impossible to find any information in the Library: a catalog of the Library is, in fact, the Library. The only way to discover the perfect copy of *Pride and Prejudice* or *Hamlet* is to stumble upon it during a random walk through $10^{1,834,097}$ books. The Library is a vast space of possibilities, but a bizarre one since there is no logical relationship between adjacent books. In the absence of any correlation in how books are shelved or placed in a room, no search strategy allows one to find progressively more perfect copies of *Hamlet*. Willfully misunderstanding the power of natural selection, creationists often employ a similar argument for the improbability of finding DNA sequences for an eye. Natural selection is defter than creationists.

A more useful metaphor for evolutionary combinatorics is found in a short article titled "A New Puzzle" by Lewis Carroll, which appeared in the British magazine *Vanity Fair* in 1879. Carroll's game connected words of the same length, each differing by a single letter (nonsense four-letter words were excluded). Carroll linked *head* and *tail* by the path *head–heal–teal–tell–tall–tail*.* A simple amusement for Carroll, but unlike the Library of Babel, Carroll's space is naturally organized so that each string of four letters is adjacent to all permutations that differ from it by only a single letter. Thus, *head* is adjacent to *heal*, as well as *heat* and *bead,* and *hxad, hejd,* and *vead.* Single changes, single mutations, enable easy navigation through the space. The 25 adjacencies for each of the four letters generate 100 four-letter strings connected to *head,* with each of these four-letter strings connected to 99 other four-letter strings by changes in a single letter. Together these adjacencies define a space of all the 456,979 (26^4) possible four-letter strings, but unlike Borges's Library, navigating Carroll's space is straightforward.[6]

A vast network of versions of *Pride and Prejudice,* each differing by a single word, illustrates two key principles about spaces. First, assuming we had some basis for choosing different versions (an evaluation function), one could move through the network finding progressively "better" versions of the book, possibly ascending toward a greatly improved version. Somewhere in this vast, multidimensional space is a peak with Jane Austen's version at the summit. Correlated spaces, such as in Carroll's game, allow exploration of progressively

* Carroll helpfully added: "It is perhaps needless to state that the links should be English words, such as might be used in good society."

better versions of a word, a gene, or a book. But there is a second intriguing aspect of this vast network. Embedded within the large network would be a vast, interconnected subnet of versions of *Pride and Prejudice* of equivalent quality, where the differences were of no import. This represents a neutral network, as described earlier for RNA.

Carroll's space represents what is technically known as a finite discrete state space, which simply means that the size of the space is limited (finite) and each word is discrete. Replace strings of four letters with longer sequences of nucleotides or amino acids, and we have vast sequence spaces of DNA, RNA, or proteins. Any possible sequence can be represented as a point in the appropriate sequence space, and one can trace an evolutionary trajectory of single mutations through a sequence space. Vast spaces of biological, cultural, and technological possibility are far more than a metaphor; since the 1930s they have been a critical analytical tool for understanding the patterns and processes of evolution. As the use of evolutionary spaces has expanded, they have made concrete the range of possibilities open to evolution and sparked continuing arguments over the nature of novelty and limits to evolution.[7]

Simple Landscapes

British evolutionary biologist John Maynard Smith recalled Lewis Carroll's word puzzle in defining a protein space in 1970. Proteins are composed of strings of 20 different amino acids. The different amino acids of the protein represent the letters, and Maynard Smith described how one protein could change into another via a path of changes to single amino acids (reflecting changes in the underlying DNA sequence). Just as the four-letter strings in Carroll's puzzle must represent real words, in Maynard Smith's version natural selection ensures that each preserved change produces a viable protein.[8]

Proteins, DNA, and RNA can each form sequence spaces. Consider a DNA sequence of six nucleotides: AATCGC. A single mutation might alter the third nucleotide from a T to an A, thus moving to an adjacent position in the space: AAACGC. In a sequence space, each sequence of nucleotides is surrounded by adjacent sequences differing by only one nucleotide. The four nucleotides of DNA (adenine, A; cytosine, C; guanine, G; and thymine, T) allow four alternative states for each of the 100 nucleotide positions in the sequence. One hundred million different DNA nucleotide sequences can be constructed for a string of 100 base pairs. Since a typical protein is about 500 amino acids long,

Maynard Smith's protein space encompasses 20^{500} different proteins, a far larger number than the number of atoms in the universe.

There is a structure to sequence spaces that influences evolution. Many of these sequences could not form viable proteins, while other sequences that might form useful proteins are simply inaccessible from most starting points. Just as some changes to a word make no sense (at least in English: *cast* to *tast*, for example) some nucleotide sequence changes might inhibit proper processing of the nucleotide sequence or folding of the resulting protein. Even small changes in a sequence can generate substantial changes in the folded structure of a protein and in its biological properties: the stability of a protein at different temperatures, or how well it connects to other proteins. Other sequences may be completely inaccessible by single-letter substitutions from a starting sequence because they are surrounded in sequence space by deleterious sequences. Consequently, the accessible universe of strings of 100 DNA nucleotides is far less than 100 million. Understanding these patterns of constraint and accessibility requires the addition of another dimension to convert the sequence space into a fitness or adaptive landscape.

Adding the evaluation function to the space of versions of *Pride and Prejudice*, each separated by changes in a single word, generates an adaptive space. Evolutionary biologists depict the relationship between combinations of genes with an additional dimension of fitness depicting how well various gene combinations are adapted to the environment. The simplest such landscape has a single, conical peak like Mount Fuji, representing a small set of well-adapted gene combinations surrounded by progressively less well-adapted combinations. In a Mount Fuji landscape, the peak represents the optimal solution to an evolutionary problem. Each step "uphill" improves fitness or adaptation (depending on whether fitness or adaptation is being measured) and will eventually lead to the peak. More complicated landscapes might have high peaks and ridges of gene combinations, each of which is well adapted, interspersed with valleys and sinkholes of poorly adapted combinations of genes. For those of a practical bent, protein spaces are essential for directed evolution of proteins to solve problems that biological evolution never had to confront, such as digesting damaging chemicals.[9]

These evolutionary landscapes were invented by geneticist Sewall Wright in 1932 as a graphical means of explaining his mathematical work to other biologists. Wright (1889–1988) was a founder of population genetics, and in 1931 he published a major, but highly quantitative, treatment of evolution. When he was invited to present a talk at the Sixth International Congress of Genetics the following year, Wright introduced graphic representations to depict different evolutionary scenarios, from a single population atop a fitness peak to the random movements of a population under drift and the division of a species into many local populations. He was reasonably sure that few of his colleagues would be able to follow the math, so he gave them pictures. Wright depicted networks of combinations of gene variants within a population. If A, B, C, and D are genes, with several different variants (alleles) for each gene, one individual might have the combination $A_1B_6C_4D_3$ (in other words, allele 1 of gene A, allele 6 of gene B, etc.), while another individual might have the combination $A_4B_6C_2D_5$. These combinations form a network upon which evolution can operate. Adding a dimension representing a value for fitness, adaptedness, or complexity converts the space from a surface into a landscape. Different combinations of genes will differ in their fitness, and across generations recombination allows a population to explore the space ($A_1B_6C_4D_3$ might become $A_1B_3C_4D_3$ and then $A_1B_3C_4D_5$). An organism with genes that function harmoniously in a particular environment would be well adapted and tend to produce more offspring (high fitness). In a Mount Fuji landscape, selection will favor gene combinations close to the peak. Gene combinations that move downslope would contribute fewer individuals to the next generation, on average, and be disfavored. Wright also recognized that changes in the environment would alter the optimal combination of genes and thus the evolutionary topography: peaks might become valleys. As Wright expected, the graphic depiction proved an enduringly popular metaphor in evolutionary biology.[10]

In the audience for Wright's 1932 talk was geneticist Theodosius Dobzhansky, who immediately recognized the power of Wright's approach. Dobzhansky (1900–1975) was born in the Ukraine and trained in Kiev and Saint Petersburg before moving to the United States in 1927 to work with geneticist Thomas Hunt Morgan at Columbia University. Recognizing that the space of possible gene combinations was immense, far larger than the collection of gene combinations represented by living organisms, Dobzhansky incorporated ecological context to represent an adaptive landscape as a sort of topographic map, with the contours corresponding to the adaptive values of combinations of genes. Unlike Wright's, Dobzhansky's landscapes included multiple species, each with its own adaptive peak. This and subsequent editions of his *Genetics and the Origin of Species* were influential in spreading Wright's ideas.[11]

When Simpson published *Tempo and Mode in Evolution* in 1944, he incorporated the adaptive landscapes of Wright and Dobzhansky but focused on the organism's phenotype rather than the underlying genotype. Thus, the landscape became a space of phenotypic possibilities. A peak on Simpson's adaptive landscape represented a morphology highly adapted to a particular environment, rather than the underlying combination of genes, and he extended Wright's adaptive landscapes to the longer timescales of paleontology. He illustrated the evolution of horses from Eocene browsers to Miocene grazers with adaptive peaks for each feeding strategy. The illustrations begin with the adaptive peak for browsing occupied and that for grazers open. The peaks move together through time, with the adaptive peak for grazing becoming occupied during the Miocene. He expanded these ideas into the concept of adaptive zones: environmental spaces with favored ecological and evolutionary strategies.[12]

In using the terms *fitness landscape* and *adaptive landscape* I have muddled two distinct issues. To clarify, a fitness landscape involves values for reproductive success (fitness) of combinations of genes or individual genomes, as in Wright's original 1932 example. In contrast, adaptive landscapes represent the mean of fitness values of all the individuals in a population. The peak in an adaptive landscape generally represents the relative fit of the population ensemble to a particular environment. One underlying assumption in much work on adaptive landscapes is that sufficient evolutionary variation exists for selection to drive a population uphill toward an adaptive peak. When population sizes are large, this is likely to be true, but the range of available variation is finite, and nature may be unable to supply what the environment demands. At smaller population sizes the available variation upon which selection can act may be limited, or may arise sometime after the opportunity forms, influencing the origins of novelties.[13]

Wright acknowledged that landscapes are a highly simplified, heuristic model. The more genes, the greater the number of dimensions in the space. The intuitive appeal of Wright's adaptive landscape metaphor also contained a trap. Multidimensional evolutionary interactions are far too complex to capture in two or three dimensions. Yet the intuitions generated by such simplified landscapes turn out to be misleading about evolutionary dynamics. The behavior of populations in a multidimensional network is often much different from the expectations from examining the two-dimensional models.

Rugged Landscapes

The earliest adaptive landscapes were graphical representations of evolutionary dynamics. Wright, Dobzhansky, and Simpson recognized that adaptive spaces were highly multidimensional, but they trusted that simpler graphic models would capture the complexities of these spaces. But do fitness and adaptive landscapes in high dimensions have the peaks and valleys as depicted by Wright? Or do other patterns emerge? Mount Fuji landscapes capture a single optimal solution, but real species are seeking to simultaneously optimize multiple features, perhaps tooth length, jaw muscles for biting, and coat color. When features are interdependent (technically known as epistasis), the resulting landscape is rugged, with many peaks rather than smooth. Rugged landscapes have many optimal or near-optimal solutions to an evolutionary problem, but imagine trying to pick out the optimal solution from among a jumble of peaks of nearly the same size, separated by deep valleys.

Concerns about rugged landscapes led theoretical biologist Stuart Kauffman to focus on a simpler model. He asked how evolution can optimize fitness when making single mutations. Pursuing this avenue led to some surprising and counterintuitive results about the complexity of organisms and the power of natural selection. His N-K model links the rugged landscapes created by the need to optimize multiple features with interaction among different components, whether DNA sequences or morphology. A key result is that a moderate degree of interaction among components creates high peaks (valuable interactions among components), but as organisms become increasingly complex the size of adaptive peaks on the landscape declines. The Alps dwindle away to Kansas. But the more the landscape looks like Kansas, the harder it is for natural selection to find an adaptive peak, if one even exists. (Kansas does have Mount Bleu, which was briefly if somewhat improbably a ski slope, but the highest point in Kansas is "Mount Sunflower," which is hard to pick out from the surrounding cornfields.) Second, even if a population reaches a peak, selection is not particularly effective at ensuring it remains there.[14]

For Kauffman, novelties arise from what has already been discovered, whether by biology or by culture. This fueled his ideas of the "adjacent possible," as introduced in the previous chapter. Recall the power of combinatorics: as the number of ingredients in a kitchen increases, both the number and complexity of the dishes increases far more rapidly. As new or longer genes are introduced, the adaptive space expands. The adjacent possible is anything one step away from something that already exists, whether that is a gene, a word, a network of ecological interactions, or an idea. The expansion of the adjacent possible will bring some novelties that were once remote closer to the occupied space, increasing the likelihood of discovery. But realizing these possibilities depends on the structure of the adaptive space, and the rugged landscapes generated by high degrees of interaction among components suggested that finding the adjacent novelties could be challenging and maintaining them equally difficult.[15]

Evolutionary landscapes involve so many dimensions that it is often difficult to develop an appropriate intuition. Mount Fuji landscapes provide a simple view of selection driving a population up a peak. Kauffman's rugged landscapes provide a different perspective. Fortunately, physicists have developed mathematical techniques for interrogating systems with many dimensions, and their application to the problem of populations in fitness spaces generates very different metaphors from Wright's or Kauffman's, and from the pictures that populate textbooks on evolutionary biology. Rather than peaks and valleys, empirical landscapes seem to have vast, connected surfaces of high fitness interspersed with deep pits of low fitness. Projected into three dimensions, such a space would look much like a block of Emmental, or "Swiss" cheese. Hence these have been nicknamed "holey landscapes." The surfaces are networks of genomes that share almost identical fitness. Small changes are likely to cause migration across this network, but occasionally a change will generate individuals with low fitness, plunging into the deep pits or holes. Ridges or plateaus connect regions of high fitness, generating a network of genotypes with essentially equivalent or neutral fitness. The problem of populations having to navigate across valleys of low fitness to reach a higher fitness peak disappears in a holey landscape, since there is likely to be a path from any viable organism to another with novel features through this neutral network.[16]

Holey landscapes provide a new framing for the distinction between microevolution and macroevolution. Much of adaptive change within a species may occur in relatively low-dimensional spaces, while macroevolutionary transitions involve higher dimensions. This is sort of the biological equivalent of wormholes in physics, which link separate parts of spacetime, and at least in *Star Trek* allow movement across the universe. One view of evolutionary novelties is that they involve such large jumps through higher-dimensional space or the addition of new dimensions. Another interesting result is the suggestion that the difference between the uncorrelated evolutionary landscapes of the Library of Mendel approach and correlated landscapes like Carroll's word game may disappear in higher dimensions. Ecology and development are missing from these models, and they are as static as those of Wright, but they provide a new way of thinking about evolutionary landscapes and how novelties might arise.[17]

Neutral Landscapes

Let us return to the vast subnet of versions of *Pride and Prejudice* differing by small changes of no consequence. Such neutral networks are deeply embedded in sequence spaces of DNA, RNA, and proteins and have been uncovered in other settings as well. Many changes to a sequence are effectively neutral, either because a change in the DNA sequence does not change the amino acid in the resulting protein or because an amino acid change does not alter the structure or properties of the final protein. While some amino acid positions in a protein are critical to the protein's function, there are other positions that can be occupied by different amino acids with no change in protein function. Therefore, many different DNA sequences generate identical proteins, a property known as many-to-one mapping (or redundancy) between the sequence and the protein. Many-to-one mapping from genotype to phenotype, whether a protein structure, the form of an organism, or the same function produced by seemingly different jaw structures in fish, allows considerable turmoil at one level (the genotype) while maintaining a functional phenotype. This is one of several techniques that effectively reduce the apparent size of evolutionary spaces.[18]

One of the best studied examples of many-to-one mapping with a neutral network is short sequences of RNA. Unlike DNA, RNA is single stranded, allowing a single RNA sequence to fold into a two-dimensional structure as adenine pairs with uracil (which replaces the thymine used by DNA) and guanine pairs with cytosine, and sometimes also with uracil. These flat, two-dimensional structures have folds, loops, and hairpin turns. Although in principle many structures could be produced from a single sequence, few of these structures achieve the lowest possible free energy (by maximizing the number of paired stretches and minimizing the number of loops in the structure). Single changes in the sequence can alter the optimal pairing in the two-dimensional structure, potentially producing very different folded structures. Thus, very small changes in the sequence can generate very divergent final structures. Yet other mutations to the sequence have no impact on either the two- or three-dimensional structure and are effectively neutral. The resulting two-dimensional structure controls further folding into a three-dimensional structure: the final phenotype. The folding of an RNA into a three-dimensional structure provides a clear mapping between the genotype (the RNA sequence), and the phenotype (the folded RNA structure).

In neutral networks, accessibility is not the same as distance, and distances cannot always be measured. Figure 4.1 illustrates four such neutral networks

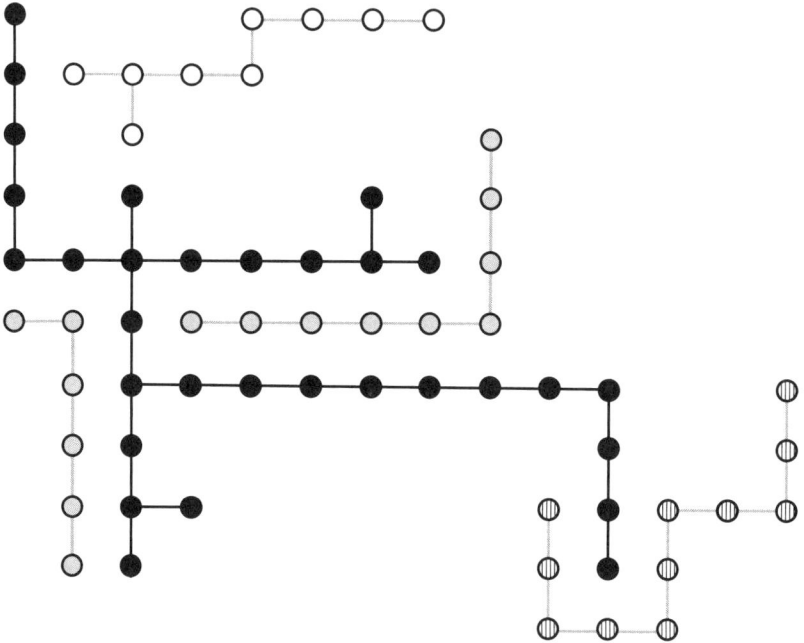

FIGURE 4.1. Neutral RNA networks and transitions to novel phenotypes. A schematic depiction of four neutral networks in sequence space. A sequence located in the upper portion of the light gray network cannot directly access the striped phenotype in the lower right, but because these are neutral networks it could diffuse across the black network until it encounters the striped network. The black network is near the dark gray one, because a random mutation to the gray network has a high probability of moving to the black network because of the long boundary between the adjacent networks. The light gray network (upper left) is not near the striped network because a random mutation to light gray has a low probability of moving to the black network. Although the light gray and striped neutral networks are of similar size, they have no borders in common, so they are not in each other's neighborhood. Thus, the probability of transition from one neutral network to another depends both on the adjacency of the networks and their size. After Fontana (2006).

in sequence space, with each network reflecting a different folded structure. Each node represents a sequence, differing from adjacent nodes by single base mutations. The neutral networks allow sequences to drift across a network within the space. There are barriers, however, to changes between adjacent networks, although these can be overcome via selection. The unoccupied nodes are unviable sequences (for example, because the sequences cannot fold correctly). The size and adjacency of the networks also determine the probability of a shift from one network to the next. Two of the networks share long boundaries (the networks with black nodes and with dark gray nodes) and thus topologically are in the same neighborhood. A random mutation in the network of black nodes is most likely to lead either to another node in the black network, or to a node in the dark gray network. On the other hand, there are two smaller networks, one in light gray nodes to the upper left and another with striped nodes to the lower right. Each of these networks shares relatively few nodes with the black network and none with any other network. Thus, while each of these small networks is accessible from the black network, the probability of achieving this transition is quite low. Crucially, these neutral networks allow evolution via the accumulation of mutations that have no immediate effect on the phenotype but may allow a later mutation to have a substantial evolutionary impact. Some recognize such impactful shifts as a phenotypic novelty.[19]

The difference between distance and accessibility has been illustrated with a map of Europe: France and Germany share a long border. If a random walk through France approaches a national boundary it is likely Germany (or the English Channel, but we will stay on land here), so any random step out of France has a fair likelihood of landing in Germany; thus Germany is highly accessible from France. France surrounds the tiny principality of Monaco, so although France is a neighbor of Monaco, Monaco is not very accessible because France's boundary with Monaco is tiny. Yet any step out of Monaco will land in France. In these examples, accessibility is not the same as distance, and one can have a sense of neighborliness without being able to compute a distance.

Dennett's Library of Mendel as the set of all possible genomes (not just sequences) shares the concern about accessibility that features in the RNA network models. Dennett distinguished between the space of all possible genomes and the space of all possible organisms, defining biological possibility as "x is possible if and only if x is an instantiation of an *accessible* genome or a feature of its phenotypic products." For Dennett his Library of Mendel is a universal design space, and he rejects the possibility of discontinuities within the space because of phylogenetic continuity. Later I will show that Dennett's definition of the Library of Mendel as the space of all possible genomes undercuts his rejection of discontinuities.[20]

Most of the spaces discussed thus far are much closer to Carroll's word space than to Borges's space. Of the 456,979 possible four-letter words available for Carroll's puzzle, only 4030 are valid words in English. Carroll's game would not work if the 4030 words were randomly distributed through the space, since the probability of one valid word being adjacent to another valid word would be small. As a writer Carroll knew that words can be correlated, with *head* adjacent to *heal*, and *heal* to *teal*, and so forth. Another way of looking at this is that the real size of the space for four-letter words is much smaller than the apparent space. The same is true of RNA spaces, so vast neutral networks allow considerable change in sequence with little impact on the folded structure of the RNA. The many-to-one mapping means that while there are 14^{15}, or 1 billion, possible strings of 15 RNA bases, many of these generate identical two-dimensional structures, so there are only 431 different two-dimensional structures. Fully half of the 1 billion different RNA sequences generate only 26 different structures (just 6 percent of the 431 possible 2D structures). Consequently, the sequence space is vastly larger than the phenotypic space.[21]

RNA sequence spaces are the key example used by those who define novelty as an issue of inaccessibility. Mutations in RNA sequences can be easily related to two-dimensional folded structures, making this one of the few cases where it is possible to cleanly map changes at the level of the genotype to the phenotype. Moreover, the presence of neutral networks and many-to-one mapping provides an explanation for the tension between the apparent discontinuous jumps represented by the apparently sudden appearance of some evolutionary novelties and the requirement for reproductive continuity between generations. Drift across through a sequence of neutral mutations in RNA followed by a shift from one network to an adjacent network allows a sudden shift in phenotype through a series of small changes in the underlying genotype. Many-to-one mappings are common; for example, with many different network structures of the genes that control a developing embryo generating a similar outcome. Redundancy is widespread in development to ensure that embryos are relatively robust to changes in the environment. Other examples include jaw structures in fish that have similar functions, and the mappings of word pronunciation to spelling. Andreas Wagner has explored this system and extended it to other evolutionary spaces, arguing this provides a general explanation for novelty (although he uses *innovation*).[22]

That discontinuity in phenotype can mask continuous changes at a lower level does not mean that there is necessarily a continuous path between any two structures in the RNA phenotypic space. Without going into the details, some changes in phenotype are very difficult to achieve, and there may not be an accessible path between two phenotypes. Some students of these RNA spaces see novelty as involving the generation of a phenotype that is not easily achievable, which requires long periods of drift in the genotype. One implication of this view is that selection may not be responsible for the generation of novelties, but only for their fate.[23]

Insights from Evolutionary Spaces

Evolutionary spaces provide insights into patterns and processes. For a Wrightian fitness landscape, gene combinations must generate an organism whose fitness can be evaluated by natural selection. This requires some correspondence between the genes and the nature of the organism. In Dobzhansky's adaptive landscapes, some relationship must exist between the frequencies of gene variants in a population and the overall fitness of the population. But often this is not the case. The gene for sickle cell anemia provides protection against malaria only if its frequency in the population is low enough that many people will have one normal variant of the gene and one sickle cell variant. Moreover, the complexities of development in plants and animals and the effects of the environment on development render any simple mapping between the genotype and the phenotype illusionary. With apologies to Richard Dawkins, the instructions for generating a new animal are not simply read off the genes but emerge from complex interactions between the developing animal and its environment.[24]

Landscapes are not fixed but dynamic, responding to changing conditions. Some changes to adaptive landscapes are due to environmental changes: drought, or a shift in the seasons. But organisms deform evolutionary landscapes, either their own or those of other species. By cutting trees and building dams beavers alter their habitat to suit, which increases their fitness as well as spilling over to constructing habitats for other species. Geneticist Richard Lewontin described such adaptive landscapes as akin to a trampoline. The math for fixed adaptive landscapes is challenging enough even in relatively few dimensions. Add large multidimensional spaces and a rubbery landscape and the math is beyond current abilities, although some interesting advances are being made.[25]

Many high-dimensionality evolutionary spaces have an interesting and rather counterintuitive property that defies our understanding of distances: everything is closer than it appears. In the introduction to this chapter, I emphasized the vast size of evolutionary spaces and the seemingly unbridgeable gaps described by paleontologists measuring morphological disparity. But several factors greatly reduce the effective size of a space. In the rugged and neutral landscapes discussed earlier, there may be a short evolutionary path between different regions of a space. Consider that infernal contraption the Rubik's Cube. For the common three by three version with nine colored squares on each side, there are 489,856,000 possible combinations (the total number of possibilities, or configuration space, is 43,252,003,274, but most of these are redundant). Yet not only can a Rubik's cube be solved relatively quickly (if not by me), in June 2023 Max Park solved one in 3.13 seconds (video is on the web). The configuration space for the Rubik's cube may be vast, but efficient algorithms yield short distances through the space. In July 2010 the "God number," or the maximum number of moves required to solve *any* configuration of the three by three Rubik's Cube was confirmed to be only 20 moves. The trick lies in the algorithm: knowing how to efficiently twist and align the cubes in 20 moves or fewer.[26]

A Rubik's Cube might seem irrelevant for the vast size of nucleotide and protein spaces but illustrates the general property that while the volume of the space increases exponentially as the length grows, the sequences' highly interconnected nature means that distances grow linearly. And the speed in which evolution can find a solution scales to sequence length. Thus, in a multidimensional space the path to the novelty may be shorter than expected. Even if a solution is "close," even under selection, the time needed to search an evolutionary landscape to reach the solution may quite lengthy, so even if selection may not necessarily find the shortest path through a space, with sufficient pressure selection will find an effective path. Despite the vast size of some of the spaces I have discussed, the *effective* size of most spaces is generally far smaller, just as in the Rubik's Cube. So many-to-one mapping reduces the effective size of a space, as do constraints such as the requirement for proteins to fold into effective three-dimensional structures. Later in this chapter I discuss other ways in which evolution has reduced the size of evolutionary spaces.[27]

Finally, and perhaps most importantly, spaces capture one of the most significant divides in understanding the dynamics of biological, cultural, and technological spaces. The idea of search across an evolutionary landscape has been embedded in discussions of evolutionary spaces since they were first used to illustrate the improvement in genes and proteins. Kauffman, with his search "through the adjacent possible," and others such as Dennett and Dawkins have viewed novelty as search through spaces. The metaphor of search has spread from biology into studies of social systems, economics, and business. At the end of this chapter, I will argue that while search is an interesting component of the generation of novelty, often the opportunities exploited by evolutionary innovation did not exist a priori but were constructed by evolution. For me, the metaphor of search misses much of the richness of the construction of novel opportunities.

Spaces for Novelty and Innovation: A Bestiary

Evolutionary spaces have been developed for gene interactions and developmental processes in animals and plants and for the shape and form of organisms. Ecological relationships have proven amenable to the construction of ecological spaces, just as the products produced by a nation's economy help define a product space, which illuminates patterns of economic growth and development. The concept of evolutionary or adaptive spaces has been extended to other areas of culture and technology as well. This section provides an overview of some of the spaces that are useful for thinking about novelty and innovation. These spaces fall into two different classes. Some are correlated spaces like Carroll's puzzle, but others are closer to Borges's Library, uncorrelated spaces where the evolutionary continuity must be provided through mapping to another space, as between genotype and phenotype.

There are other features of sequence spaces of DNA, RNA, and proteins worth describing. Each sequence is adjacent to all permutations that differ from it by only a single change. In the case of protein spaces, the distance between two spaces is the sum of the fewest changes required to transform one protein into the other (or, more accurately, the number of changes required to reach the sequence of each protein from an ancestral sequence). The distance between two proteins found in related organisms serves as a metric of the amount of evolutionary change between them. Such measurements have long been used to build evolutionary trees from molecular sequence data, although the actual evolutionary path between two sequences may be longer than the shortest possible distance.

Moreover, many directions in a protein space may not be readily accessible. As with RNA and DNA, proteins fold into two- and three-dimensional structures, and similar structures are found in proteins with very different functions. For example, the hemoglobin proteins, which transport oxygen in our blood, are constructed from many helical units joined by shorter connecting segments. Maintaining a functional globin protein during migration through a protein space requires preserving these long helical motifs. Some changes in the amino acid sequence preserve the links required to fold the protein into the correct shape, but many others would not. Maynard Smith's protein space was later extended to a space of the variety of folded structures that can be generated from repeats of different structural motifs. As expected, when actual evolutionary spaces are compared against the possible, known repeat proteins form only a small part of this potential space.[28]

Such functional interference between mutations has a critical implication for searching sequence spaces: the order in which mutations arise often influences evolutionary possibility. It is hard to show this experimentally in elephants, but time, money, and people are all that are required to do the experiments in yeast. Evolutionary biologist Joseph Thornton has tracked the evolutionary history of heat shock protein 90 (Hsp90), a protein that ensures proper folding of other proteins. Hsp90 is found across eukaryotes, which allowed Thornton and his team to reconstruct the evolutionary trajectory of the protein and the order in which new mutations were added. Most mutations were dependent upon earlier mutations, so that the order in which mutations arose was critical. Once introduced, genetic changes often became entrenched by subsequent changes. To the extent that this is a general result (and it probably is), the number of viable evolutionary trajectories through an evolutionary space may be many fewer than anticipated from the dimensionality of the space.[29]

The models of novelty in sequence space that I have described thus far center on single base mutations and may be efficient at optimizing the function of a protein, but this is a highly inefficient means of exploring a protein space. More efficient exploration of sequence spaces requires incorporating a broader range of changes in sequence space, such as deletions and rearrangements of larger blocks of sequence, as well as recognizing that DNA sequences can have many functions, including regulatory activities.[30]

The RNA model has been extended to other systems, including regulatory interactions and metabolic networks. A regulatory space encompasses gene regulatory circuits, with neighboring circuits differing by one interaction. Andreas Wagner has argued that, as with the RNA genotype space, many regulatory circuits would produce the same gene expression pattern, forming a vast network of viable regulatory circuits. But if Wagner is correct about the general structure of such networks, neighborhoods of different circuits should express different phenotypes. Thus, novelty arises through single changes in the structure of a regulatory network that generate a different phenotype.[31]

I will have more to say about the networks of interactions among genes that control the development of organisms from egg to embryo to adult in chapter 6, but these interaction networks are embedded in a large space of gene regulatory networks, or GRNs.

A principal function of gene regulatory networks is to control development, ensuring that similar adults result despite variations in food and temperature, and other environmental vagaries during growth. The importance of limiting variation as development proceeds led developmental biologist Conrad Waddington to invert fitness landscapes so that multiple developmental trajectories form valleys leading to a common outcome. He envisions specific cell types (heart muscle, or specific types of neurons, for example) as marbles funneled down the valleys to achieve the desired final cell type. Increased temperature or some other environmental shift might bounce a marble slightly up the side of a metaphorical valley, but the canalization of development would ensure it returned to course. Waddington used his developmental landscape to illustrate how the outcomes of cellular differentiation could be constrained. The topography of the landscape reflects fitness values, although because Waddington inverted the topography, low spots have the highest fitness. Waddington's insights have remained enormously influential for those interested in developmental evolution.[32]

Developmental spaces lack the structure of sequences spaces, but there is an underlying structure nonetheless, and developmental abnormalities provide insights into these regularities. Abnormalities illustrate underlying constraints on the range of potential forms. Additional digits—polydactyly—occur in about one in a thousand live births, common enough that individuals with more than five toes seemingly had a special status among the ancestral Puebloans in Chaco Canyon, New Mexico, over a thousand years ago. The constraint of five digits did not arise with the origin of vertebrate appendages, however, for early tetrapods had a variety of different numbers of digits, and only later did five become fixed. Additional limbs are far less common, indicating a much tighter developmental constraint on limb development.[33]

Counting the number of species or clades provides little information about the elaboration of form. The various animal phyla and classes recognized by biologists are descended from Cuvier's embranchments and recognize gaps and discontinuities. Taxonomic ranks in Linnean classification reflect some intuitive sense of differences in form and thus distance in a shape space. For example, snails within the class Gastropoda are more similar to one another than they are to clams and scallops in the class Bivalvia, much less to the arthropods or echinoderms. Such intuitions may have been sufficient for many decades, but intuition is not data. Spaces built from shapes of organisms were inherent in Simpson's discussion of horses, and their mathematization began with D'Arcy Thompson's book *On Growth and Form.* He laid a grid of points over skulls, bones, and shells. Stretching and deforming the grid (largely by hand) illustrated how one skull could be deformed into another. Some novelties can be incorporated in such a scheme, such as the growth of horns, but the number of novelties that can be captured by deformations of a grid is relatively small. In the early 1960s paleontologist David Raup used Thompson's insights about logarithmically coiled shells to devise three coiling parameters, which jointly define a space for virtually all snails, bivalves, and similar organisms. Within such a morphospace one can plot not only all living mollusks but also the great diversity of extinct fossil forms, and forms that have never existed. Since Raup's pioneering work, this approach has been extended to a variety of plants, animals, and even ice crystals, from needles to snowflakes.[34]

There is a critical difference among these spaces. Raup's original morphospace of coiled shells is a theoretical morphospace, in principle capturing the full potential range of shapes. Many morphospaces, however, are constructed of measurements of fossil or living organisms. In this case statistical techniques reduce the dimensionality of the space to something manageable. This is the second of the three ways of reducing dimensionality, but it is one that we impose on the data. An interesting issue that needs more exploration is whether statistical treatments may in some cases reflect biological constraints on how organisms evolve.

Morphospaces became more useful when they illuminated how organisms occupied such spaces over time. Doing so required a means of measuring distances, akin to using the number of single changes to measure distance in a sequence space, and a phylogenetic tree of the relationships among taxa. Developing these techniques was spurred by reaction to Gould's *Wonderful Life.* *Wonderful Life* was a creative argument that the gaps between the major animal groups arose during the Cambrian Explosion at the dawn of animal life. To support his argument Gould deployed the wonderful fossils of the middle Cambrian Burgess Shale fauna, many of them preserving in fine detail the anatomy of eyes, guts, gills, and other rarely preserved anatomical details. Gould argued that evolutionary theory implicitly (and sometimes explicitly) expected morphological disparity to increase gradually—what he termed "the cone of increasing disparity." But the sudden appearance of so many disparate morphologies in the Burgess Shale required a different view of evolution. Gould's argument was compelling to many (and infuriating to others), but it was not data, and data win. Michael Foote was an undergraduate at Harvard while Gould was writing *Wonderful Life* (and is now a professor at the University of Chicago). Foote's early career as a paleontologist turned Gould's ruminations into science by devising means for measuring the structure of morphological diversity (known in the trade as disparity) and quantifying the differences between species. Since Foote's early work, morphospaces have been developed for tiny microfossils and sea urchins to leaves and trees. Combining morphospaces with adaptive landscapes in a study of the evolution of early vertebrate limbs identified distinct adaptive peaks for fish and terrestrial vertebrates and showed that early vertebrates already occupied the base of the peak for tetrapods. Analyzing patterns of disparity provides a more rigorous basis for the arguments about apparent discontinuities in evolutionary history that were discussed previously.[35]

Some morphospaces have been constructed by measuring both living and fossil representatives of a clade. Others, like Raup's coiling morphospace, are generated by absolute parameters, rather than measurements, and hence can capture organisms that have never existed. These theoretical morphospaces are particularly useful in examining how much of a possibility space is occupied, how this occupation has changed through time, and what parts of a space have never been explored by evolution. As with the sequence spaces described earlier in this chapter, the utilized portion of most theoretical morphospaces is much less than the possible space.

An illustration of the insights into novelty and innovation available from a well-constructed morphospace is provided by a recent study of plants, from their closest relatives among green algae through flowering plants. Each of the major divisions of plants, from the green algae through liverworts and mosses to ferns to gymnosperms (pines and cycads) to flowering plants, occupies a discrete region of the morphospace. Moreover, unlike animals, the increases in phenotypic disparity were episodic, extending into the Cenozoic, and are tied to introductions of novel means of reproduction.[36]

Three important facts have emerged from studies of morphospaces: First, the distribution of organisms within morphospaces is lumpy: clusters of cats are distinct from those of dogs, elephants, or beavers. A general property of most evolutionary spaces is that the distribution of actual genes, proteins, shapes, or phenotypes is clumpy (technically, underdispersed). Such lumpy distributions are one of the more perplexing facts about the living world. If evolution primarily involves many small changes, why do quantitative analyses of morphology find so little evidence of continuous gradation (Gould's "cone of increasing disparity" or Kaufman's "adjacent possible")? The dichotomy between transformationist and saltational approaches to novelty discussed in chapter 2 arises from a division over whether lumpy distributions reflect the disappearance of intermediate forms (which Darwin argued *must* have existed), or whether it is a real evolutionary signal requiring additional explanation. This is fundamentally a disagreement about the lessons we can draw from an evolutionary space.

The lumpy distributions of forms coexist with repeated discoveries of the same solutions to evolutionary problems, from toxins and antifreeze proteins in the blood of fish to overall morphology, and this is the second critical finding. In a space of forms, the marsupial thylacine "wolf" and modern wolves would be close together, as would the Jurassic "beaver" *Castorocauda* and modern beavers, and ichthyosaurs and dolphins. The saber-toothed cat of the La Brea tar pits in Los Angeles would wind up close to the many other clades of saber-toothed cats that evolved during the late Cenozoic, as well as Permian-aged gorgonopsids and even some Cretaceous forms, which are more closely related to modern marsupials like kangaroos than they are to placental mammals. In other words, saber-toothed predators repeatedly evolved very similar morphologies, and similar patterns of convergence have been found in many other clades. Consequently, proximity in morphospace may provide little insight into evolutionary history. This problem is alleviated by projecting evolutionary relationships onto the morphospace. Combining morphology with an evolutionary tree reveals whether taxa close together in morphospace are also close relatives or distant relatives that have converged on a similar shape.

The ubiquity of convergence and repeated, or iterative, evolutions has led several paleontologists to argue that the possibilities available to evolution are limited: similar solutions turn up time and time again. The ability to "spit" venom has evolved independently in three different lineages of cobras, for example. In each case, existing cobra venoms have been modified in the same fashion to increase the resulting pain in their mammalian adversaries. Returning to saber-tooths, long, scimitar-shaped canines evolved in gorgonopsians, or "mammal-like" reptiles, 260 million years ago in the Permian, 40 million years ago in cousins of the cats (the Nimravidae), and multiple times among different groups of cats (family Felidae) beginning about 25 million years ago. The repeated evolution of hyper-carnivores with long canines seems to be a solution to the presence of lots of large prey on which to feast. In the context of evolutionary spaces, this suggests that the "space" for such predators has appeared repeatedly, with different clades taking advantage of the opportunity.[37]

Furthermore, morphospaces are often hierarchically structured, or path dependent, such that accessing a region of a space allows access to new opportunities. This has been shown beautifully for the design space of flower petals, but seems likely to be a general feature of many phenomenological evolutionary spaces. Convergence within such structured evolutionary spaces, as with cobra venoms, is of less significance than that in more distant cases. Moreover, even when evolution generates convergent forms, the underlying genetic or developmental pathways may be very different. In a study of closely related butterfly species where color patterns have evolved to mimic each other, distinct regulatory pathways are involved. The ubiquity of convergence has implications for ongoing disputes over whether evolution is open-ended or deterministic, and thus whether novel forms are truly novel, an issue examined in chapter 10.[38]

Finally, and perhaps most significantly for understanding novelty and innovation, building a morphospace requires some common properties among the organisms. As the disparity between organisms increases, it becomes increasingly difficult to find common characters. That is why a space of a coffee cup and a mushroom makes no sense. Insects, snails, and saber-toothed cats share few common morphological features. So, while I could build a morphospace for insects, one for snails, and another for vertebrate skulls, for example, no single space could incorporate all three clades. And this is a critical clue to the nature of evolutionary novelty: some of the most interesting novelties are those that generate new evolutionary spaces.[39]

Sabre-tooth cats fed almost exclusively on large prey. Morphology reflects how an organism interacts with the environment, including other organisms. Does it burrow? Is it a predator or a grazer? How easily can it move around? Among marine organisms over the past 600 million years predation and burrowing deep into sediment have each become more common. During the Paleozoic most marine gastropods, the group I study, filtered food out of seawater or found it in the muck on the bottom. A few species were parasites, but even fewer species were predators. Yet today about 90 percent of all marine snails are predatory, including some with harpoons and wonderful toxins. (Well, biochemically wonderful. Not so wonderful for people who unsuspectingly pick up the wrong shell.)

These ecological roles permit construction of a variety of ecospaces, which reveal that species from very distant clades have similar functions. Ecospaces have been defined for fossilized marine animals based on differences in feeding, mobility, and position above or below the sea bottom. Shallow-burrowing filter feeders such as clams and worms group together, while another group includes active predators along the seafloor, such as arthropods and modern snails. These three axes of feeding, mobility, and position yield 216 possible modes of life among marine animals. Just as a theoretical morphospace includes some potential shapes that are biologically implausible, ecospaces can include ecologically meaningless combinations. Baleen whales are great at filtering krill out of seawater, but no flying organism could filter insects out of the air. Bats come close, but they target specific prey. Some ecological combinations are plausible but seem to have never been tried. We found that only 92 of the 216 possible ecospaces had been occupied in the past 600 million years. Similar studies have examined Mesozoic terrestrial ecosystems, dominated by dinosaurs, and the mammalian ecosystems of the late Cenozoic. About 19 of the ecological roles invented by dinosaurs were reoccupied by placental mammals, again suggesting a certain conserved structure to ecological communities.[40]

Ecospaces illustrate two different facets of innovation: The first arises when a group discovers an entirely new way of making a living, as did many clades during the Cambrian Explosion when 30 of these strategies were first filled. But innovations may also occur when a new group of animals develops the ability to access resources already used by another group. For example, gastropod mollusks, crabs and other arthropods, starfish, skates and rays, and fish are all mobile predators on the seafloor. These patterns of ecological convergence in life habits represent innovations for a clade but not global innovations. There are interesting questions about whether similar evolutionary novelties have been required for accessing the same part of an ecospace, and about differences in the impact of innovations by different clades accessing similar ecological resources.

Neither fitness nor adaptation are incorporated in most ecospaces, so they are not evolutionary landscapes. Moreover, since these spaces are categorical and thus discontinuous, it is not even possible to develop an evolutionary landscape. But morphology, ecological function, and similar spaces each are special cases of a phenotypic space. When Dennett proposed his Library of Mendel, a space of all possible genomes, he argued that it approximated a "Universal Design Space." While Dennett recognizes that some parts of his design space may be difficult or impossible to access, he dismisses the possibility of evolutionary discontinuities. Yet emergent features and discontinuities in developmental, morphological, and functional spaces suggest that phenotypic spaces can be largely decoupled from underlying genotypic spaces.

Later in this chapter we will examine cases in which the mapping from geno-type to phenotype is imperfect.[41]

Darwin was aware of gaps in the fossil record when he wrote the *Origin* but explained this as an inevitable consequence of the inadequacies of preserva-tion. Most individuals of any species are not preserved as fossils, nor are most species. It was not terribly surprising that these problems had turned what initially *must* have been a continuous distribution of form into a fossil record that is, even at best, rather sketchy. Indeed, as an undergraduate I learned the ritual self-flagellation of paleontologists ("Oh, our record is so poor . . . we are unworthy to discuss evolutionary processes").

Dobzhansky's adaptive landscapes provided another explanation for dis-continuities, as he viewed the structure of opportunity as patchy for functional and ecological reasons. For example, the distribution of seed sizes may be clumpy because small seeds come from grasses, larger seeds from bushes, and even larger seeds from trees. An uneven distribution of seed sizes will influ-ence the distribution of the sizes and shapes of the beaks of the birds that eat the seeds. The repeated evolution of saber-tooth cats could be interpreted as a persistent peak in adaptive space. If the distribution of resources and other opportunities was patchy then the phenotypic responses would necessarily be patchy as well. For Dobzhansky, the adaptive landscape of opportunities, with peaks and valleys, presented a scaffold upon which evolution could play out. Some peaks might never be reached, but the landscape reflected the possibili-ties on offer.

The final explanation is that underdispersed patterns reflect abrupt pheno-typic transitions. Depending on how development translates the genome into a phenotype, gradual genetic change might result in abrupt and discontinuous morphology, as was the case with RNA folding. Evolutionary novelties, in this view, reflect a fundamentally distinct type of evolutionary dynamic from adap-tive and neutral evolutionary changes. This scenario is consistent with the saltational perspectives of Geoffroy, Huxley, Schindewolf, Goldschmidt, and many others, but is now more firmly rooted in our understanding of develop-mental mechanisms, as taken up in chapter 6. The second and third possibili-ties differ from Darwin's explanation by acknowledging that the pattern of form is lumpy but differ from each other in whether the source of the lumpi-ness lies in ecological structure or in the internal dynamics of development. Schumpeter's definition of technological innovation as quoted in chapter 3 falls in this category, particularly his insistence that innovations "cannot be decomposed into infinitesimal steps."[42]

SPACES FOR CULTURE AND TECHNOLOGY

The utility of evolutionary spaces for understanding the challenges facing organisms carries over to the social, cultural, and technological realms, so it should not be surprising that similar approaches have been applied. Economists César Hidalgo and Ricardo Hausmann introduced product spaces to understand the development trajectories and potentials of different countries. A product space is a network of products linked by the probability that they are both exported from the same country, which is likely to indicate that they require similar capabilities. A commercial jet airplane requires thousands of different products and thus lies in the densely connected core of a product space, along with other metal products, machinery, and chemicals. The peripheral regions are occupied by simpler, easier to produce, and less well-connected items, such as agricultural and forest products, clothing, and fisheries. Sophisticated products require considerable human and physical capital, as well as institutions (banks, engineering firms, even lawyers) and infrastructure, while less sophisticated products rely upon land (farming) or less-skilled labor. The economic development of a country involves diffusion through the product space as it acquires the capability to produce new and more complicated products. These new capabilities are likely to be constrained by existing capabilities: a country that produces bicycles is unlikely to begin producing jet aircraft, for example.[43]

Hidalgo and colleagues capture the utility of the space and the importance of accessibility with this analogy:

> Think of a product as a tree and the set of all products as a forest. A country is composed of a collection of firms, i.e., of monkeys that live on different trees and exploit those products. The process of growth implies moving from a poorer part of the forest, where trees have little fruit, to better parts of the forest. This implies that monkeys would have to jump distances, that is, redeploy (human, physical, and institutional) capital toward goods that are different from those currently under production. Traditional growth theory assumes there is always a tree within reach; hence, the structure of this forest is unimportant. However, if this forest is heterogeneous, with some dense areas and other more-deserted ones, and if monkeys can jump only limited distances, then monkeys may be unable to move through the forest. If this is the case, the structure of this space and a country's orientation within it become of great importance to the development of countries.[44]

Subsequent work explicitly tied products to the capabilities required to generate them and led to the introduction of an economic complexity index as a predictor of future development. Curiously, the product space analyzed is modular, with clusters of highly connected products separated from other clusters by poorly occupied regions, suggesting that it may be difficult to access different modules within the space. The difficulties countries have in moving through this product space add a new consideration for those interested in economic development. While a product space captures relationships at a point in time, the product space evolves as new products are invented and as innovations (from banking to multinational supply networks) increase the capacity to generate novel products.[45]

Design or evolutionary spaces have been applied to other areas of culture and technology. Linguists have long been interested in the total design space for elements of language, with Greenberg taking an early step toward this goal by describing the ordering of the primary components of grammar: subjects (S), objects (O), and verbs (V). These three components generate six possible orders: SOV, SVO, OSV, OVS, VSO, VOS. In English, "Becky hit the ball" follows SVO word order. Like Raup's morphospace, parts of this design space were initially thought to be vacant but subsequent studies have uncovered examples of each of the six possibilities. The occupation of these spaces is highly uneven, however, with about 96 percent of all the world's languages falling into the SOV, SVO, and VSO categories. Greenberg also identified several "universals" associated with different structures. As with spaces described earlier, linguists have explored how languages have evolved through these design spaces and how social structures, population size, and other factors correlate with word order.[46]

Greenberg's focus on grammar captured a significant but still small aspect of linguistic diversity. A more recent effort encompassing 195 grammatical features across 2400 languages allowed a more refined analysis. Here the possible space is more than 10^{34} possible grammars (if the features are independent). Nonetheless, the distribution of languages within this space revealed what the authors described as "enduring constraints" influencing linguistic evolution.[47]

Raup's theoretical morphospace for coiled shells has been repurposed to construct a design space of network architectures including food webs, electrical circuits, metabolic networks, and neural circuits. There is potential to develop this approach for both the analysis of actual networks and exploring possibilities of new types of networks. An alternative cultural design space is more structured than Raup's, attempting to encode behaviors as a hierarchically structured recipe. This approach captures both decisions and the actions based on those decisions. For example, knapping a stone tool can be described as a series of actions needed to detach flakes and eventually produce the desired form. One advantage of this approach is that it may more closely capture teaching and learning, yet at the cost of a broader view of cultural or technological design.[48]

Perhaps the most widespread use of adaptive evolutionary spaces in culture and technology has been fitness or adaptive landscapes, commonly the N-K model popularized by Kauffman. This has been used both metaphorically and empirically in production rules and management strategies in economics and in organization management, in anthropology and sociology, in cognitive sciences, and in political science. The idea of the fitness landscape often serves as a metaphor or conceptual framework rather than as a basis for modeling or empirical study, reflecting the frequent use of "search" as a metaphor in business strategy. Studies of culture have identified many recurrent features, such as public acknowledgement of marriages and supernatural agents that seek to do harm (rather than the ubiquity of bad luck). But are these features culturally based, or do they reflect the way we think? These have been described as "cognitive attractors," or basins in a cultural space that are found in many different cultures.[49]

This bestiary of spaces relates to novelty and innovation in different ways. Novelties arise in sequence spaces (DNA, RNA, and proteins), as well as in regulatory and developmental spaces. Many of the phenotypic spaces, where novelties are expressed, are phenomenological. And as described earlier, many-to-one mapping and other discontinuities between different types of spaces mean that a novelty at one level, perhaps a new regulatory process, need not translate into novelties at a phenotypic level. Similarly, novelties in development or morphology need not be driven by novelties at the sequence or genomic level. To preview the definition of innovation developed in the next chapter, innovations are recognized as changes in the structure and function of ecological, cultural, or technological networks. Consequently, they are expressed as changes in ecological, functional, cultural, or economic spaces. But to understand how both novelty and innovation arise in these spaces, we need a firmer grounding on the nature of spaces. Many evolutionary spaces have high dimensionality, but understanding how such spaces function is not easy with the limitations of our "Euclidean intuitions."[50]

A Problem of Topology

A space might be defined as a continuous area that is available or unoccupied. A space might be full of people, as in a crowded train station, or as empty as a desert or outer space. To a mathematician a space is defined by a set of points with a specified structure or relationship that establish the boundaries or nature of the space. Thus, the space of a city or a desert is defined by geographic coordinates and the relationships between them. We live in a world governed by three spatial dimensions and one of time. We think of the spatial dimensions as having regular and symmetric properties: a foot or meter in one direction is the same length as a foot or meter in a different direction. These are the properties of Euclidean vector spaces, and many evolutionary spaces have been implicitly Euclidean, or at least metric, spaces.

Consider a theater. An "empty" theater is not empty; it simply does not have bodies in seats. But the regular arrangement of seats around the theater provides a sense of the distribution of bodies in a full theater. If we know how the theater is used, we are even able to predict how the seats will be filled: In a movie cinema seats in the middle of a row about halfway back from the screen will fill first, followed by those along the aisles. For live theater the pricy seats are close to the stage. And in a college lecture hall the prime seats are along the aisles toward the back. In most geographic space one can often move simply from one point to an adjacent position, barring obstacles. These are Euclidean spaces. But spaces may be non-Euclidean, including some where even the concept of a distance between points is meaningless.

Lewis Carroll's space of four-letter words has the happy property that we can easily compute the distance between different sequences. In the example used earlier, moving from *head* to *tail* required five changes, which is the distance between these sequences (there may be a shorter distance using nonsense strings). Even though *cast* and *fame* share only a single letter, there is a simple path from one word to the other. The same logic applies to computing distances for sequences of DNA, RNA, and proteins. Things become more difficult with other spaces. By describing the space of animal morphologies as clumpy, we are implicitly assuming that there is some meaningful sense of distance between the morphologies: trilobites are all closer to one another in some space than they are to a clump of butterflies. But the concept of distance requires that a space has specific characteristics. If I ask you to compute the distance between a mushroom and a coffee cup you would, with some justification, question my sanity. If I enquired about the distance between an orchid and an aardvark, the question might seem just as silly, until upon a moment's reflection you realized that it was only silly in a phenotypic or developmental space (where aardvarks and orchids have nothing in common), but if we drop into a sequence space the distance is easily calculated, and this is the basis for constructing large-scale trees of life.

Back to "The Library of Babel." The narrator of Borges's story provides some clues to the nature of the Library: it is spherical, the center can be anywhere, in any hexagonal room, and as there are no boundaries the circumference is unattainable; it is limitless and periodic, in the sense that if you travel infinitely in any direction, you will find the same books in the same order. Although Borges probably did not intend this, there are sufficient clues to reveal the large-scale structure of the Library. Within the Library we can move through hexagonal rooms on one level, or up and down between levels (so it appears to be a Euclidean space). But with no circumference, no boundaries and no limits, the space must eventually become non-Euclidean. In mathematical terms, the Library must be a manifold: a space that is locally Euclidean but on a larger scale becomes non-Euclidean. I will skip the details, but the Library is likely what is known as a 3-torus, which you can think of, or try to think of, as a four-dimensional donut where the donut is not in a larger space but *is* the space.[51]

Even the metaphors used to describe patterns of innovation are influenced by implicit assumptions about the topology of these spaces, how lineages can explore them, and whether the spaces are constructed. The topology of spaces can influence whether phenotypes are readily accessible or difficult for evolution to discover. This section explores the topology of evolutionary spaces because the assumption that most spaces are Euclidean is often not correct, yet our intuitions remain Euclidean. It is generally assumed that the higher-dimensional adaptive spaces discussed earlier are Euclidean, or that we can use appropriate statistical approaches to generate a Euclidean space. As it happens, many morphospaces are manifolds, just like Borges's Library. And lest it appear that I am indulging again in my interest in Borges, the concept of non-Euclidean spaces, and of manifolds in particular, turns out to be of great significance for understanding the topology of evolutionary spaces and the nature of novelty and innovation.[52]

For our purposes we need only consider the types of spaces shown in figure 4.2. Euclidean spaces are a special case of metric spaces in which distances can be computed, but the dimensions of the space may not be at right angles (orthogonal) to one another. Topological spaces have the property that one

Euclidean vector space

Metric space

Topological space

Pre-topological space

Set

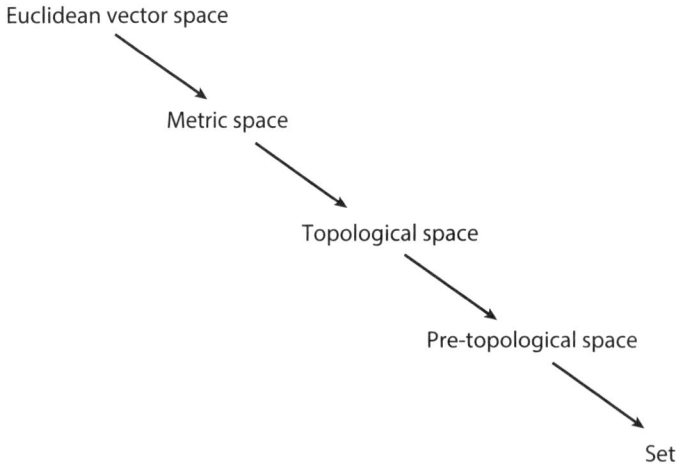

FIGURE 4.2. Relationships between topological spaces. We are most familiar with a Euclidian vector space, in which the dimensions of the space are orthogonal, but there are many non-Euclidean spaces. In metric spaces the distances between entities can be measured, as in a Euclidean space, but the axes may not be orthogonal to one another. Topological spaces have a sense of "nearness" or "adjacency" of objects, but distances between objects cannot be quantified. Topological, metric, and Euclidean spaces are all bounded, but pre-topological spaces and sets are not bounded. In mathematical terms, the set is not closed. In pre-topological spaces there is a sense that objects are within a neighborhood, unlike a set.

can determine whether objects are near or far, but it is not possible to quantify the distances between the objects. Euclidean, metric, and topological spaces are bounded, meaning that a limit or bound of the space can be defined. In unbounded, pre-topological spaces there is still a sense that objects may be in the same neighborhood, but without any bounds on the space. Finally, a set is a group of objects (a pinecone, a coffee cup, and a snowball) with no relationship among them.

Returning to the RNA folding space described earlier, while distances can be determined between different RNA sequences, in the phenotypic space there was only a sense that the black and dark gray networks were neighbors while the light gray network was not in the neighborhood of any of the other neutral networks. Because there is no meaningful measurement of distance in these spaces, folded RNA shapes do not exist in a Euclidean or metric space, but in a pre-topological space. The genius of using RNA folding to

TABLE 4.1. Features of Evolutionary Spaces

Evolutionary Space	Space or Landscape	Topology	Novelty or Innovation
Genotype—Wright	Both	Metric	Novelty
Genotype—NK	Both	Metric	Novelty
Protein	Both	Metric	Novelty
Developmental	Both	Variable	Novelty
Wagner-reg	Space	Euclidean	Novelty
Wagner-met	Space	Euclidean	Novelty
Adaptive (Simpsonian)	Landscape	Variable	Variable
Morphospaces, landmark	Space	Locally Euclidean grading to pre-topology with increasing morphologic scope	Variable
No landmarks	Space	Topology to pre-topology	Variable
Skeletal design	Space	Set	Novelty
Phenotypic trait	Variable	Variable	Innovation
Functional	Variable	Variable	Innovation
Ecospace	Space	Set	Innovation

Protein spaces have been viewed as the mapping from a sequence space, but in a broader context both protein and developmental spaces are intermediaries between genetic and phenotypic spaces, with novelties and innovations largely occurring in different types of spaces. Most of the spaces discussed here are variants of phenotypic spaces. Wagner-reg and Wagner-met are the regulatory and metabolic spaces described by A. Wagner (2014). In general, morphospaces are not landscapes, but see S. Arnold et al. (2001) for a discussion of morphospaces as adaptive landscapes. "Variable" indicates that topology of the space or involvement in novelty or innovation may vary depending on the taxonomic breadth under study. After Erwin 2017b.

explore evolution comes from the many-to-one mapping between the genotype (the sequence) and the phenotype (the two-dimensional folded shape), which illustrates how changes in a Euclidean space can play out as disjunct changes in a phenotypic space. Consequently, even though the folded RNAs exist in a pre-topological space, we can still determine the distances between them through the genotypic space. If this genotype-to-phenotype mapping breaks down, whether because developmental processes have decoupled the phenotype from the underlying genes or because the environment has a significant influence on the form of an organism, establishing the distance between forms in topological or pre-topological spaces is not possible.

The topology of the evolutionary spaces described here differs greatly. A morphospace for closely related species with similar morphologies may be Euclidean, or at least metric. But as the disparity increases, the nature of the space changes and it becomes increasingly non-Euclidean. Most shelled gastropods (snails) and cephalopods (there are a few exceptions) can be encompassed within a metric space, and this is often true within groups that in Linnean taxonomy are described as classes or orders. But comparing morphology between classes, as with different classes of echinoderms, or between phyla, is more fraught. Table 4.1 summarizes the features of the evolutionary spaces described here and their expected topology. Given the challenge of non-Euclidean spaces, my colleague at the Santa Fe Institute, David Krakauer, suggested focusing on density rather than distance might be more useful. Some regions of a space may have a greater density of novelties or innovations than others, even if we cannot readily identify the distances among these new possibilities. In the concluding section of this chapter, I will argue that the RNA case is often misleading, and a fuller understanding of these processes requires a different approach.[53]

Search versus Construction

Is mathematics discovered or created? Things differ in quantity without human intervention, but counting requires a concept of numbers. Real numbers are fine, but then come rational, irrational, and transcendental numbers. Dividing the circumference of a circle by its diameter will yield 3.14 . . . anywhere in the universe, even if Earth is the only planet that knows this as π. Adding imaginary numbers to real numbers generates complex numbers. Some have argued that abstract mathematical objects exist in some eternal space, independent of humans. Andreas Wagner encapsulates this view of sequence novelty: "Unlike galaxies, which self-assemble through the gravitational attraction of cosmic matter, or biological membranes, which self-organize through the love-hate relationship of lipid molecules with water; genotype networks do not emerge over time. They exist in the timeless eternal realm of nature's libraries."[54] There is an undeniable attraction to this view of universal libraries that exist, like Euclid's postulates, in some uncorrupted space of possibilities. But a more convincing argument (at least to me) is that math is a language that follows rules we defined. This debate over the objectivity of mathematics illustrates differing perspectives on how evolution explores evolutionary spaces, and whether the origins of evolutionary novelty lie in search through a space of options, or in the expansion and construction of evolutionary spaces.

If an evolutionary space defines all possible entities, whether sequences or books, then novelty becomes an issue of discovery or accessibility of unexplored regions. In this sense both sequence spaces and the combinatorics of the Rubik's Cube are bounded. But the complexity of the genomes of real organisms is far greater than that of the sequence space. Complexity is challenging to define, but most of us would agree that flies are more complex than sponges, and humans more complex than flies. But building almost any animal requires no more than about 20,000 genes. Sponges usually have fewer, humans a few thousand more. But to a first approximation the complexity of an animal is not a function of the size of its genome. Rather, it reflects the complexity of the interactions between genes and how they are controlled during development. This decoupling of organismal complexity from genome size, or the phenotypic space from the sequence spaces, seems strong evidence that evolution is not bounded. Logically, therefore, some evolutionary spaces must be constructed through the history of life.[55]

Conceiving of evolution as a process of sequential search driven by adaptation was part of the fitness and adaptive landscapes of Wright, Fisher, and Simpson. As these ideas have been elaborated and Mount Fuji landscapes relegated to introductory textbooks, the complexities of the search process have become more apparent. The recognition of the importance of neutral evolution beginning in the late 1960s eventually led to the RNA model described earlier, which has been central to combinatoric views of novelty, particularly the recent work of Andreas Wagner. Human problem solving has been described as a process of search for solutions as far back as Adam Smith, and this was articulated by Herbert Simon in the late 1950s, whose work influenced later generations in economics, psychology, and management.[56]

Earlier I noted the impossibility of searching Borges's Library; for some biologists, evolutionary change in most spaces (except morphospaces) occurs via search or diffusion within an existing space. DNA, RNA, and protein spaces can be defined a priori for sequences of any length, as in the previous example of a DNA sequence of 100 nucleotides. The version of Lewis Carroll's game described in the introduction to this chapter generates a search via single nucleotide changes. If the transitions are relatively easy, this may eventually lead to new evolutionary opportunities. For example, in the RNA sequence space, most of these changes are small-scale changes. Because the two-dimensional folded state is predictable from the RNA sequence, search in the

Euclidean sequence space may result in novel forms in the pre-topological phenotypic space. Highly multidimensional fitness spaces facilitate adaptive changes within a few dimensions, with macroevolution and novelty being generated in other dimensions.

Kauffman encapsulated his view of evolution as a search process as the adjacent possible:

> Adaptive evolution is a search process—driven by mutation, recombination, and selection—on fixed or deformed fitness landscapes. An adapting population flows over the landscape under these forces. The structure of such landscapes, smooth or rugged, governs both the evolvability of populations and the sustained fitness of their members. The structure of fitness landscapes inevitably imposes limitation on adaptive search.[57]

Search is integral to applying Kauffmann's N-K model to management strategy: a review of recent studies concluded that "important insights can be gained by linking the notion of search in rugged performance landscapes to practitioner-oriented practices and frameworks."[58]

As the adjacent possible expands through time, new opportunities become accessible simply because they lap up against existing species. A remarkable diversity of fossil trilobites is found in Cambrian to Devonian rocks (520 to 359 million years ago): small, blind agnostids; *Isotelus* was tens of centimeters long; others had long spines curving back over the head. After trilobites originated, these forms were accessible via Kauffman's adjacent possible as steps through trilobite developmental and morphological spaces. Search is a powerful metaphor for adaptive evolution, and there are many cases where this perspective is appropriate. In fact, most evolutionary changes, whether adaptive, deleterious, or neutral, represent small, incremental searches within the appropriate space. An example that illustrates many-to-one mapping in morphology comes from studies of wrasses, small, often brightly colored fish found in oceans around the globe. Wrasses have a variety of jaw structures, but functional analysis revealed decoupling of the structure of the jaw from its function. In other words, very different wrasse jaws are functionally equivalent. One interpretation of this result is that neutral search through the variety of jaw structures may have preserved function, as with the RNA model described previously.[59]

The adjacent possible may be infinitely expandable, although extinction can purge whole regions (as happened when trilobites became extinct 251 million years ago). There are lots of large, fat pigeons around today, so presumably the possibility space for something like a passenger pigeon is accessible, even if the result might not have the reddish-pink breast of the original or share its highly social behavior.

Search can be more expansive than the sorts of changes that feature in Borges's Library. In principle, we can expand the idea of search to incorporate other genomic changes, including insertions and deletions of varying lengths within a gene, the fusion of two genes together, and changes in the regulatory sequences that control when genes are expressed during development, which increases the effectiveness of the search. Inclusion of such alters the space and pathways through it, likely generating anisotropies that limit the path of search within a space. Other search strategies include jumps of various sizes through a space. A Lévy flight is a random walk where the size of each jump or step is drawn from a heavy-tailed distribution; other types of flights draw the jump sizes from different distributions. Lévy flights are common in animal search patterns but can also arise in molecular evolution. For example, insertion of a new regulatory element in a gene can trigger expression of the gene in a new context (the jump), followed by adaptive changes to improve the functioning of the gene.[60]

But is search the only appropriate metaphor for the generation of novelty? One of my primary arguments is that novelty and innovation are often best understood as the construction of new opportunities, possibilities that could not have been envisioned, and that invoking search as the dominant metaphor precludes understanding this dynamic. Construction is related to Kauffman's adjacent possible in that the possibilities of construction expand through time, but in contrast to Kauffman's view, novelties need not be adjacent to realized genomes, developmental patterns, organisms, or other phenomena. This section contrasts the assumptions of search with the possibilities of construction, including such questions as: How stable is genotype-to-phenotype mapping? Is evolution incremental to the adjacent possible in an evolutionary space? How might new opportunities and new spaces be constructed through novelty? First, though, what do I mean by construction?

Consider something as simple as the personal computer. The "space" for computing was present during the time of ancient Greece in the sense that mathematicians had developed a rich and diverse array of mathematical tools. But until 1902 there was no evidence for sophisticated computing devices. Discovery of the remarkable Antikythera mechanism from a shipwreck revealed that by 100 BCE the ancient Greeks had both the collective knowledge and technical capacity to design and construct remarkably sophisticated analog computers. In the sense of this book, this was a technological novelty, with no evidence for any innovation. The complexity of the Antikythera mechanism suggests it must have been preceded by earlier, cruder devices, none of which are known. Slide rules date to the early 1600s, analog computers began appearing in the late nineteenth century, and digital computing is less than a century old. The personal computer required a technological and economic scaffold that arose only in the 1980s. Advocates of search as the source of innovation argue that economic and technological developments progressively allowed access to new possibilities—possibilities that always existed, an argument reminiscent of early nineteenth-century German idealistic morphologists (chapter 2). If evolutionary novelty is chiefly a process of search, then opportunities exist in some Platonic space but are simply inaccessible. Evidently the Apple Macintosh was there all along. Steve Jobs just had to rummage through the creations of the Homebrew Computer Club, discovering the Apple II and stumbling over the Lisa before eventually finding the Mac.

A concept of "eternal" spaces may be useful in some contexts, but suggesting that the "space" of personal computers existed in the time of ancient Greece and was simply "inaccessible" obscures exactly what we seek to understand about cultural and technological innovation: the social, economic, and technological changes necessary to put an iMac on my desk and a smartphone in your pocket. Whether the first state was ancient Egypt or Mesopotamia, this form of organizing human society had no precedent. Characterizing early states as a consequence of searching through the space of possible organizations of human society is neither helpful nor illuminating. Similarly, in biology, it is the processes that allowed the origin of animals or the early vertebrates to crawl onto land that we seek to understand. To revise Kauffman's view, what is interesting is not adjacency, but possibility. A critical component of novelty and innovation in biology, culture, or technology is generation of new evolutionary spaces and thus new opportunities and how new means of exploring evolutionary spaces arise. This is a central focus of chapters 6–9.

Several features of evolution challenge a microevolutionary, search-dominated view of novelty. Some models of search assume that change occurs via a single nucleotide or amino acid change, or a single change in a regulatory interaction. Cumulatively, such changes allow exploration of sequence spaces. But even for microbes the variety of evolutionary operators is greater than single base-pair substitution. Sequences may have deletions or insertions of variable length. Proteins assemble into three-dimensional structures of columnar helices and pleated sheets that correspond to discrete domains in the corresponding genes. The evolution of gene and protein sequences involves factors beyond those captured by a protein space. Some of these include the position of the gene within the genome, pleiotropic effects such as a protein's position within a biological network, or how readily organisms can dispense with its activity. New genes are assembled by the formation of new domains or the rearrangement of existing domains. Other genes have components that can be assembled into many different proteins, as sort of a mix-and-match process of alternative splicing. Over evolutionary time spans, genes may be transferred from other species. Thus, evolution operates as much by a diversity of combinatoric operators as it does the traditional conception of search.

Regulatory interactions are even more difficult to capture as search. Before genome sequencing was possible and molecular biologists could only measure the total amount of DNA in a cell, many biologists expected some relationship between the number of genes in a genome and the complexity of an organism (leaving aside for the moment the intractable problems of defining complexity). Humans should have far more genes than mice, and mice more than sponges. So much for theory. Although animals generally have more genes than their single-celled relatives, genome sequencing has shown that variations in gene number or genome size are largely unrelated to any measure of developmental or phenotypic complexity. The complexity of animals reflects variation not in sequence space but in the regulatory interactions that enable development, as discussed in chapter 6.

The case of short RNA sequences mapping from the genotype to their folded secondary structures described earlier is so valuable because it is so rare. Few biological processes exhibit such linear mappings, particularly in clades like plants and animals with complex development. This complexity of genotype-to-phenotype mapping includes responses to physiological and environmental information that require flexible responses. Over evolutionary time, either within a species or across related species within a lineage, genomes, proteins, and regulatory information are subject to a broader range of changes than envisioned in the RNA folding model. In a sense, these processes add entropy to the relationship between the genotype and phenotype.[61]

One might rescue search and the insights from the RNA and related models by enlarging them, expanding the number of dimensions, including sequences of varying lengths, and adding operators such as deletions or rearrangements. Wagner's regulatory space could be extended from the set of all five-gene interactions to six genes, or some higher number. We could expand the space to cover all conceivable traits. But such an argument fails, because many evolved traits could not have been defined a priori and are historically contingent.

The sequence spaces for DNA, RNA, and proteins of all conceivable lengths and permutations would be a truly vast space in which realized genes and proteins were embedded. Stuart Kauffman describes these as "pre-statable" spaces: spaces for which all possible states can be defined. Borges's Library is pre-statable as it includes all books of a given length. Much of classical physics operates within pre-statable spaces. Yet few of the spaces described here are pre-statable, among them sequence spaces, Bambach's ecospace, morphospaces for logarithmically coiled organisms like clams and bivalves, and the skeletal design space. But only the sequence spaces are directly searchable. The others are phenotypic spaces or spaces for innovation rather than novelty. In general, biological, social, and technological novelties arise in non-pre-statable spaces. This issue resolves the question posed in chapter 1 about the possibility of a general, quantitative model for novelty and innovation. Consequently, general "laws" of novelty and innovation are probably possible only by defining novelty as accessibility in a pre-statable space. To the extent that other types of novelty are of interest (and they are), we must be concerned with the evolution and construction of new spaces.[62]

Evolutionary spaces can evolve in three ways. First, new evolutionary operators may allow new ways of exploring an existing space. Alternative splicing of RNA sequences, lateral transfer of genes from one clade to another, and fusion of protein domains each change the exploration of a sequence space. These changes allow lineages to move into new regions of a space. Second, spaces may expand or contract. Longer proteins expand protein spaces, while the duplication of segments in an arthropod expands the arthropod morphospace. The developmental and morphological spaces for trilobites and non-avian dinosaurs vanished when they went extinct 251 and 66 million years ago, respectively. Disappearance of a clade through extinction may free up ecospace and create new opportunities for other clades. New spaces might be constructed via genetic or developmental novelties. Finally, spaces interact, in the sense that organisms create opportunities for other organisms, whether of the same species or of very different species. Both coral reefs and tropical forests are classic examples of widespread ecosystem engineering, but this is common in many other settings as well.

If the extension and modification of existing spaces and generation of new spaces are key components of novelty, this raises intriguing questions about how spaces evolve. Comparative studies of plant and animal development have revealed how developmental spaces have evolved over time, and that developmental capacity for novelty may long precede the ecological possibility of successful innovation. My emphasis here on the construction of new spaces does not diminish the importance of search. Search involves the exploration of combinatoric solutions once a space has been generated, adapting those that are viable and avoiding those with pitfalls.

Novelty: Opportunity and Limits

One of the most intriguing aspects of evolutionary spaces is how they expose the tension between the possible and the actual—what could be generated versus what evolution has achieved. Building theoretical morphospaces poses questions about the distribution of living and fossil forms, and why some forms have never appeared. That something has not evolved does not mean that it could not evolve (despite the remarkable number of papers that continue to make this error). Regions of an evolutionary space can be unattainable or inaccessible. Unattainable parts of a space are not possible for either physical or biological reasons: A DNA triple helix; sequences of amino acids that code for proteins that misfold or cause fatal defects; most of the creatures inhabiting the bar scene of *Star Wars: A New Hope.* Some theoretically possible ecological strategies are just tough ways to make a living, such as a filter-feeding flyer. Inaccessible regions are more interesting. These are areas of an evolutionary space that reflect viable design options if only evolution could access them. For example, an island of viable proteins marooned in a sea of nonviable sequences. For the RNA sequence networks, novelty represents accessing some difficult-to-reach sequences, but there may be many useful RNA sequences that cannot be reached by any plausible route of changes in single amino acids, nor even by recombination of sequences. In more technical terms, these constraints mean that the spaces are not isotropic, but anisotropic.[63]

Unattainable or inaccessible regions remind us that novelty and innovation have always been constrained, and such regions effectively limit their dimensionality. Limits or constraints may represent an absence of genetic or developmental variation in a particular direction. But some limitations can be overcome, and regions of a space that might have been unviable 500 million years ago can become an evolutionary opportunity. That we have tetrapods, with two pairs of limbs, and no vertebrate hexapods, with three pairs, appears to reflect a developmental constraint. There is no physical reason why such a vertebrate could not exist (and there seem to be no end of robotic hexapods), but the variation upon which selection could act did not arise early in vertebrate history and probably could not today. Physical constraints are harder to overcome, however, as they often reflect limits to optimal performance.

Thus, evolutionary constraints play a crucial role in evolutionary novelty and innovation by illustrating the bounds of evolutionary change and where real opportunity might lie. Here I raise three general issues for thinking about limits to the range of novelties that might be generated, or their likelihood of success as innovations: standards; scaling relationships, or the influence that size and energy have on possibilities; and the ubiquity of convergence, which some have interpreted as evidence that the possibilities open to evolution are limited.[64]

Each new computer I acquire generates contributions for the Smithsonian's National Collection of Unusable Computer Cables, reflecting the persistent replacement of technology standards. Sudoku puzzles have some squares already filled, constraining possible entries to other squares. Standards may limit a space of possibilities while also opening new opportunities by ensuring that components fit together. The four bases of adenine, cytosine, guanine, and thymine are the standards for building DNA, just as a set of 20 amino acids constitutes the building blocks of proteins. Among many similar examples are the fact that virtually all amino acids in proteins are of the L form rather than its mirror image, the D form; the voltage gradient that allows communication between neurons; lipid cell membranes; and aspects of metabolism. Technological standards surround us, from the QWERTY keyboard and how cell phones divide frequencies to weights and the synchronization of time. When standards go awry the result can be disaster, as happened in 1999 with the Mars Climate Orbiter. Software at NASA was sending signals to the engines in units of pound-seconds, while the software on the probe was expecting input in newton-seconds. The Orbiter should have gracefully checked its speed and entered Martian orbit but instead disintegrated in the Martian atmosphere. Standards can be critical to the success or failure of innovations. The absence of a global standard for electrical plugs requires me to fish out the right adapter from my shoebox of plugs or run the challenge of finding the correct adaptor in Windhoek, Namibia. Standards are ubiquitous in culture as well.[65]

A second limitation on the range of potential novelties arises from basic physics. Over your life-span your heart will beat about the same number of times as that of a mouse or a whale over their life-spans. As the mass of an animal increases, so too does the energy it consumes (the metabolic rate), but the metabolic rate grows only about three-quarters of the increase in body mass.

Since the life-span of a mouse is far less than that of a whale, there must also be a relationship between heart rate and body weight—for mammals, heart rate declines with increasing body weight. These are just a few examples of relationships between the size of organisms and the rate of physiological activities. There are dozens of these interlinked relationships in everything from pico-plankton to the biggest trees and the largest animals, reflecting optimal biological solutions to energy expenditures. A consequence of these scaling relationships is that if one knows the size of a mammal, the average values of brain size, heart rate, and the likely number of offspring can easily be determined. Similar relationships obtain for plants as well.[66]

These scaling relationships were first uncovered in the early twentieth century, but their causes have been controversial. One interpretation is that they reflect a linkage between surface area and volume. Absorption of nutrients in the gut is a function of surface area, as are exchange of oxygen and carbon dioxide in the lungs and photosynthetic capacity of leaves. In contrast, the energy required by a plant or animal is a function of its volume. When length doubles, surface area increases by the square of the length, and volume by the cube. In other words, the relationship between surface area and volume follows an exponential relationship, known as a power law. Consequently, as size increases there is a growing disconnect between supply, controlled by surface area, and demand, controlled by volume, and for many years this tension between surface area and volume was assumed to influence scaling relationships. Plotting scaling relationships on a logarithmic scale (where increments of size, volume, or other features increase by a factor of 10) produces the general scaling (or allometric) equation $y = ax^b$, where x is usually some measure of body size. If $b = 1$ there is a linear relationship, but in biology b is usually between 0 and 1. Consider the oxygen consumption of a house cat and a lion. Lions are much larger, but their oxygen demand is less than we might expect based simply on the difference in mass of cats and lions. This means that lions use oxygen more efficiently than cats, and the factor b tells us how oxygen demand scales to increases in body size. The ratio between increases in surface area and volume is 0.66, and biologists expected this would be the value of b.

An alternative interpretation of scaling relationships originated in 1932 with the work of Max Kleiber, an iconoclastic Swiss biologist who measured respiration in cattle at the University of California, Davis. Kleiber found that the ratio of metabolic rate to body bass was 0.75, not the expected 0.66, which became known as Kleiber's law. In the late 1990s a wonderful collaboration at the Santa Fe Institute among physicist Geoff West, ecologist Jim Brown, and

Brown's then graduate student Brian Enquist generated a new explanation for Kleiber's law. Surface area and volume play roles as factors in the distribution of resources through the network of arteries and veins in mammals, or the vascular structures of plants, which have similar branching patterns that connect water and nutrients in the soil with the products of photosynthesis. What West, Brown, and Enquist described as the metabolic theory of ecology focused on these distributional networks and proved capable of explaining a huge range of scaling relationships, from life-spans of animals to life-spans of companies. The model predicted a scaling relationship of 0.75 between metabolic rate and size, and a host of other scaling relationships that were variants of quarter-power scaling.[67]

Under the West-Brown-Enquist model the optimal function of many traits will be limited to a range of sizes. Once a lineage accesses part of this range, scaling relationships ease access to the rest of the range, often without other major novelties. Organisms above and below this size range have different functional optima to which they would adapt. Scaling relationships effectively reduce the potential size of spaces, limiting the possibilities to organisms reasonably close to the mean. Yet changes in body size as such have not featured prominently in discussions of novelty and innovation. For example, the three-quarter-power scaling relationship is more variable in mammals than other organisms, illustrating that novelties can restructure even such fundamental relationships. The importance of body size is discussed further in the next chapter, and the importance of changes in body size feature prominently in discussions later in the book, particularly in association with the rapid increase in body size with the appearance of animals during the Cambrian Explosion, and a long-term reduction in body size leading to birds.[68]

Allometric scaling relationships are found in aspects of human societies as well. Just as the metabolic rate for animals scales with body mass, the number of patents issued scales with population size, and across companies income and assets scale with the number of employees. The metabolic theory of scaling has been extended from ecology to cities, to show that networks of exchange of energy, resources, and information within a city exhibit scaling relationships like those in biology, but with a critical difference. The number of gas stations relative to the size of a city scales sublinearly, so the larger the population the fewer gas stations per capita. The same scaling relationship holds true for other supply and transportation networks and infrastructure in general, such as roads, water and gas lines, and the length of electrical lines. But in a wonderfully counterintuitive result, there are many features of cities that scale supralinearly. These increasing returns to scale have been documented for wages, the number of professionals, crime, restaurants, and patents. This means that the larger the city the greater the per capita social capital created. Just as organisms are constrained to follow certain relationships, so are cities. One implication of this finding is that cities of the same population size will share many features, whether they are in South Africa, the Netherlands, China, or the United States. Later work found similar relationships among cities in the Basin of Mexico (modern-day Mexico City) before the arrival of the Spanish, suggesting that the characteristics of human social networks that generate these increasing returns to scale are a universal of human settlements.[69]

Supralinear growth of the sort observed in cities cannot persist forever. Eventually the energetic demands will outstrip the capacity of supply networks, and West has argued that their persistence is dependent upon a supply of innovations that essentially changes the rules and allows growth to continue. The energetics of Western societies seems to confirm this argument, as we have moved from wood to coal to gas and oil, and now increasingly to wind and solar power. General-purpose technologies may be a source of innovations that allow supralinear growth.[70]

A final concern with the structure of evolutionary spaces is the ubiquity of similar or identical solutions to evolutionary problems discovered by very different groups of organisms. Some of these convergences are broad, as with birds, beetles, and bats. In each case, wings allow access to a new habitat and generate new ecological possibilities. Tuna, lamnoid sharks, dolphins, and some Jurassic ichthyosaurs independently adopted similarly teardrop-shaped bodies, fin shapes, and even a characteristic structure of body-wall musculature. But all these groups are fast, open-ocean predators, and the requirements for this lifestyle channel the evolution of any vertebrate along similar paths to maximize propulsive power.

The more interesting convergences are more specific, where structures and function are almost identical. Earlier I noted the example of virtually identical glycoproteins protecting northern cod in the Arctic and the notothenoid fish of the Antarctic from freezing. Although these fish are only distantly related, their antifreeze proteins were recruited from the same family of genes and evolved similar structures to provide the same functionality. Convergences among animal toxins used for defense or deterrence and to capture prey are widespread among snakes, scorpions, bees, spiders, sea anemones and jellyfish, cone shells, echinoderms, and even the platypus. Some venoms are unique: the brown recluse spider of the south-central United States contains a very potent venom that causes necrosis (cell death), but the active enzyme is found only among sicariid spiders. Although there is a very large library of potential venoms, very different animal groups often recruit the same proteins to generate almost identical venoms, for example by co-opting neurotransmitters from the brain. Despite the many independent origins of venom secretion across animals, a similar suite of genes has been recruited, while there is much less convergence in how lineage-specific cells are regulated.[71]

Novelty, as used here, applies to characters within clades. Similar novelties may appear in different clades or within the same larger clade, as with the Jurassic and modern beavers or some novel adaptations to xeric (dry) habitats among flowering plants.

Much of the interest in convergence, however, has arisen from more philosophical questions about whether the general space of evolutionary possibility is so limited that convergences are fated to arise repeatedly. Convergences are monuments to the power of natural selection, but some have gone further, arguing that the ubiquity of convergence indicates the number of evolutionary trajectories is limited, so that the histories of even very different organisms will play out in roughly similar ways. In this deterministic perspective scaling, physics, and the demands of different modes of life constrain the nature of evolutionary novelty and of evolution more generally. If the number of solutions to some evolutionary problems is limited, this has important implications for understanding the space in which evolution operates, and the role of innovation in generating new spaces. We return to this topic in the final chapter.

Discontinuities, Real and Apparent

There are obvious gaps between different animal groups. Huxley viewed these as a challenge to Darwin's gradual view of evolution, and apparent discontinuities have continued to fuel disagreements over discontinuities as a metric of novelty and innovation. The distribution of shapes of plants and animals is neither evenly nor randomly distributed. Earlier I argued that the picture is even more complex, as the topology of evolutionary spaces is Euclidean at small scales but decays to non-Euclidean. The challenge of discontinuities extends to humans and our societies as well, whether it is curiosity over how geniuses are different from the rest of us, the origins of human consciousness, or the almost simultaneous appearance of complex states in Mesopotamia, China, the Indus River, Egypt, Peru, and Mexico. Over the past three chapters

I have described five different explanations for these discontinuities and il-lustrated how they support wildly discordant views of the evolutionary reality of novelty and innovation.

Discontinuities as an artifact of the paucity of the fossil record and the ex-tinction of intermediate forms was Darwin's response to Huxley and is the first explanation for discontinuity. This perspective assumes that if we had a com-plete history of any branch of the tree of life, we would see a continuous dis-tribution of morphologies. While some of my paleontological colleagues have shown that the architecture of sedimentation can generate apparent jumps in morphology (often because sediments are not preserved), the sort of quantita-tive assessments of morphological disparity discussed earlier reveal that Dar-win's explanation for discontinuity can apply in some cases but is insufficient as a general explanation.

The second potential explanation for discontinuity was proposed by Dobzhansky and others in the 1930s and 1940s. Dobzhansky suggested that the clumpy distributions of morphologies are real but reflect the distributions of ecological opportunities within an adaptive space. Not all morphological con-figurations yield viable organisms, and those we do find represent peaks in an adaptive space. Even if the delivery of genetic and morphological variation were continuous, ecological opportunity would essentially form a scaffold within an adaptive landscape defining where species (or larger clades) might exist. If the opportunities are widely distributed, so too will be the resulting forms. In this scenario the transitions between ecological opportunities will be so rapid and infrequent that they are unlikely to be captured in the fossil record. As described in chapter 2, Simpson viewed all large-scale evolutionary patterns, including evolutionary novelty, as arising from adaptive radiations. So, any discontinuities in novelties reflected the structure of evolutionary spaces and the underlying ecological and evolutionary opportunities.

A macroevolutionary third alternative extends the ideas of Eldredge and Gould's punctuated equilibrium to broader taxonomic scales, with differential success of species and clades. Novelties still reflect ecological opportunities, as in the second scenario, but some of these arise after mass extinctions or other biotic crises. Simpson was heading this in direction in his books of 1944 and 1953 (or at least paleontologists of the past few decades have read him as doing so). Much of the work on higher taxa (new orders, classes, and phyla) falls within this scenario.

A fourth way of looking at discontinuities emerges from work on sequence spaces, and particularly the RNA spaces described in this chapter. Here novelty reflects the limited accessibility of configurations. As mentioned earlier, this view of novelty is distinguished from many other discussions by the focus on the *process* as a critical factor in generating the novelty, rather than in the product or the consequences in terms of a new morphology, developmental process, or behavior. This approach seems limited to pre-statable evolutionary spaces.

Finally, discontinuities may reflect the genomic and developmental processes generating evolutionary variation and thus underpin novelty. If the first, second, and fourth scenarios are largely microevolutionary, invoking the steady accumulation of small changes, this view is more willing to accept the possibility of the discontinuous *origins* of novel features. The views of G. Wagner, Müller, Moczek, and others are firmly within this approach, with Wagner emphasizing specific modifications to gene regulatory networks as a driving force for the generation of new homologous characters.

One of the intriguing features of these five explanations is the wide variation in presumed linkage between novelty and ecological or evolutionary impact. Darwin, Mayr, and others accorded primacy to ecological opportunity in the generation of novelty but downplayed the significance of variation as an important factor. As I discussed in chapter 2, I read their work as assuming that the supply of relevant new genetic variation was sufficiently frequent that what mattered was the ecological opportunity. In contrast, comparative developmental biologists such as Müller and G. Wagner have been more interested in the delivery of novel variation (and correspondingly have often largely ignored the factors involved in the persistence and success of a novelty once it arises).

Evolution has been envisioned as the exploration of vast spaces of possibilities, which generates new opportunities even as it forecloses other possibilities. The historical trajectories of lineages can be described as movement through spaces of genes, proteins, development, form, and ecology, among others. Some trajectories may be advantageous paths to increased fitness, while others may be effectively neutral. Here I want to reemphasize the insight from the RNA studies that novelty reflects the relative inaccessibility of a folded structure and is "discovered" through a population wandering through a series of neutral changes. Although some spaces may be eternal libraries of possibilities, as Andreas Wagner suggests, the picture is complicated when one considers the variety of operators in evolutionary spaces. Novelties arise when search encounters very different phenotypes from the position at the beginning of the search. Single nucleotide changes have occurred since the solidification of the genetic code perhaps four billion years ago. But other operators have appeared more recently, including those that change the length of a sequence, insert or delete regulatory controls, or allow the formation of new sorts of institutions. This challenges the idea of search as the appropriate metaphor for evolutionary novelties. In many cases, the generation of new patterns of regulatory control itself generates new opportunities, whether these opportunities are biological, cultural, or technological. Consequently, in many cases the metaphor of search is inappropriate, and we need to consider how evolution constructs new opportunities that may subsequently be exploited by a pattern of search.

Our natural inclination to assume spaces are Euclidean is often flawed. Objects in phenotypic spaces may be in the same neighborhood, but it may not be possible to specify a distance between them. If this is so, what implications does this have for the controversy over discontinuities? If there is no reasonable notion of "distance" between *Anomalocaris* and the lobopod *Aysheaia* (at least in terms of adult form, if not genetically), then how can one claim there is a discontinuity between them? An interesting issue for quantitatively adept paleontologists will be how issues of continuity, discontinuity, and disparity reflect the topology of the evolutionary space.

The past three chapters have examined how novelty and innovation have been approached over the past two centuries. In Gould's gedankenexperiment we would run life again from the Cambrian to see whether the same novelties appeared in the same clades, and whether they transformed into innovations in similar ways. As Gould realized, that will remain a thought experiment. But experimental evolutionists led by microbiologist Richard Lenski pioneered another way of exploring the importance of contingency—by tracing the history of microbial lineages through tens of thousands of generations. They discovered an important novelty in their cultures and the insights it revealed provide the conceptual foundation for the framework for novelty and innovation that forms the next chapter.

5

A Conceptual Framework
for Novelty and Innovation

ON FEBRUARY 24, 1988, microbiologist Richard Lenski took a sample of the common human gut microbe *Escherichia coli* and divided it into 12 identical populations. Every single day since then, Lenski or another member of his laboratory at Michigan State University has taken 0.1 milliliters of each sample and inoculated a fresh sample bottle full of growth medium. With each generation of *E. coli* taking about 20 minutes, Lenski's group has been able to track the history of these 12 independent lineages for more than 75,000 generations and counting. Each 500 generations they save a sample, preserving a frozen fossil record of each lineage. This long-term evolution experiment (LTEE) has yielded important insights into evolution and is a favorite of most evolutionary biologists.

E. coli is a frequently used experimental species, which feeds off the glucose present in the growth medium. Since it takes about a day for a new culture to exhaust the available glucose, that controls the cycle of the experiment. But standard laboratory growth medium contains more than glucose. Another carbon source is present, citrate, but *E. coli* does not feed off citrate (in fact, there is more citrate than glucose in the growth medium). About 31,000 generations into the experiment, Lenski and one of his students, Zak Blount, discovered a new strain had evolved in one lineage with a mutation that enabled it to feed off citrate. Voilá, a novelty.

Interrogation of this *Cit+* mutant led to the discovery that the mutation alone was insufficient for *E. coli* to metabolize citrate. An earlier, potentiating mutation was required before the *Cit+* mutant could arise. Consulting the samples in freezers revealed that this potentiating mutation had arisen in two other lineages, but the *Cit+* mutation had not occurred in those lineages, so the microbes were unable to exploit the citrate in the growth medium. The *Cit+* mutant was not sufficient on its own, either. Further mutations were required to stabilize the *Cit+* mutant and allow the strain to take full advantage of the new opportunity. With these changes, the mutant *Cit+* strains reach much larger population sizes than strains unable metabolize citrate.[1]

From this work Blount, Lenski, and their colleagues proposed a model for the generation of novelty involving potentiation, or the establishment of the necessary preconditions for the novelty, actualization of the novelty, and finally refinement, involving the adaptive changes required to fully utilize the novelty. Critically, these steps are independent. The acquisition of potentiating mutations provides no guarantee that novelty will ever arise, and a novelty without the potentiating mutations is likely to fail.

The conceptual framework for novelty and innovation developed here builds off the foundation of the Lenski-Blount model, for it provides a way of understanding the phases in the generation of phenotypic novelty and the mechanisms responsible for their eventual ecological and evolutionary success (or failure) as innovations. This framework involves distinct, if potentially overlapping, phases: Potentiation encompasses the suite of necessary environmental, genetic, cultural, economic, or technological preconditions for generation of a novelty. Here I am *not* thinking of anything that happened before a novelty. The Berlin Wall fell a bit over nine years before the first iPhone appeared, but there was no causal connection between the two. Potentiation covers necessary preconditions for the generation of a specific novelty. Potentiating events may be followed by the formation of a novelty (which may occur over some length of time), which melds into the subsequent phase of adaptive refinement. Hopeful monsters do not exist, and never have. Any novelty is likely to require a period of subsequent adaptive adjustment for the clade that possesses the novelty. Novelties may also trigger or precipitate further novelties before the clade can fully take advantage of the potential provided by them. If further environmental or ecological changes are required before clades can capitalize on these opportunities, lags may occur. Fossil evidence shows lags of tens to hundreds of millions of years before a change in the structure of ecological networks occurs to generate an innovation. Where novelties contributed to founding a new clade, this generally occurs during this potentiation phase.

This chapter begins by summarizing evidence from preceding chapters establishing novelty and innovation as real evolutionary problems requiring explanations beyond those provided by microevolutionary studies of adaptation and speciation, and indeed beyond macroevolutionary approaches from pale-ontologists focusing on the relative success of various clades. The previous dis-cussions also allow enumeration of the requirements for a useful conceptual framework. Given the array of different types of novelties and innovations through the history of life, claims for something as grand as a theory seem pre-sumptuous (or foolhardy). Rather, the framework developed in this chapter is a way of thinking about the processes of generating novelty and innovation in particular cases and trying to uncover generalities. At the conclusion of this chapter, I broach the topic of what work one might want such a conceptual framework or model to do. Such goals might seem obvious, but conceptual scaffolds can serve a variety of purposes, and different scientists have different goals. The right model for one use may be quite different from the approach required to address a different set of questions. All this also keeps philosophers of science busy sorting out what people mean or should have meant.

Distinguishing Novelty and Innovation

Four lines of evidence support a distinction between novelty and innovation. Each of these has been mentioned in the preceding three chapters, and this section summarizes this evidence, although we will return to several of these issues in latter sections.[2]

First, episodes of rapid increase in morphological disparity decoupled from increases in the number of species have been documented with robust quantitative techniques for several dozen clades, ranging from trilobites to the Triassic radiations of ichthyosaurs, dinosaurs, and other "reptilian" clades. Al-though very controversial among some evolutionary biologists and paleon-tologists when first proposed by Stephen Jay Gould in his book *Wonderful Life* in 1989, subsequent studies with a variety of different approaches have estab-lished that rapid disparity increases early in the history of a clade are quite common, although certainly not universal. Many of these events involve the appearance of significant novelties, as well as the establishment of new clades, followed by the ecological impact of innovation. The key point here is the link between novelties and early disparity, decoupled from the diversification of species.[3]

Second, one of the more contentious arguments for novelty as a distinct evolutionary process began with Mivart's challenge to Darwin over the origins of complex adaptations such as a wing, as described in chapter 2—what is the selective advantage of the earliest forms of a complex adaptation? How do the new functions associated with a novelty arise, particularly if the new function requires a complex structure? To some, such complex adaptations challenged the explanatory power of natural selection, and they have long been used by creationists as well as by those advancing alternative evolutionary theories, including orthogenesis and mutation-driven theories. Yet over the past few decades, the apparent problem presented by complex adaptations as a counter to natural selection has weakened. Phylogenetic analyses have decomposed many of these features into suites of characters and character states. Developmental studies have similarly dissected complex adaptations and shed light on the likely evolutionary trajectories through which they were constructed. Together, such a comparative approach to adaptations of varying complexity has shown that in many cases adaptive explanations are sufficient, while in others the individuation of new characters or suites of characters has been distinguished.[4]

The third line of evidence is the poverty of adaptive radiations as a general explanation for novelty and innovation. Adaptive radiations have been deployed to explain the number of taxa, the origin of novelty, and much of the history of life. As discussed in chapter 7, evolutionary biologists now recognize a broader range of evolutionary diversification events, as well as the challenges of rigorously identifying adaptive radiations. Among this diversity, however, there are certainly events that Simpson would recognize as meeting his criteria, including Darwin's finches in the Galapagos Islands. The challenge that I and others have confronted is the paucity of evidence among well-studied cases that they generate much phenotypic novelty, and this is true even if one focuses on adaptive radiations associated with putative key innovations. Moreover, identification of macroevolutionary lags, as discussed in the preceding chapters, requires acknowledging a distinction between origin of a novelty and spread of the clade containing the novelty to ecological and evolutionary importance. Although macroevolutionary lags were first recognized by paleontologists based on morphological evidence, comparative genetic and developmental studies have revealed two interesting patterns, which are discussed in greater detail in the next chapter: First, potentiating mutations may be necessary before a novelty can arise, as in Lenski's long-term evolution experiment with the *Cit+* mutant in *E. coli*. Second, I will note some of the many cases in which developmental novelties arise well before their expression in morphology. Thus, developmental capacity is decoupled from morphological novelty, a pattern that occurs with the origin of Bilateria and the origin of tetrapods (figure 5.1). Together, these discoveries raise insurmountable problems for

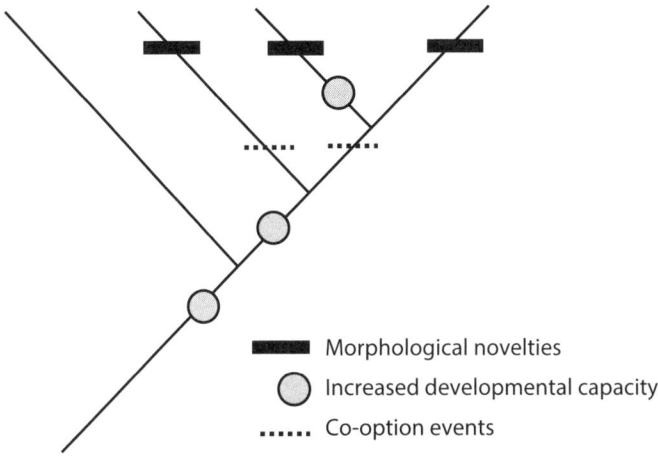

FIGURE 5.1. A hypothetical phylogenetic tree with an example of potential relationships between increases in developmental capacity (circles), which allow subsequent co-option of gene regulatory subcircuits (dotted lines) with later morphological novelties (bars). In this example, similar gene co-option events happen independently in different lineages, as suggested in the text for the origins of similar patterning systems in bilaterian animals.

Simpson's claim that the history of life can be considered a series of adaptive radiations, or that they are the foundation of evolutionary novelty.

Finally, the origins of the eukaryotic cell, multicellularity, and the generation of colonies each require cooperation between previously competing units and create new, higher levels of selection. Initially described as major evolutionary transitions, these are some of the most extreme examples of novelty via individuation, because they represent the generation of new types of evolutionary individuals. As described in chapter 3, most of these events are also associated with fundamental innovations: the origins of life, eukaryotes, multicellularity, and human culture. As Maynard Smith and Szathmáry recognized, these represent nonuniformitarian evolutionary events: rare and in some cases unique evolutionary episodes where studies of evolution today provide limited insight. Moreover, the new evolutionary individuals generated during these transitions fundamentally change the evolutionary dynamic in ways akin to general-purpose technologies. Although the conditions leading to major evolutionary transitions remain disputed, this class of novelties creates challenges for any adaptive scenario.

Whether taken individually or together, studies of morphological disparity, many complex adaptations, integration, and major evolutionary transitions, together with evidence that adaptive radiation scenarios are insufficient to explain the origins of novelties and the success of innovation, support the argument that understanding them requires moving beyond classic, adaptationist evolutionary approaches.

Requirements for a Conceptual Framework

Models involve tradeoffs. Generality comes at the cost of the details that enrich specific cases, but models tuned to a few specific cases may lack sufficient breadth to be interesting. So enumerating desiderata for any model illustrates the complexity of any conceptual scaffold for a broad understanding of novelty and innovation. *Requirements* may be a misnomer here, for some of these features can be rejected for other sorts of models, as discussed later.

Early models of technological growth envisioned a linear pathway from scientific research through invention of a new technology and eventual diffusion of the technology to consumers. While such a linear path generates a straightforward model, empirical studies provided little support. In biology, a linear pathway from novel forms through establishment of a new clade to ecological success is similarly unlikely. Although there may be some examples of such trajectories, more commonly feedback, lags, and other complexities are interposed, complicating the reliability of any clear path of causality.

Evolutionary novelties arise at many different levels. New genes form, the regulatory tools that control how these genes are expressed increase (or, occasionally, decrease) in complexity, mutations cobble together new proteins from pieces of old ones, and new physiological traits arise. As we saw in chapter 3, novelties are found in morphology, in behavior, and in ecological interactions. While most recent discussions of novelty have focused on new structures and their underlying developmental processes, more fully understanding novelties requires recognizing that they arise at many levels. This raises the question of how these different sorts of novelty interact.

For example, novelty at one level may facilitate but not necessarily entail novelty at another level. When Hox genes were thought to underpin much of animal development in the 1990s, their divergence into multiple copies was viewed as a trigger for the early diversification of animals. But as Hox genes were found in cnidarians, it became clear that the gene duplications were likely necessary for animal diversification but did not directly generate morphological novelties. In chapter 3 I introduced the idea that behavioral novelties may provide a pathway for morphological change, and chapter 8 provides several examples. The relationship between novelties of different kinds remains a promising arena for future work.

Success as an innovation may depend upon environmental and ecological conditions decoupled from the formation of a novelty, as was the case with the expansion of grasslands long after the origin of grasses. Lucinid bivalves originated in the Silurian, but did not become ecologically significant, or taxonomically diverse, until the Late Cretaceous rise of seagrasses and mangroves. Some novelties and innovations may facilitate further novelties, while others will require prior novelty or innovation to generate the opportunity space for them to form.[5]

These macroevolutionary lags reflect a more general requirement for necessary preconditions, which can influence both novelty and innovation. Just as the *Cit*+ mutant required prior, potentiating mutations before it could arise, potentiation has played a powerful role for other novelties. Here are three brief examples. New regulatory devices sited distant from the genes they control (called distal enhancers) appear to have been a necessary potentiating feature for complex spatial and temporal control of animal development. The pea family (legumes) includes beans, chickpeas, lentils, and peanuts and covers plants like clover, which farmers plant to help rejuvenate the soil. Symbiotic bacteria among their rhizomes convert nitrogen in the air into ammonia, a critical nutrient for plant growth. But whole-genome duplications (polyploidy) appear to have been required for the formation of these nitrogen-fixing root nodules among legumes. Finally, genes in many eukaryotic cells are divided into pieces, with protein-coding regions (exons) separated by noncoding introns. The distribution of introns across eukaryotes suggests intron-richness potentiates complex multicellularity. Identifying potentiating factors is challenging for unique novelties and innovations.[6]

The Cardinal Virtues: Radical, Generative, Consequential

Definitions of novelty are so disparate because of the discordant valences assigned to different aspects of the problem. Dictionary definitions are "new," "original," or "unusual," capturing the view that novelties are a structural or functional departure from an earlier condition. My emphasis earlier in this chapter on the importance of disparity in recognizing novelties associated with the Cambrian Explosion of animals reflects my concern over the magnitude of the change between an ancestral form and the novel descendant. Indeed, the contrast I drew between saltational and transformational views of novelty in chapter 2 implicitly valued radical change as a cardinal attribute of novelty, and the radicality of a novelty may be the most obvious definition. There is little doubt that beaver-like Jurassic mammals were radical, for example, but considerable doubt about whether they had the ecological impact on rivers and streams of modern beavers.[7]

Novelty has also been viewed as "unanticipated knowledge":

> This is not the problem of observing the "impossible", that is an event whose possibility we have considered but whose probability we judge to be 0. Rather, the problem arises when we observe an event whose existence we did not even previously suspect; this is the so-called problem of "unanticipated knowledge."[8]

Others who work on novelty (including a few friends) will disagree with emphasizing radical transformations. Simpson's invocation of adaptive radiations relies on claims about *generativity* as a critical component of evolutionary novelties: their importance derives at least in part from the generation of new species or clades entering a new adaptive zone. Just as phyla or classes are something we can count as an imprecise metric of radicalness, species are things that one can count as a measure of the productivity or generativity of an evolutionary diversification. Following from Simpson's emphasis on the central role of adaptive radiations, many evolutionary biologists have focused on generativity as a critical component of novelty. In economics, "general-purpose technologies" can impact large segments of an economy and enable many different sorts of technology. They include language, pottery, the generation of energy by fire, the steam engine, and electric motors. Vacuum tubes, transistors, and semiconductors are all found in many products. Not surprisingly, some historians of technology have focused on general-purpose technologies as "engines of growth." Novelties can also be generative by sparking further novelties, and several examples of these will be presented in later chapters.[9]

Finally, novelties may have ecological and evolutionary impacts on their clade or, through ecological interactions, on other species with whom they share an environment. The origin of oxygen-producing photosynthesis about

three billion years ago eventually led to the generation of the oxygen-rich atmosphere in which animals thrive, an innovation with pervasive effects on
most organisms and on the geochemistry of the oceans and atmosphere. Novelties may be recognized because they were consequential, as in cases where
the novelty was access to a new neutral network. The major evolutionary transitions of Maynard Smith and Szathmáry, from the origins of life, eukaryotes,
and multicellularity to human language, were consequential. They enabled life
to pursue opportunities that had not previously existed and generated new
evolutionary individuals. Many discussions of innovations, following the definition used here, are really about the consequences of events.

Paleontologists have developed a range of tools to quantify the differences
between ancestor and descendant. Yet sometimes the things one most wants
to measure are those most refractory to measurement. As discussed in chapter 4,
the more radical a novelty the greater the difficulty in computing a distance.
Some evolutionary biologists relied upon counting the number of phyla,
classes, or orders as a metric for novelty. Until the rise of phylogenetic systematics, few worried about whether trilobites, as a class of arthropods, were
somehow equivalent to bivalves (clams), a class of mollusks. There they were,
each different from other classes of arthropods or mollusks, and that allows us
to count them. Also recall from chapter 4 that even when we can measure a
distance between two forms, there is little reason to think that evolution followed a direct path. The historical evolutionary distance can be vastly different
from the shortest apparent route between two forms. The idea of culture is
equally tricky. Many species exhibit culture and learning, and chimpanzees
have recently been shown to exhibit the accumulation of cultural traditions.
What distinguishes humans is the extent of our cumulative culture. But is it
possible to measure the extent of change in cumulative culture? Combinatorics, the number of different possible combinations, has been one approach,
but readers are likely to reach different conclusions over how satisfactory this
is as a measure of radicalness.[10]

Few writers have attempted to disentangle these three aspects of novelty.
Simpson is perhaps the most obvious example, for his emphasis on new
higher taxa and new adaptive zones combined elements of each. Today no
approach to novelty can be successful without addressing, and perhaps reconciling, these different virtues of radical, generative, or consequential. A single
episode of novelty and innovation can incorporate all three attributes. Examples abound: the origins of eukaryotes, oxygenic photosynthesis, the angiosperm flower, and insects with complete metamorphosis would qualify as

novelties by any of these criteria. But only radicality can be assessed soon after the generation of a novelty. Both generativity and consequentiality depend on whether a novelty becomes successfully incorporated within a clade and how successful that clade becomes (whether in terms of new species, ecological impact, or by some other criteria). In other words, generativity and consequentiality are ecological and evolutionary attributes of innovations, not of novelties, and can be assessed only historically. Many environmental and ecological factors will influence whether a novelty succeeds, so using them as part of the definition of novelty is problematic. From this I conclude that definitions of novelty must focus on radicality as a metric. At least in principle this can be assessed directly rather than retrospectively.

For a particular research program, focusing on the radical nature of a novelty, its generative potential or its ecological, evolutionary, cultural, or economic consequences may reflect different epistemic values. Problems arise, however, when these different aspects of novelty are conflated. If we recognize a novelty by its radical nature, but then argue that it must necessarily have generated many new species or had downstream evolutionary consequences, we have made a logical error. The most important reason for keeping clear about these attributes is simply clarity of discussion about novelty and innovation. I have attended conferences that degenerated into days of confusion simply because well-meaning organizers wanted to avoid the issue of adequately defining novelty at the outset.

A Conceptual Framework

The framework presented here is not a linear model for turning novelties into innovation. Rather, it recognizes that contingencies, positive and negative feedback, and temporal lags may be involved at various stages of the process.[11]

Potentiation

Potentiation encompasses a range of earlier events that are required to establish the conditions for the novelty to arise, such as the prior mutation for the Cit+ mutation in E. coli, and mutational order. As mentioned in the previous chapter, the order in which events occur plays a decisive role in the outcome. The order of mutations is important in genetics, where geneticists have described enhancing, enabling, and permissive mutations that facilitate subsequent evolutionary changes. The same phenomenon has also been identified in the structuring of gene regulatory networks and in the evolution of technology, providing support for the significance of potentiating events in evolution more generally.[12]

The potentiating mutations for the *Cit+* mutation are hardly unique, and similar patterns are known in other systems, often as "pre-adaptation." Snapping shrimp are crustaceans with asymmetric claws, found on coral reefs and in other shallow marine habitats. They make quite a ruckus, as they snap the larger claw to create a cavitating bubble. But this ability to snap requires several prior changes to the claw, including the tooth-cavity system. As was the case with the *Cit+* mutation, tooth-cavity systems arose many times, but the final novelty of the functional snapping claw evolved only once and led to the explosive radiation of at least 500 species of snapping shrimp (so even I might be willing to call this a "key novelty"). A similar example is found in ants, where most social ants have a variety of different castes that differ in size and shape, allowing for a division of labor among the castes. One of the most remarkable castes is the large super-soldiers of the genus *Pheidole*. *Pheidole* is a large genus with hundreds of species, and super-soldiers have evolved multiple times among distantly related species. But these are not independent productions of this novel morphology. Rather, the potential to generate this morphology apparently arose some 35 to 60 million years ago and has been independently triggered in different species by environmental conditions.[13]

One objection might be the challenges of reliably identifying potentiating changes. Certainly, the potential for exploring the mutations for *Cit+* in *E. coli* were almost unique: Lenski and Blount had a record of all 12 independent lineages every 500 generations extending back to the beginning of the long-term evolution experiment. Having identified the apparent potentiating mutations in the *Cit+* lineage, they could go back and check for the same change in other lineages. In addition, the *Cit+* change was sufficiently straightforward to allow them to identify the potentiating mutations. But evidence for potentiation already exists in three areas: behavior; variability in response to environmental changes, known as phenotypic plasticity; and changes in the function of genes or other structures, particularly through co-option.

Earlier I described studies suggesting that the origin of new structures can sometimes be found in behavioral changes in response to new circumstances. The spread of grasslands in many temperate areas of the world during the Miocene triggered changes in the behavior of grazing animals. Grasses contain small silica bodies—unlike the leaves of shrubs or trees—and horses, elephants, and other animals who graze on grasses have evolved teeth that continue to grow. These high-crowned teeth allow for tooth wear from the silica in grasses, so that an animal's teeth will last rather than be worn down to a nubbin. Pale-ontologists can establish when animals began eating grasses from differences in carbon isotopes, which are incorporated as teeth grow. As early elephants were confronted with spreading grasslands 8–10 million years ago, carbon isotopes show that they shifted from browsing on shrubs and trees to eating substantial quantities of grass. But the shift to grasses occurred millions of years before the adaptive response of growing longer (high-crowned) teeth. The evolution of high-crowned teeth in many herbivores was not a morphological novelty, although one could argue that the shift from browsing to grazing, which opened a vast new suite of ecological opportunities, was a behavioral novelty. But we have better data for the relationship between behavioral changes and evolutionary response than exists in other cases where pheno-typic novelties arise, including changes in locomotion associated with walking on land, flight, and bipedal locomotion in humans.[14]

The shift from browsing to grazing with the spread of grasslands is just one example of how behavioral changes can be induced by changes in environmental factors. The shift from living in trees to walking across African grasslands led to upright posture and bipedality in humans, a shift in behavior leading to a shift in morphology. Such changes are easier in species where environmental cues change development. If growing mussels sense the presence of crabs or other predators, they invest more resources in a thicker shell, stronger byssal threads attaching the shell to a rock, and stronger muscles. Plants have similar responses to insects or herbivores, and some crustaceans produce spines in response to predators. Recall that in chapter 3 I introduced the "plasticity first" hypothesis, in which such developmentally plastic responses establish the conditions for the successful introduction of novelties. An interesting argument can be made that species with greater variability in their responses to the environment might be more likely to generate novelties. In such cases the phenotypic plasticity acts as a potentiating factor for subsequent novelty.[15]

The legs of centipedes and millipedes and the enveloping wings of skates and rays arose through repurposing of existing networks of genes to build appendages in a new location. The co-option of regulatory circuits established for one function into a new locality or a new function is one of the most potent forces of evolutionary novelty. The larval skeleton of all living sea urchins and a lobe on the posterior of a group of flies, novelties by any definition, also reflect the redeployment of existing networks of regulatory genes into a new developmental address. Thus, the original gene regulatory networks were potentiating events for the later novelties. As discussed in greater detail in chapter 6, such co-options are extraordinarily widespread foundations of evolutionary novelty.[16]

Decades of evolutionary biologists knew such changes of function as pre-adaptation. In 1982 Stephen Jay Gould and Elisabeth Vrba suggested the term *exaptation* would be more appropriate, as "pre-adaptation" had the connotation of a structure having been formed in advance of need, in contravention of all that we know about natural selection. Gould and Vrba pointed out that what were described as pre-adaptations were all fine adaptations to their original use, but they were also pre-positioned to be co-opted into new uses, and often new novelties. Feathers, the malleus and incus that allow hearing in mammals, and the origin of vertebrate lungs from the swim bladders of fish have each been proposed as an exaptation. Most cases of exaptation do not result in the formation of an evolutionary novelty, so while not all exaptations are novelties, the frequency of co-option suggests that many potentiating factors may be exaptations.[17]

Evaluating potentiation becomes increasingly challenging the deeper we go into the past, as it can become difficult to disentangle causality. Comparative studies of the evolution of developmental processes provides a framework for such analysis, however, as it can establish possible potentiating events, which can later be evaluated by experimental approaches. Most importantly, however, the challenges of identifying potentiation should not blind us to the importance of such events. A final caveat. Applying potentiation to *any* historically prior event would render the concept meaningless. To be useful, the idea of potentiation applies only to events that were directly necessary for the generation of a novelty.

Novelty

The central motivating issues for this book are investigating commonalities in novelty across biology and across biological, cultural, and technological domains, and evaluating the importance of novelty and innovation in the generation of diversity and complexity in the history life, culture, and technology. One approach would be enumerating exemplars or case studies based upon some intuitive belief in what constitutes a novelty, but this avenue could be justifiably criticized as ad hoc. I began this research program intending to articulate a single definition for evolutionary novelty, in the belief that only by rigorously circumscribing the phenomena of interest could we make progress.

But confronted with the complexity of the problem I have considerable sympathy for the views of Brigandt and Love in accepting definitional diversity, limiting the range of problems considered yet not so narrowly as to exclude important events. But their interest as philosophers of biology was principally about what biologists viewed as within the remit of novelty, rather than the comparative questions with which I am concerned.[18]

In discussing radicalness, consequentiality, and generativity I concluded that a character-based definition of novelty is the most workable. Definitions based on process are inherently limited to the results of those processes and hence are too restrictive to encompass the scope of novelties of interest. Definitions based on the consequences of a novelty depend upon a wide range of events beyond the origin of novelty, conflate novelty with innovation (as used here), and preclude inquiry into why some novelties never generate innovations.

Building off the discussions in chapter 3 of novelties as new, homologous, individuated characters, I recognize three different types of novelties. First are new character identities, such as the insect wing, a new gene without homologous components, or a new form of regulatory structure, such as the regulatory roles of small RNAs in developing animals. This is equivalent to Günter Wagner's type 1 novelty. The second type of novelty involves deep modifications to an existing character such that the character becomes newly individuated. This is a common form of novelty for appendages, particularly in arthropods, and includes the butterfly wing; the hard, outer wing (elytra) of beetles; antennae; feeding structures; walking legs and swimming legs—all of which are modified appendages. These novelties are homologous structures that have been so deeply differentiated as to become effectively newly individuated structures and are equivalent to Wagner's type 2 novelties. Finally, new characters arising through combinations are one of the most effective strategies for generating novelties, forming the third class. Whether combining domains of different proteins to form new proteins, or technological components for an invention, recombination has long been a potent source of novelty.[19]

Focusing on individuation is applicable across a wide range of phenomena, from the construction of new genes or the wiring of gene regulatory subcircuits to new phenotypic characters. It can also be applied to cultural and technological features, from language to the formation of cultural, political, and economic institutions and the invention of new technology. Günter Wagner's discussion of feathers as novelties exemplifies why individuation is so central to the concept. Feathers evolved from vertebrate placodes, simple thickenings in the skin. As a reminder, feathers are a novelty, but the myriad variety of feathers constitutes alternative character states, from the downy undercoat of geese to the strongly asymmetric flight feathers of a hawk. Wagner initially linked morphological novelty to the formation of recursively wired gene regulatory networks, but I decline to follow this part of the definition because I view the issue of the underlying structures as a hypothesis to be tested, rather than an intrinsic part of the definition (more on this in the next chapter).[20]

Focusing on individuation alone, however, has drawbacks. One of the hallmarks of humans is the ability to generate things beyond the capacity of one person, or even a small group of people. Such cumulative cultural evolution is perhaps the defining characteristic of humans and has been intensively studied by cultural evolutionists. No other species accumulates an expanding stock of cultural attributes or has proven capable of using them for such pervasive remodeling of its environment. But this also illustrates the complexity of characterizing novelties by individuation. The evolution of human culture required cognitive changes among individual early humans. But cumulative culture is a social phenomenon that allows generations to build upon a body of knowledge and generate long-lasting institutions such as universities, but requires high-fidelity methods of transmission. But characterizing the novelties as individuated characters is challenging: is cumulative culture, as a collective property, the novelty, or are the specific changes that allow cumulative cultural evolution the suite of novelties? I will return to these issues in chapter 9, where cultural and technological novelty and innovation are discussed in greater detail.[21]

I make no claim that novelties necessarily arise suddenly, although some may. Goldschmidt's emphasis on "hopeful monsters" and wholesale genetic revolutions poisoned subsequent discussions of novelty by linking them to rate, rather than mechanism. Some evolutionary biologists happily wield the cudgel of hopeful monsters against those whom they view as transgressors against evolutionary orthodoxy, but this is inappropriate. Some of the novelties I discuss in the following chapter were likely rapid, simply because it is as hard to co-opt half a gene regulatory circuit as it is to understand the function of 15 percent of a wing. But apparently sudden discontinuities observed in the fossil record still represent the accumulation of change over some interval of time.

My contention here is that most of the interesting novelties (an issue of consequentiality here, not individuation) did not arise through adaptive search as described at the conclusion of chapter 4, but through novel combinations that generate new spaces, or new means of searching a space. A pressing issue discussed in later chapters is to what extent novelties arise from available variability or through other processes that can generate novelty. Any means of generating novelties necessarily requires that viable organisms exist throughout the process. Evolutionary continuity is not negotiable.

Adaptive Refinement

As with the *Cit+* mutation in *E. coli*, novelties will require an interval of subsequent adaptive evolution. Morphological novelties may require adjustments in developmental processes and in other aspects of morphology. Behavioral novelties may generate adaptive changes in morphology, and novelties in development may need accommodations in regulatory pathways or in the complex of proteins and DNA that structures gene expression. The cleanest sort of developmental novelty involves co-option and redeployment of existing regulatory circuits. But even here adjustments occur. An example discussed in more detail in the next chapter is the long pectoral fin of skates and rays, which extends onto their head and allows their novel method of swimming. All other vertebrates have a single growth center controlling limb development, but in skates and rays the ancestral regulatory module for limb development was co-opted to form a second growth center. The patterns of expression of developmental genes are very different, presumably as the new, anterior module was modified to generate the unique fin morphology. The complex features we recognize as morphological novelties are unlikely to appear fully-formed and without needing further refinement.[22]

Exploitation

An interval of further potentiation and the generation of adaptive changes may arise to fully exploit the initial potential of a novelty. Macroevolutionary lags occur because of the contingent nature of changes in the physical environment or ecological interactions required for innovation.

The importance of further potentiation underlies the distinction between invention and innovation and has long been evident to military historians. The British invented tanks during World War I as a response to the long stalemate in the trenches of Europe. But they were first used as mobile platforms for machine guns to protect infantry advances. World War I ended before either Allied or Axis military had sufficient experience to recognize the potential for tanks to operate independently. If tanks were the invention, the development of appropriate tactics and military doctrine in the interwar years was the potentiation needed before the military use of tanks reached an apogee during the great tank battles of World War II. Military history is a long pattern of inventions and then generation of appropriate tactics, followed by counter-technology and tactics, from English longbowmen against French cavalry at Crecy in 1346 to Ukrainian use of antitank weapons and drones against invading Russian forces in 2022. Some military inventions had more limited success. The Byzantines invented Greek fire, an early flammable liquid somewhat like napalm. Yet their success spraying it through nozzles as a flamethrower was short-lived. The recipe was such a closely guarded secret that its use had ended by the tenth century.[23]

Innovation

The exploitation of a novelty is an ecological process, and thus is dependent upon the environmental and ecological setting. When novelties have been linked to adaptive radiations, the diversification of taxa is the signal for exploitation of an ecological opportunity. But Simpson articulated several other triggers for adaptive radiations. And as explored further in later chapters, there are many classes of evolutionary diversification that are not adaptive radiations. Hence the need for a metric other than adaptive radiation for the success of a novelty. Using the same logic that Schumpeter did in distinguishing economic invention and innovation, I regard changes to the structure of ecological networks as the hallmark of innovation. And by ecological network I mean more than traditional food webs constructed by ecologists.

Food webs are networks of trophic relationships among interacting species. In other words, who eats whom, as well as parasitism, the detritivores that complete the flow of nutrients from dead organisms back into the ocean or the soil, and other, more specialized forms of interactions. A parrotfish nibbling on coral is a link in a network for a coral reef, as are connections between a host coral and its photosynthesizing symbionts. What food webs miss are the nontrophic interactions among organisms, some of which may be just as critical to the functioning of a community. Reefs are a habitat for dozens or hundreds of species with which the corals and coralline algae that construct the reef may have no direct trophic interaction. Similarly, sponges, oysters, and other animals that filter bacteria and other organisms out of the water have a tremendous impact on the overall ecological communities in which they live. These are all ecosystem engineers, and to capture the full impact of an innovation we need to examine the full network of ecological interactions.

In chapter 7 I will discuss several approaches to more rigorously capturing the dynamics of innovation. One employs the rapidly growing suite of tools to investigate network dynamics, to assess when a new species containing a novelty is sufficiently critical to the network that removing the species changes its dynamics. Fortunately, removing the species can be done computationally if our understanding of the network is sufficiently robust. Another approach relies upon a long-standing interest among ecologists in ecosystem novelty (which encompasses innovation here), an increasingly important issue as human-induced changes in climate and habitat generate new combinations of species. One group of ecologists defined ecosystem novelty as "the degree of dissimilarity of a system, measured in one or more dimensions relative to a reference baseline," while others have focused more on new species combinations and changes in ecosystem functioning. But ecological communities are rarely static, and may have different species in different locations, and change through time as well, limiting the utility of some of these approaches for understanding the impact of evolutionary novelties. Extensions to these approaches that focus on ecosystem functions rather than just new species are promising and may provide a useful metric for innovation.[24]

A significant research question for the future is identifying similar changes in ecological networks and asking how often they are linked to novelty, the inverse question. My intuitive expectation is that most innovations will arise independently of evolutionary novelty, largely driven by adaptive changes in species, extinctions, or local extirpation, or by environmental changes.

NOVELTY, ADAPTATION, AND DIVERSIFICATION

How do novelty, innovation, and adaptation differ from one another? These terms are often used interchangeably because most biologists do not distinguish between novelty and innovation, and many believe that novelty and innovation are the result of adaptive change. Others, weary of other terminological battles, avoid rigorous definition in the belief that science is best served by being intellectually inclusive in the hope that a consensus will emerge. No research program can succeed without rigorously circumscribing the domain of study. The probability of reaching false conclusions about the underlying mechanisms is greatly increased by failing to specify the objects of interest. In this chapter I have argued that novelty is associated with the formation of newly individuated characters. There is a crucial difference between the formation of new individuated characters and the generation of new character states.

The origins of new feathers and of arthropod appendages are each an example of a new homologous character, whereas the formation of different kinds of feathers or appendages (character states) does not represent novelties but adaptive evolutionary change. I have acknowledged that this is a highly restrictive definition of novelty relative to many other proposals, but it is relatively unambiguous. The hypothesis advanced by Wagner, that individuated characters reflect the construction of gene regulatory subcircuits, is consistent with other studies by Eric Davidson and me. But this does remain a hypothesis, as the number of morphological or phenotypic characters for which we have detailed knowledge of the underlying developmental gene regulatory networks remains quite limited. But the available data are consistent with this view.

I think a larger issue is whether Wagner's definition of novelty is too restrictive: Does it eliminate some evolutionary changes we think should be included as novelties? For example, changes in body size often allow access to otherwise unavailable ecological opportunities. The remarkable decrease in the size of maniraptoran dinosaurs leading to the origin of birds and the steady increase in body size of whales greatly expanded the ecological and evolutionary opportunities for these clades. But identifiable morphological novelties are not associated with the origin of either clade (feathers evolved before birds, while some of the novelties associated with whales, particularly feeding structures, evolved after the origin of whales). If novelty is correctly based on characters, then changes in size, however important they may have been, are rarely associated with diagnostic changes in particular characters.

I do not view either of these cases as a reason for expanding the definition of novelty, however. These are examples where changes in size are associated with the origin of a clade, but not the generation of new phenotypic characters. I will return to these examples in chapter 9, but in these cases the origins of birds and whales (Cetacea) represent both the origin of a clade and an evolutionary innovation.

CAVEATS

The limitations and difficulties with this conceptual framework are worth stating explicitly. I do not believe that these issues are insurmountable, although resolution of some of them will require further research.

- The most critical point is that the framework suggested here *does not* require events to occur in the order listed. Adaptive changes may arise

before phenotypic novelties, an issue that has long been discussed as
"pre-adaptation," and which Gould and Vrba described as exaptation,
and neutral changes may provide opportunities for the generation of
subsequent novelties. While Lenski's work motivated this scenario,
other models of protein evolution emphasize factors such as compen-
sating mutations, which might enrich this view of novelty.

- Macroevolutionary lags reflect the importance of environmental and
ecological events or factors, but there is no reason to expect any clear
relationship between them and novelties. Indeed, the potential for
decoupling is expected and differentiates this framework from adaptive
evolution models.

- As I noted in chapter 3, as novelty has become the hip thing to study,
many scientists have adopted a definition of novelty sufficiently flexible
to incorporate their favorite organism/system/question. But too flexible
a definition of novelty makes the question so squishy that rigorous study
becomes almost impossible. Wagner's definition of novelty as formation
of individuated, homologous phenotypic characters usefully restricts
the scope of what is considered as evolutionary novelty.

- From the foregoing it may appear that phenotypic novelties are a
necessary precondition for the ecological transformations recognized
as evolutionary innovations. In other words, that there is no microevolutionary, adaptive trajectory to innovation. This certainly appears to
have been the view of saltationists, for example, but there is neither
logical nor empirical support for such a claim. Ecological and evolutionary innovations may occur without novelty. Evolutionary transformations and the origins of major clades do not necessarily require
novelty. Indeed, as I discuss later, there are many episodes in the history
of life that are transformative but are due entirely to adaptive evolution
and drift.

- Novelty and innovation differ significantly in this conceptual framework. Novelties affect various attributes of organisms, from genes to
culture. Since novelties are character based they will cluster phylogenetically, while innovations may cluster ecologically. Environmental
factors may generate clusters of innovations. In other words, those
novelties that facilitate other novelties will tend to cluster within a
clade. In contrast, since innovations are an ecological rather than a
phylogenetic phenomenon, generative innovations will cluster ecologically or geographically.

WHAT'S A MODEL TO DO?

I have described this as a conceptual framework rather than a model because it is primarily a way of thinking about the problems of novelty and innovation, and not because events will necessarily follow a particular trajectory. The cases presented in chapters 6 to 9 illustrate such variability. The nature of models is a fraught issue in both science and the philosophy of science. Models of novelty and innovation come in a wide range of flavors to serve different purposes, depending upon the questions being asked.

Models are employed for many reasons, but Scott Page's book *The Model Thinker* identifies seven: to reason, explain, design, communicate, act, predict, or explore (which helpfully form the acronym REDCAPE). Models help us reason about the circumstances that we seek to understand and to think logically about how different components of a system interact. Ideally, in explaining phenomena of interest models help us understand mechanisms and generate testable hypotheses. Models necessarily simplify reality, so the act of designing a model requires the model maker to identify some features to include in the model, while excluding others. Sewell Wright invented fitness landscapes largely to communicate with his less quantitative colleagues, as we saw in chapter 4. As should be clear from chapters 2 and 3 we currently have a failure to communicate what we mean by novelty and innovation, often because of buried assumptions about the nature of evolution. Such assumptions may canalize models along a particular course. Often this canalization is not evident to the reader. Fisher and Wright laid the foundations for quantitative population genetics with a suite of interlinked models that allow geneticists to do everything from understanding adaptation to predicting the amount of milk produced by crossing different breeds of dairy cattle. But the power of quantitative population genetics comes from a certain agnosticism about the nature of a gene and the complexity of regulatory and developmental interactions. Some approaches to evolutionary novelty have remained similarly agnostic. Combinatoric models based on sequence spaces of DNA or proteins, for example, assume that novelty is defined by new structures with new functions, and thus have an implicitly consequential model of novelty. This is not necessarily a false model of novelty and innovation, but it does contain within it, as any model will, certain assumptions about issues such as search. In economics, the models used in policy studies and other areas often serve to guide action (how can a country increase rates of invention and innovation, for example?) and generate predications. Making predictions (technically, retrodictions) about the past might seem even more treacherous than New York Yankees' manager Yogi Berra's observation that "it is tough to make predictions, especially about the future." The history of life is a rich test bed for evaluating evolutionary models. By allowing us to play with our intuition and test different scenarios, models are a tool for exploration and refining our understanding of the complexity of our natural, cultural, and technological world.[25]

The intellectual framework of this book is guided by the goals of the project. One objective is to explore the possibility of identifying whether a such an intellectual scaffold or a modification of it might apply to novelty, invention, and innovation across biological, cultural, and technological domains. There are reasons for thinking that this is not a wholly quixotic task. Insights from the evolutionary theory of the day have frequently been invoked by historians, economists, and others interested in technological evolution, while cultural evolution has drawn from work on other aspects of biological evolution since Darwin's works.[26]

I have been less explicit, however, about a second objective, and that is to evaluate the role that novelty and innovation have played in the expansion of diversity. Most economists have now concluded that economic growth (with gross domestic product, or GDP, as the metric) is underpinned by technological innovation. Is the same true of cultural evolution more broadly, and of biological evolution? Economists have a fairly straightforward metric in GDP (although one that turns out to be more complicated in practice, as we will see). The obvious analog in biology is the number of species, or species diversity. But many species are ecologically redundant: they serve similar or identical ecological roles. To understand the role of novelty and innovation we need a broader view of biodiversity than simply the number of species. Fortunately, biologists have recognized that biodiversity encompasses a far richer scope than simply the number of species. Morphological diversity, or disparity, has already been discussed. Phylogenetic diversity depends upon the structure of phylogenetic trees and has become an important metric in conservation biology. Other aspects of biodiversity include the architectural structure of ecological communities, the variety of ecological functions, genomic diversity, the range of developmental processes, and the richness of behaviors. If these other aspects of biodiversity closely tracked the number of species, that would be a useful proxy for biodiversity, but in fact there is relatively little correlation between species diversity and these other metrics, as already discussed for disparity in chapter 4.[27]

The argument here is that distinctions among potentiation, novelty, refinement through adaptation, and innovation (possibly preceded by further potentiation) provide an intellectual scaffold for exploring how a wide range of novelties and innovations in the biological, cultural, and technological domains have been constructed. This framework raises many questions: How wide a range of events can generate potentiation? Is novelty required for innovation? (It depends, at least in part, on one's definition of novelty.) Is it possible to cleanly separate novelty and innovation, as suggested here? To return to the questions posed in chapter 1, can we identify commonalities in the different aspects of this framework across biological, cultural, and technological domains? Not all these elements need necessarily be present in each of the cases presented, but they provide a set of questions to guide investigations. In chapter 6 the focus is on the genetic and developmental processes that generate novelties, before chapter 7 turns to the broader ecological and evolutionary dynamics of innovations, and chapter 8 presents a series of cases. As humans we possess unique traits, of which language and cumulative culture feature prominently in the discussion of cultural and technological novelty and innovation in chapter 9. In the concluding chapter I return to the possibility of comprehensive models of novelty and innovation and similarities that may exist across these domains.

6

The Origins of Novelties

VERTEBRATE LIMBS, flowers, and the wings of insects each represent classic novelties. But do they represent novelties all the way down? New genes, new tissues, new types of cells, and new developmental processes underlying novel morphologies? One might anticipate that organisms with greater apparent complexity (squid, orchids, or perhaps even humans) would have more genes and more cell types. But therein lies a paradox at the heart of evolutionary novelty: new genes and new cell types have accumulated in many lineages of plants and animals, but it has long been evident that novel morphologies rarely depend on new genes or new networks of interaction among genes. Evolution is frugal, repurposing existing genomes and interactions to generate novel structures. As French biologist Francois Jacob recognized, the principle of evolution is *bricolage*, tinkering, not engineering.[1]

In 2000 the genome of the fly *Drosophila melanogaster* was found to have about 14,000 genes distributed across 144 megabases (a measure of the length of the DNA), while the human genome is about 3100 megabases. Humans were also known to have multiple copies of many genes that existed in only single copies in the fly (such as the Hox genes). And of course, we are far more complex than flies. Yet in one of the most stunning insults to the human psyche since Darwin published the *Origin*, the initial sequencing of the human genome in 2003 identified only 25,000 genes (the number of genes continues to be revised, most recently to about 20,300). Among sponges, extensive duplication of genes has produced genomes of some 40,000 genes, while sea anemones and other cnidarians have 18,000–20,000 genes. Genome size roughly doubled with the origin of animals, but only about 20,000 genes are needed to build any animal, from a sponge to a lobster or a human. This should not be a surprise following the discussions of combinatoric complexity in chapter 4. Humans do have new genes and more cell types than a fly, and similar expansions in genome size are known from the origin of vascular plants, but the number of new genes is insufficient to generate underlying novel developmental pathways. Rather, novelties rely on reuse, co-option, and repurposing of existing genes and genomic capacity for new functions, and in finding new forms of controlling those functions to generate novel structures.[2]

This chapter examines the sources underlying each of these three types of novelties within the biological domain: newly individuated characters, deep transformations, and novel combinations. The critical role played by potentiating factors was introduced in the previous chapter, but since they are often context specific here they are treated within case studies of novelties, but this is not intended as exhaustive analysis. Rather, these are illustrative applications of the conceptual framework. Following consideration of the nature of innovation in chapter 7, several of the studies introduced here will carry through to chapter 8 on innovation, including the origin of the eukaryotic cell, the early diversification of animals and the Ediacaran–Cambrian Explosion, and the spread of flowering plants.

These case studies illustrate different mechanisms for generating novelties. The examination of the origins of animal multicellularity and the elaboration of arthropods during the Cambrian radiation provides a general introduction to the role of developmental interactions. The success of arthropods has arisen in part because of their frequent individualization of different appendages to generate mouthparts, feeding structures, gills, and walking legs. The co-option of networks of gene interactions has been involved in many novelties, including insect wings, beetle horns, and the helmets of treehoppers that are examined here. These cases also suggest that the magnitude of a novelty may depend on where it arises in development. Plant novelties provide informative similarities to and differences from those in animals, and emphasize the potentiation delivered by duplication of entire genomes (which also occurred at the base of vertebrates). Although new cell types were critical to the origin and early diversification of animals, one of the best studied examples of the generation of new cell types and a new organ involves the placenta of eutherian (or placental) mammals, an example that also provides an opportunity to illustrate the role of transposable genetic elements. One of the best studied examples of a deep transformation of existing morphology is the extension of pectoral fins into the sweeping wings of skates and rays. Finally, the origin of the eukaryotic cell provides a paradigmatic example of combinations as a route to novelty.

A particularly intriguing discovery from comparative developmental studies is that novelties may generate the capacity and potential for changes at other levels, and often this capacity may arise well before subsequent novelties. This pattern of early potentiation has been associated with expansion of regulatory capacity in lineages leading to the origin of animals, from patterns of genome duplication that provided the raw material for novelties among various plant clades, and even from the genetic underpinnings associated with the shift from fins to limbs in shallow seaways some 400 million years ago. This provides a new perspective on the apparently abrupt appearance of novelties in the fossil record, as potentiation in the genome and in developmental processes may be largely unobservable.

The Origins of Animal Multicellularity

The Ediacaran–Cambrian radiation of animals involves increased genomic, developmental, and ecological complexity in one of the most concentrated and far-reaching episodes of novelty and innovation in the history of life. Every major clade of animals appeared between 570 and 520 million years ago: sponges to vertebrates and a plethora of early, enigmatic forms. Hundreds of new cell types and dozens of tissues and organ systems were required, along with the necessary developmental patterning systems. By combining studies of living animals with fossil evidence we can peer back through time and establish surprisingly robust hypotheses about the tempo and mode of evolutionary changes, including insights into the origin of fundamental animal novelties. Fully understanding these innovations requires consideration of ecological interactions and changes in climate and geochemical cycles—notably, a substantial increase in oxygen levels in the oceans and atmosphere. Unraveling not just the biological events but the complex chains of causality in which they are embedded continues to attract attention from molecular and developmental biologists, geochemists, and others.

These animal architectures have also been remarkably enduring. How have these novelties become resistant to subsequent change? The many aspects of basic animal architectures (eye and gut morphology, how limbs or shells are generated, cell and organ types) have remained essentially stable for the past half-billion years, despite considerable environmental churn and extraordinary evolutionary fecundity. Beginning in the 1990s, this conservation of morphology and developmental patterning was extended to the genome, with the discovery of highly conserved developmental genes and gene patterning systems, as illustrated by eyes.

Vertebrate eyes share little similarity with those of a fly or an octopus, and each of these is very different from the much simpler eye spots found among clams or cnidarians. A 1977 paper suggested that eyes had evolved perhaps 65 times across animals. Underlying this incredible structural diversity, however, is a shared suite of opsin proteins and a common regulatory machinery. Some evolutionary biologists argue that natural selection has preserved these architectures through stabilizing selection. To others this seems an inadequate explanation. My late colleague Eric Davidson, a developmental biologist, proposed that this architectural stability was achieved by recursively wired gene regulatory networks, in which feedback locked a subcircuit of genes into a particular regulatory state that ensures a specific developmental and morphological outcome. Through years of experimental work, Davidson and his colleagues identified such a recursively wired subcircuit at the core of gut formation in developing sea urchins, and similar circuits are involved in heart formation. Davidson showed that perturbing these subcircuits by knocking out the function of any of the five core genes killed off the developing embryo, showing that these interactions locked cells into a gene expression pattern. This and other evidence suggested that once these expression patterns evolved, they would be resistant to subsequent evolutionary change. Davidson and I called these kernels, by analogy to the computer program at the core of an operating system. Evolutionary changes still occurred, but successful changes were more likely either upstream or downstream of the kernel within the regulatory network.[3]

The same year we published our ideas, Günter Wagner identified gene regulatory networks with a similar topology as the stabilizing factor underlying cell types. With his long-standing interests in homology, Wagner christened these character homology interaction networks, or ChINs, which figure prominently in his 2014 book on evolutionary novelty. Kernels and ChINs involve the same gene subcircuit structure, but what Davidson and I did not appreciate was that if kernels originated to ensure the stable generation of cell types, they were ideally positioned to serve as the core of larger networks to ensure stable developmental generation of regions of the developing embryo such as the heart or gut. Recognizing that the persistent stability of novelty was a fundamental problem, Wagner and his colleagues subsequently broadened their concept to character identity mechanisms (or ChIMs), to include cell types, tissues, and organs. This encapsulated Wagner's focus on novelty as a means of generating new individuated characters. As our understanding of the processes of the regulatory genome has expanded, however, it has become clear that evolution has discovered a wider range of mechanisms to ensure the conservation of developmental and morphological architectures, while allowing continued evolvability. A key feature of novelties associated with the origin and early evolution of animals is the generation of conserved patterning mechanisms and their subsequent co-option to generate many aspects of morphology.[4]

The Ediacaran–Cambrian radiation began during the Ediacaran Period, with the oldest animal fossils dating to 570 million years ago, and extends into the Early Cambrian. The earliest macroscopic animal fossils are found among rocks in the eastern Canadian province of Newfoundland, where long fronds, disks, and more ornate shapes make up the Ediacaran fauna, named after the Ediacaran Hills in South Australia, where such fossils were first characterized. The number of different forms found in the Ediacaran fauna increases in younger rocks in Russia, China, Australia, Namibia, and elsewhere. The organisms represented by these fossils can be a half a meter or more long, but all were soft-bodied with no evidence for a shell, carapace, or skeleton until close to the Cambrian boundary. *Dickinsonia* is pancake shaped with a central furrow and appears superficially segmented. Careful study of hundreds of *Dickinsonia* fossils, many with the preservation of fine detail, reveals evidence of tissues, muscles, and a primitive nervous system but no evidence for a gut or regional differentiation, much less a mouth, appendages, eyes, or similar features.[5]

By 540 million years ago the earliest skeletonized tubes, spines, and shells appeared. Soon after this, all the major clades of animals appear as fossils, from sponges through various mollusks, trilobites and a huge variety of other arthropods, and even the earliest vertebrates. Sponges and arthropods had already split into distinct major lineages when they first appear as fossils, confirming that there was a gap in the record of their earliest history. Some fossils represent skeletonized organisms, but paleontologists are enormously fortunate to have a dozen or more places with exquisite preservation of animals with soft parts. In such cases legs, eyes, and traces of the gut can be found, and a few fossil arthropods reveal the structure of nerves and the brain. Together this rich fossil record documents the sudden appearance of representatives of extinct clades and stem groups to living clades (figure 6.1). The absolute number of Cambrian species was far lower than in modern oceans, but a striking and fundamental feature of this Ediacaran–Cambrian radiation was that the earliest members of a large clade staked out the morphological boundaries occupied by the clade for the following tens or hundreds of millions of years. In his book about the mid-Cambrian Burgess Shale fossils, Stephen Jay Gould described this pattern as early maximal disparity, a claim that has largely been confirmed by subsequent quantitative studies of disparity.[6]

The apparent suddenness of the Cambrian Explosion was evident when Darwin wrote the *Origin*, but he explained this as an artifact, a lack of preservation

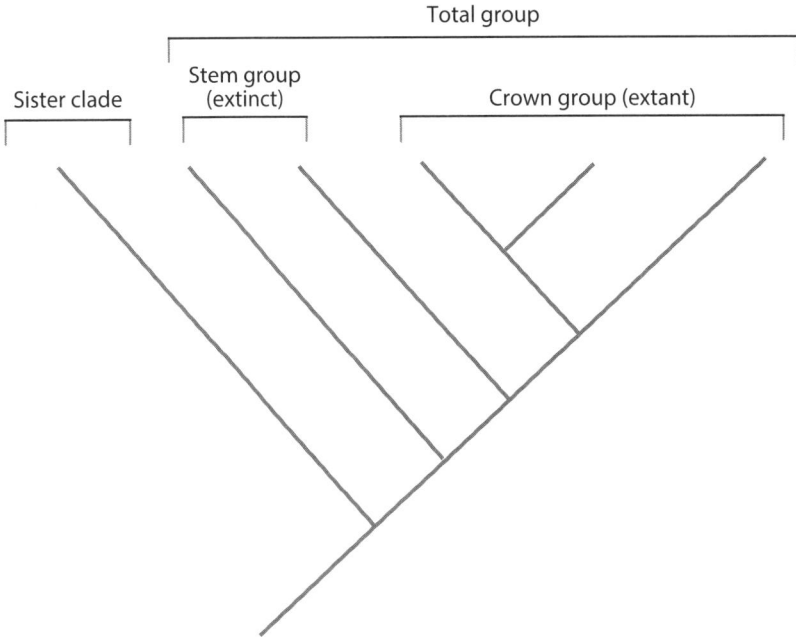

FIGURE 6.1. Sister, stem, crown, and total group clades.

of a continuous and gradual divergence of animals. By the late nineteenth century this view had become more sophisticated, with geologists such as Charles Walcott, a noted Cambrian stratigrapher and then secretary of the Smithsonian Institution, invoking a lengthy gap known as the Lipalian interval, during which no sediments were preserved. Although the idea of a Lipalian interval vanished long ago, early in my career paleontologists had a poor understanding of time through the Ediacaran and Cambrian, with many believing that the Ediacaran fossils predated the Cambrian by tens of millions of years. We have now established a fairly continuous record of fossils from the Ediacaran into the Cambrian (see chapter 8). Despite this, one still might argue that the suddenness of the Cambrian event was created by poor fossilization or some other artifact. A strong reason for rejecting such a view is provided by examining the sediments of this time. Burrows, trails, and trackways are also preserved in the rock record as trace fossils. There is a robust trace fossil record through this interval, which exhibits a profound transformation to well-burrowed sediment near the base of the Cambrian, with a trackway from an organism with paired appendages described from latest

Ediacaran-aged rocks in south China. Through the early Cambrian there is an increase in burrows that penetrate vertically into the sediment and in the size and complexity of burrows. From this we conclude that the body size of many animals increased, and that some had the nervous and muscular systems to form vertical burrows and sensory systems to track their movements. So marked is this change that geologists would easily recognize that a significant evolutionary event had occurred even if no skeletal or soft-bodied fossils were preserved.[7]

Comparisons of molecular sequences through a molecular clock to infer when clades diverged provides another avenue to insight into early animal evolution. This requires knowledge of the evolutionary relationships among the taxa, careful calibration of the clock using well-dated fossils, and appropriate statistical methods. The most recent consensus phylogeny for animals is shown in figure 6.2, with the structure of the tree calibrated using molecular clock results. Animals are part of a larger clade, the Holozoa, with the Fungi as the next closest, or sister group, to the Holozoa. At the base of the tree lie the other holozoan clades, to which we will turn shortly. Sponges (Porifera) are the most basal animal group, and molecular clocks estimate that they originated about 700 million years ago (with considerable uncertainty on this estimate), followed by the split of cnidarians (jellyfish, sea anemones, and *Hydra*) about 650 million years ago. The divergence of the three great bilaterian animal clades at about 630 million years ago generated the deuterostomes (echinoderms such as starfish and sea urchins, chordates including vertebrates, and other groups) and the protostomes; the Lophotrochozoa (mollusks and annelids); and the Ecdysozoa, which includes the arthropods and related clades.[8]

The central finding from these studies is a gap between the inferred divergence points and the first appearance of fossils representing these clades, as shown in figure 6.2. When two species diverge at a node, we expect that they will be virtually indistinguishable and only later acquire derived features that characterize the newly independent lineages. Thus, we should expect that divergences as measured with molecular clocks will precede the first appearance of characteristic fossils. But the gaps are tens of millions of years, and many interpret this result as indicating a macroevolutionary lag between the acquisition of important genomic and developmental novelties and the expression of these as morphological novelties, leading to evolutionary innovations. From this one concludes that the origin of animals and the divergence of many of the basal clades were effectively decoupled from the morphological novelties that now characterize these groups, the acquisition of larger body

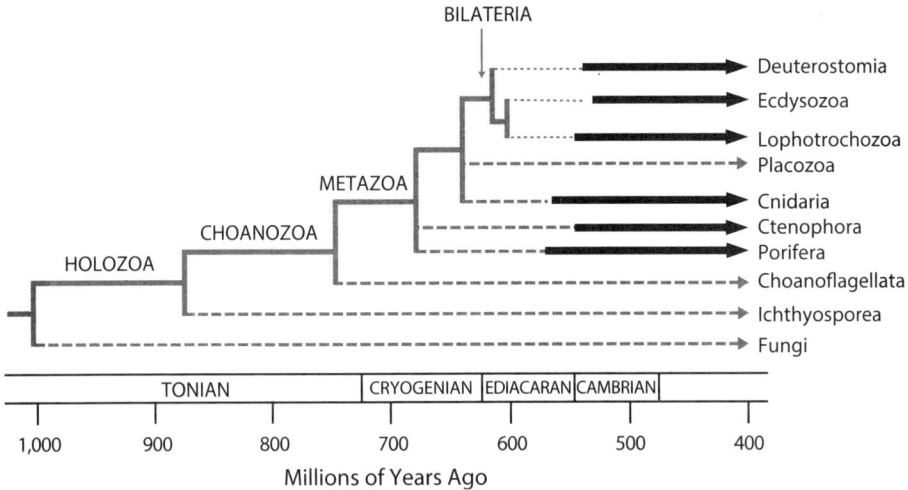

FIGURE 6.2. Recent consensus animal phylogeny based on molecular data, with estimated divergences based on molecular clock studies. Considerable uncertainties remain. Key nodes for the discussion in the text are the origin of animals, the divergence among the Bilateria, and divergences at the base of the Panarthropoda, within the Ecdysozoa. The heavy black lines indicate the known fossil records for each clade, while the dotted lines are the inferred history of each clade based on divergence times inferred from molecular clock estimates.

size, and the onset of appearance of skeletons, which is the defining feature of the Cambrian Explosion. The remainder of this section will examine some of the novelties associated with three key nodes (shown in figure 6.2): the origin of animals; the split among the Bilateria into Lophotrochozoa, Ecdysozoa, and Deuterostomia; and within the Ecdysozoa, the initial novelties toward arthropods. We will return to this story in chapter 8 in addressing the associated morphological changes and ecological innovations.[9]

Each of these transitions required multitudes of cell types, which must be formed at the correct time and in the correct place in the developing embryo. Precisely assembling these cell types is the task of gene regulatory networks, and a brief introduction into how these networks coordinate gene activities helps to appreciate the novelties introduced during the early evolution of animals. A schematic of a representative animal gene is shown in figure 6.3. Promoters, enhancers, and insulators are significant regulatory components. Promoters define where transcription begins, enhancers control transcription of a gene by interacting with promoters, while insulators limit the activity of

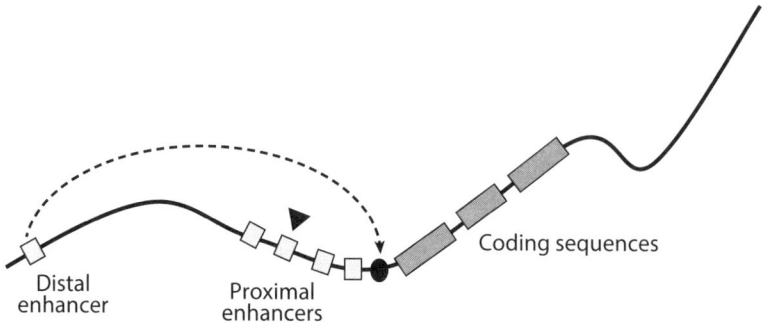

FIGURE 6.3. A representative metazoan gene, with the distal and proximal enhancers involved in gene regulatory activity to the left of the gene that contains the coding sequences. In the middle is the promoter, where transcription activity begins, a proximal transcription-factor binding site (black circle), and then a string of other transcription-factor binding sites are strung out to the left, forming a proximal enhancer (gray boxes, with one transcription factor (triangle) bound). The distal enhancer is the grey box to the far left. Insulators, other important components of the regulatory genome, are not apparent at this scale and would be further to the right and left of this sequence.

enhancers to specific regions. Transcription factors are short pieces of protein with a segment that readily binds to a specific region of DNA, known as a transcription-factor binding site. The specificity of this DNA binding ability allows transcription factors to bind to a promoter or enhancer to control gene activity. Transcription factors often act combinatorically, meaning that several must bind to a single gene to either activate or repress the activity of the gene. Further away from the gene may be distal enhancers, with additional transcription-factor binding sites. At a broader scale, genes that are expressed at the same time may be clustered into a region of DNA bounded by insulators and other regulatory sequences that allow the entire domain to become active. Finally, the activity of transcription factors and other regulators forms networks of interaction that regulate gene expression, with the suite of expressed genes defining a specific type of cell, such as a muscle or nerve cell. There are more actors involved in the regulatory genome than I have sketched here, but these will suffice to illustrate some of the critical novelties associated with the origin and early diversification of animals.

Until about 2015, cell types were defined by their function and what they looked like under a microscope: skin (epidermis), nerve, muscle, and so forth, each broken down into subtypes based on studies extending back into the nineteenth century. New molecular techniques that allow examination of the RNA expressed within individual cells has upended this scheme in favor of identifying the combination of genes expressed in each cell type (more specifically, by their RNA transcriptome). Fortunately, there is considerable overlap between these approaches, but the RNA-based method distinguishes cell types that are not evident with traditional methods. This has been a tremendous boon for health care, particularly in the ability to distinguish different types of cancer cells, but when applied to evolutionary questions it has revealed that sponges have almost twice as many cell types as previously thought, and that sea anemones possess a greater variety of nerve cells than one ever imagined. From such studies one can also infer the evolutionary history of cell types, revealing a substantial increase in the number of cell types in the ancestor of all animals. While each of these new cell types represents a novelty, the true novelty was the capacity to generate many cell types and control their expression in a developing organism.

From Holozoa to Metazoa

The cousins of animals are a suite of largely single-celled eukaryotes with relatively complex life cycles. One clade of these, the choanoflagellates, has been identified for more than a century as the immediate relatives of animals because its members look like the feeding cells of a sponge. It was long assumed that sponges arose as a sort of multicellular choanoflagellate. Today, however, we have far greater understanding of choanoflagellates and the related lineages shown on figure 6.2, largely because of the work of developmental biologists Nicole King and Iñaki Ruiz-Trillo. With their research groups and collaborators, King and Ruiz-Trillo revealed new insights into the non-metazoan holozoans and the transition to animals.

Along with their complex life cycles, each holozoan clade contains some species that are multicellular. As an example, Nicole King's group reported a sheetlike, colonial choanoflagellate that responds to light by inverting the sheet using a primitive version of a system also employed by animals for movement. The ecology and life cycles of these organisms encouraged the evolution of multicellularity, which required tools for signaling between cells, cell adhesion, and a limited suite of transcription factors. This tool kit is sufficient to generate distinct cell types through the life cycle and across an organism, with some cell types having multiple functions. The transition to animals built on this capacity to generate different cell types during a complex life cycle as the foundation for different cell types in early animals. Many of the regulatory genes that would eventually have prominent roles in animal development are found today among other holozoan clades as well as animals and thus must have existed in their common ancestor (the most basal holozoan node in figure 6.2). In other words, holozoans possessed capacity for complex development that was only realized in the animal lineage.[10]

Many of the genomic changes at the transition to animals were not novelties, although some facilitated later developmental and morphological novelties. Among such changes were increases in the number of genes, in the size of families of transcription factors, in the number of transcription-factor binding sites per gene, and in the number of enhancers per gene. In most eukaryotes transcription factors bind immediately adjacent to the promoter, as shown with the proximal enhancer in figure 6.3. But animals also have enhancers located far from the promoter, as shown, and they may even lie in other genes. Since enhancers operate by binding multiple transcription factors, expanding the number of enhancers greatly increases the regulatory code, enabling new cell types and expression of genes in different contexts and permitting more combinations of genes to be expressed in a cell type. Yet sponges, ctenophores, and other non-bilaterian clades appear to largely rely on the proximal gene regulation typical of other holozoans, rather than distal enhancers. The changes identified at the base of Metazoa, other than distal enhancers, new types of promoters, and some other regulatory changes, seem to be changes of degree (more genes, more transcription factors, greater cell–cell signaling, and more transcription-factor binding sites per enhancer) rather than wholescale novelties.[11]

With the origin of animals, the regulatory tools from holozoans were repurposed for increasingly complex developmental patterning. If we interpret this information in light of the timing of early animal evolution, this suggests an initial phase perhaps 600 million years ago when development was dominated by proximal gene regulation, as in the other holozoan clades, but with larger genome sizes and greater capacity to generate cell types. As regulatory complexity expanded with increases in the number of transcription-factor binding sites per gene, more cell types could be generated, which we see in living sponges and cnidarians. Nerve and muscle cells are distinct, but they share an evolutionary history dating to a cell that engaged in both functions, which in turn appears linked to stem cells. Thus, the earliest phase of animal evolution involved division of cells with multiple functions into separate cell types, which formed simple tissues with some degree of patterning from front to back and top to bottom (anterior–posterior and dorsal–ventral). The gene regulatory networks underlying initial cell types and simple tissues were relatively flat, but in the next section I will argue that complex developmental patterning to generate more complex tissues and organs largely occurred independently in different lineages closer to their appearance in the fossil record.[12]

Sponges are not simply a collection of cell types. Their epidermis, or the outer covering of cell, is a simple tissue that also has sensory and contractile functions, and some tissue-level coordination between cells. Cnidarians, the next major node on the evolutionary tree, possess more cell types than sponges and proper metazoan tissues but lack organs. Cnidarians have greater developmental patterning than sponges, with numerous morphological novelties, but this seems to have been largely accomplished with the capacity already available to sponges. Organs normally arise through interactions between adjacent tissues, and the various bilaterian clades have multiple tissues and organs. Thus, the generation of so many novel morphologies during the Ediacaran–Cambrian radiation required not just cell types but also establishment of processes to generate tissues and organs, as well as the novel tissue and organ types.[13]

Origins of Bilateria

Sponges, cnidarians, ctenophores, and placozoans are fascinating clades of animals. Comparative developmental studies have made ctenophores a favorite of those investigating the origin of nervous systems, and placozoans, minute sheets of cells lacking tissues or organs, are also having a bit of notoriety for they shed light on the earliest stages of animal evolution. But the most fundamental event after the origin of animals was the diversification of the bilaterian animals into three great clades: deuterostomes (including echinoderms and vertebrates), ecdysozoans (arthropods and nematodes), and lophotrochozoans (including mollusks and annelid worms) (see figure 6.2). The appearance of large-bodied representatives of these groups, often skeletonized, is the signal event of the Cambrian Explosion.

A fundamental problem in understanding the origin of the bilaterian clades is distinguishing between three possible explanations, for which segmentation serves as a useful and relatively uncontroversial example. Segmentation is the serial repetition of body parts along the main axis, with complete segmentation in arthropods, annelids, and chordates. But other clades such as mollusks exhibit partial segmentation. In principle, segmentation could have arisen once in the ancestor of all bilaterian animals and then have been lost in some lineages, which would make it a homologous character (or, more accurately, a complex of characters), or it might have arisen independently in different lineages and thus be convergent. But there is a third, slightly more complex, alternative. The developmental processes that form segments are controlled by the same network of genes, but the phenotypic expression, here segmentation, arose independently in different lineages. In this case the morphology is not homologous, but the underlying developmental processes are at least partly homologous, and this is indeed the case, which illustrates a problem with relying on homology as a defining feature of evolutionary novelty. Later, I will argue that many other bilaterian features are most plausibly interpreted as cases of deeply conserved developmental mechanisms independently co-opted into controlling similar features in different clades.[14]

Segmentation requires sophisticated developmental processes and more hierarchical gene regulatory networks, which in turn requires the intercalation of transcription factors and new promoters into the networks. Gene regulatory networks that initially specified types of cells often seem to have been embedded in larger networks when they were co-opted for new functions. Other regulatory capacities expanded as well, particularly increases in control over the architecture of chromosomes, allowing greater deployment of the distal enhancers mentioned earlier. For example, in the invertebrate lancelet *Amphioxus* the Hox gene cluster is a single domain, with most of the regulatory control elements immediately upstream. In fish and other jawed vertebrates some Hox clusters are divided into two domains, with regulatory elements both upstream and downstream of the genes. This allows these genes to be expressed in more settings.[15]

Co-option of existing gene regulatory networks adds enhancers, so the circuit can be active in a new location in the developing embryo, or at a different time. The subcircuit is already operational, and recycling an already-functional module is more straightforward than constructing a new regulatory subcircuit, just as computer programmers reuse bits of code. Indeed, some developmental subcircuits have been so commonly reused, and are such a common source of developmental novelty, that Davidson and I described them as "plug-ins."[16]

How would such co-option have impacted the early history of bilaterians? I envision the last common ancestor of protostomes and deuterostomes (node X) to have been relatively simple, with tens of cell types and simple tissues. Its developmental patterning would have included anterior–posterior and

dorsal–ventral systems, as well as the beginnings of the proximo-distal patterning now widely employed in appendages. In contrast to suggestions from
others, I do not believe the evidence available supports such organisms having
segmentation (which would make segmentation homologous), image-forming
eyes, or a complex brain. Such a relatively simple protostome-deuterostome
ancestor necessitates widespread later co-option to explain the distribution of
highly conserved developmental genes across bilaterians. Such co-option helped
generate segmentation, the brain and nervous system, sensory systems, appendages, and a gut with different regions. That sophisticated developmental
patterning systems would be independently built on common systems is not
particularly surprising, given the extent of co-option documented in gene
regulatory networks.[17]

Panarthropod Novelties

The beautifully preserved fossil arthropods of the middle Cambrian Burgess
Shale were Stephen Jay Gould's primary evidence for extensive disparity. And
Gould wrote *Wonderful Life* before the full reveal of disparity among Early
Cambrian lobopodians and arthropods from the Chengjiang fauna in South
China, and other similar localities (figure 6.4). The arthropods with their
jointed, segmented appendages belong to a broader panarthropod clade with
the minute tardigrades, or water bears, and onychophorans, a group now found
among leaf litter in tropical forests. Onychophorans and tardigrades have
stumpy, unsegmented legs attached to a wormlike body. The remarkable diversity of lobopodians in the Burgess Shale and Chengjiang fauna has revealed
that living tardigrades and onychophorans are a mere remanent of these once-
thriving groups. Some of Gould's "weird wonders," such as *Anomalocaris* and
Opabinia, have now been comfortably placed within this mélange.[18]

Arthropods are segmented from front to back, patterned by Hox genes,
and have such a diversity of appendages growing out of these segments that
they have been described as the "Swiss Army knives" of the animal world. Each
segment has an independent identity but possesses information about adjacent segments. The nineteenth-century anatomist Richard Owen described
this as serial homology: "representative or repetitive relation in the segments
of the same organism." Other examples of serial homology are vertebrae; the
paired fins of fish, which led to vertebrate forelimbs and hindlimbs; and
the segments of earthworms. In millipedes and some other groups of arthropods the pairs of legs are replicated but not individuated, although those in

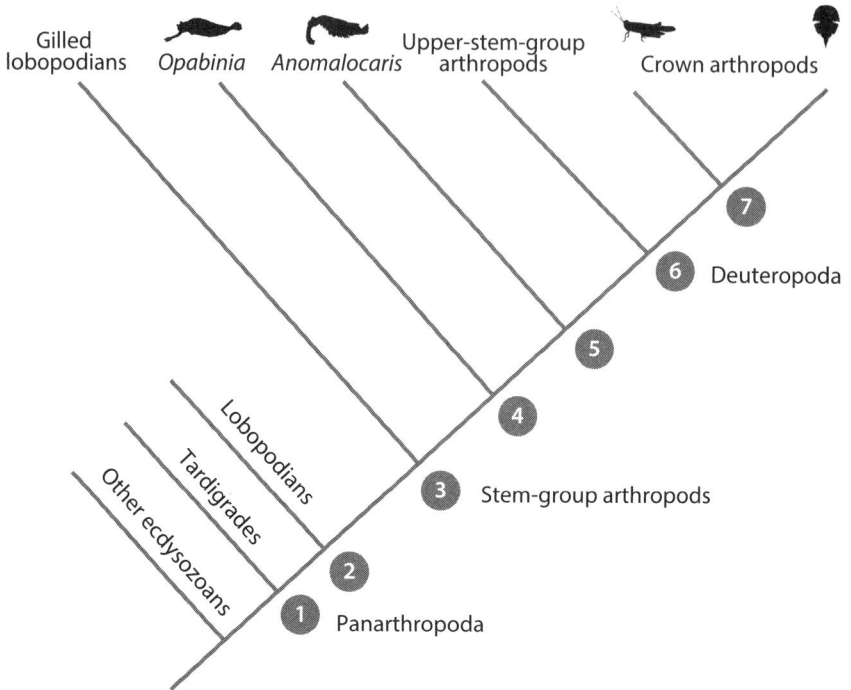

FIGURE 6.4. Simplified phylogeny for early arthropods with key evolutionary novelties. Node 1 is the base of Panarthropoda, node 2 is the divergence between lobopodians and stem-group arthropods, node 3 is the base of stem arthropods, node 4 is the node leading to *Opabinia* between gilled lobopodians and radiodonts (*Anomalocaris* and allies), with *Anomalocaris* at node 5, Deutero-poda at node 6, and crown arthropods at node 7. Images from Phylopic.org; *Opabinia* image by Nobu Tamura (vectorized by T. Michael Keesey) cc by 3.0.

the head may be. Since developmental patterning systems within each segment can be redeployed to a new purpose, this facilitates the rapid generation of novel, individuated appendages. This latent developmental capacity for novel appendages enabled arthropods to generate structures as diverse as the mouthparts of trilobites and the raptorial appendages of shrimp and crabs.[19]

The arrival of arthropods in the Cambrian is signaled by trackways in latest Ediacaran and earliest Cambrian rocks, but no body fossils are known until after 520 million years ago, when a bonanza of both hard- and soft-bodied arthropods appear in the Burgess Shale–type deposits, as well as many different lobopodians. Resolving the evolutionary relationships among these fossils and the living panarthropod clades has been challenging. Lobopodians and arthropods were already so diverse when they first appeared that the order of appearance is of no use in reconstructing the sequence of evolutionary events. Fortunately, the phylogenetic relationship among these fossils is sufficiently well resolved (see figure 6.4) to track the acquisition of key morphological novelties. Moreover, and despite the great antiquity of these events, a comparative approach with living descendants allows many inferences to be made about the underlying developmental novelties.[20]

These morphological novelties provide a classic example of the importance of serial homology for individuation of new characters. This discussion also provides a foundation for examination of other arthropod novelties later in this chapter.

At the base of the Panarthropoda (node 1 on figure 6.4), the body of the animal was not segmented, and the limbs were unsegmented and unjointed. A primitive axis of proximal-to-distal differentiation was present and may have been co-opted from anterior–posterior patterning of the nervous system. The many lobopodians overturned a lot of long-held assumptions about the evolution of arthropods, providing combinations of characters that had never been imagined by earlier invertebrate zoologists. By the ancestor of lobopodians and stem-group arthropods (node 2), some animals had dispersed protective plates (sclerites), and simple eyes were present. The next node (node 3) is the base of stem-group arthropods, and we find that lobopodians were not a distinct clade but rather a disparate assemblage arrayed along a lineage leading to arthropods. These organisms feature swimming flaps along the sides, unskeletonized front appendages, and a segmented gut.

How these animals are related remains difficult to sort out, but *Opabinia*, one of the most iconic of the "weird wonders," lies between the "gilled lobopods" and the radiodonts (node 4). *Opabinia* had a more developed head than other lobopodians, mushroom-shaped eyes, and a long, grasping, claw. *Anomalocaris*, the giant predator of Early and middle Cambrian seas, is the best known of the radiodonts (node 5). Among the dozen or more radiodontid Early Cambrian genera are species with adaptation for filter feeding and others that were specialized predators. The appearance of radiodonts marks significant new novelties and adaptations, including many features characteristically associated with the "true" arthropods (or Euarthropoda). Among these is a pair of specialized, jointed (arthropodized) appendages attaching ahead of the mouth (protocerebral). Compound eyes appear, with over 16,000 simple eyes, or ocelli, making up a single eye of some species; these eyes are paired and perch on protruding stalks. At the next node, for the Deuteropoda (node 6), the protocerebral appendage of the radiodonts has become reduced to specialized structures around the mouth. Both the head and body are segmented, all appendages have become jointed and segmented, and the head is composed of three segments with specialized, differentiated appendages. Among the different clades arising here are bivalved arthropods and two different clades each with large appendages emerging from the head. To preview the innovations discussed in chapter 8, by this node these proto-arthropod lineages had sufficient capacity to generate a wide variety of different morphologies, responding to ecological opportunities in Cambrian seas.

There are two major clades of crown arthropods (node 9), the chelicerates (horseshoe crabs, spiders, and allies) and the mandibulates, including crabs and insects. A plausible case could be made that trilobites were designed by malevolent beings to plague seekers of arthropod phylogeny, for they include characters associated with both chelicerates and mandibulates; they lie somewhere near node 6, although just where remains a mystery. A few decades ago, it was unclear whether there were any chelicerates or mandibulates among the Cambrian fossils, but both stem chelicerates and stem mandibulates have since been discovered.

This overview provides a scaffold to highlight significant novelties. In appendages individuation began with the division of an unsegmented appendage into segments (podomeres) separated by joints (the process of arthropodization). Like all bilaterian animals, ancestral panarthropods possessed machinery for differentiating cells along a gradient, probably initially for patterning the nervous system. This gene regulatory system was co-opted to pattern appendages along the body to the distal end of the appendages. The jointed appendages first appeared at the front of the body, and the underlying gene regulatory network was co-opted to establish the joints between segments. Initially each appendage segment was undifferentiated, but some segments acquired specialized features, while other segments lost appendages entirely. Developmental studies have identified shared patterns among the anterior segments from Onychophora to crown-group arthropods, indicating that these segments were established before they became specialized. In general, arthropod appendages were individuated from head to tail; and in some clades appendages toward the rear remain unspecialized. From this we might expect that the core regulatory processes are conserved across appendages (for example, in joint formation), while unique patterning mechanisms developed for each appendage, and indeed that is just the pattern identified through comparative developmental studies. One reason that arthropods have been so successful is that their developmental system possesses a remarkable capacity to generate novelties via serial homology and for co-option of regulatory novelties. This modular structure enables specialization and individuation of appendages. Such evolvability of arthropods returns later in this chapter with the horns of beetles and treehopper helmets.

The Cambrian Explosion was one of the most extensive single episodes of novelty generation in the history of life. Here I have focused on the genetic, developmental, and morphological novelties of the Ediacaran and Cambrian not just because I know it well, but because the depth of comparative developmental studies allows us to infer something about how these novelties arose and how novelties in genes and developmental patterning may be decoupled from the construction of new morphologies, particularly those expressed in the fossil record. But just as any general view of novelty and innovation must address these events, such a view would be profoundly myopic if it could *only* address these events. Early bursts of morphological novelty and of morphological disparity have been identified for other clades. As illustrated by other cases examined in this chapter, novelties continued to arise through much of the remainder of the Phanerozoic in animals as well as plants.

The Greening of Land

Some paleontologists have argued that early plant evolution was similar to the Cambrian Explosion, but most recognize that the origin of flowering plants involved a fundamentally different, largely sequential pattern of appearance of morphological novelties. Conifers are the exception to this pattern, as their disparity remained low for their first hundred million years and only began to increase in the Triassic, before reaching a peak in the Late Jurassic, about 150 million years ago. Nonetheless, there are also interesting similarities in patterns of novelty and innovation between plants and animals. In the late 1980s this was described as "dancing to a different beat," with plant evolution responsive to the appearance of novelties and less influenced by environmental factors (implying that the distinction between novelty and innovation may be less relevant for plants). Whatever the driving forces underlying plant evolution, their spread transformed the Earth's geochemical cycles, influenced the weathering of the continents and the transfer of nutrients and sediments to the oceans, and created evolutionary opportunities for terrestrial microbes, fungi, animals, and other plants. Comparative studies of plant development and the underlying developmental patterning systems have been advancing rapidly, but our understanding is not yet as rich as for the early history of animals.[21]

Land plants descended from freshwater green algae. Green algae were already multicellular, but moving onto land required plants to solve problems of support, reproduction in air rather than water, resistance to desiccation, the ability to acquire and transfer water, and photosynthesizing in air. As with animals, this required the generation of new cell types, tissues, and organs, as well as the processes to coordinate their development and function. Green algae, like the unicellular ancestors of animals, already possessed many of the necessary developmental and morphological novelties. The simple freshwater alga *Chara* plays the same role for land plants as unicellular holozoans do for animals. The genome of *Chara* includes more plant hormones, signaling pathways, and transcription factors than those of other algae, as well as changes necessary for development in land plants. All these changes occurred before fossil bryophytes, the hornworts, liverworts, and mosses, appeared in the mid-Silurian about 430 million years ago. Some evidence suggests that bryophytes may have been on land as early as the late Cambrian (~500 million years ago). Bryophytes possessed many of the key traits needed to survive on land, but they lacked lignin, which stiffened cell walls to support plants and allow water transport.[22]

Vascular plants include ferns and seed plants, with the seed plants encompassing cycads, conifers, and flowering plants, as well as many extinct groups beloved only by paleobotanists. Successful adaptation to life on land required roots, lignin, a waxy cuticle, the capacity for three-dimensional growth (in contrast to the mat-like or filamentous habitat of algal lineages). Some of these developed independently in different lineages. For example, leaves evolved independently among ferns, seed plants, and two other clades, and each time the clade rapidly diversified into new forms. But except for the much later flowering plants, each of these clades seems to have exhausted the variety of accessible leaf types within a few tens of millions of years.[23]

As with animals, molecular clock analyses suggest that vascular plants originated well before they first appear in the fossil record, followed by a gap between the earliest appearance of vascular plants as fossils and their later diversification in the Devonian. These results date the origin of land plants to between 600 and 500 million years ago. The oldest vascular plants date to about 425 million years ago, not far off the oldest fossils at 420 million years ago, during the Silurian.[24]

Two pulses of morphological novelty among vascular plants were associated with shifts in reproduction: first in the Devonian some 380 million years ago, followed by the Early Cretaceous explosion of flowering plants beginning about 130 million years ago. The first pulse happened as plants freed themselves from the need for water to aid reproduction, shifting to airborne pollination, while during the second flowering plants recruited targeted pollinators among insects, birds, and other animals. The productivity of ecosystems may also have been limited before the rise of flowering plants, constraining evolutionary opportunities.[25]

The morphological novelties in early plants were built upon a suite of genetic, developmental, and cellular potential that arose in groups of green algae. Major bursts in the number of genes occurred perhaps 850 million years ago among the green algal ancestors of plants, and later before the origin of land plants. Each of these episodes involved far more new genes than arose with the origin of animals, but as with animals, what drove the changes leading to land plants was primarily new pathways of regulating the existing genes. Transcription factors regulate plant development, as in animals, and an expansion of transcription-factor families played an important role in their new developmental pathways. Ferns are closely related to seed plants, with cycads being the most basal living clade of seed plants. Recent sequencing of the genomes of several ferns and cycads has provided new insights into key novelties in seed plants, including co-option of many genes and gene regulatory networks for new functions. Genes involved in seed formation, pollination, and plant architecture in seed plants first appear in ferns, where they play similar roles. Unlike in animals, however, some genes were also transferred between clades through horizontal gene transfer, a process common among microbes but infrequent among animals. The fern and cycad genomes also contain many genes derived from bacteria and fungi. One of the more surprising examples comes from the cycad, which contains four copies of a fungal gene providing resistance to herbivory in seed plants.[26]

The strongest similarity with the morphological explosion among animals is found among the architectures of trees. Modern forests are largely composed of conifers and dicots, which share a common architecture. Cycads and their unique architectures persist in some regions. But early Carboniferous forests displayed at least six fundamentally different ways of constructing a tree. So at least for tree architectures, modern forests represent a greatly diminished sample.[27]

Many new developmental novelties accompany the appearance of flowering plants (angiosperms) during the Cretaceous. Angiosperms and gymnosperms may have split as far back as the Carboniferous, with the last common ancestor of living angiosperms dating to perhaps 200 million years ago. By about

135 million years ago, during the Early Cretaceous, all major clades of flowering plants were present.[28]

I will focus on carpels, the characteristic female reproductive structures of flowering plants. Carpels protect ovules and allow targeted fertilization, creating opportunities for flowering plants to coevolve with insect, bird, and other pollinators. Such enclosed fertilization contrasts with gymnosperms, where the ovules are exposed, and fertilization is often from airborne pollen. Carpels also permitted plants to develop protection against self-fertilization, thus reducing inbreeding and enhancing variability. In many plants carpels provide the tissue from which fruits develop, proving a new means of seed dispersal. Loath as I am to identify "key innovations," I would certainly allow that carpels have been a significant evolutionary novelty. They facilitated an amazing diversity of pollination and seed dispersal systems, as flowering plants have become dominant features of most terrestrial ecosystems. Studies of modern plants and their development have allowed reconstruction of the evolutionary trajectory to carpels. Many of the regulatory genes have been identified, but not at the level of detail known for animals. The genes involved in carpel development were assembled sequentially through vascular plant evolution. Many of the key genes arose before land plants or with the origin of seed plants, hundreds of millions of years before they were co-opted for new roles. Carpels are another example of a novelty helping to construct a new evolutionary space.[29]

But note the problem. Even if carpels in the last common ancestor of all angiosperms were as simple as thought, molecular clock studies still place the origin of this novelty well before the appearance of fossil angiosperms or later innovations in terrestrial ecosystems. Carpels were not a key innovation. They were an important, even a critical, novelty for flowering plants, but their success involved other ecological and environmental factors, as taken up in chapter 8.

Genome doubling and even quadrupling has occurred frequently in the history of plants, and such whole-genome duplications are often invoked as a driver of evolutionary novelty. A possible twofold genome duplication near the origin of vertebrates generated four sets of Hox genes (although many were subsequently lost), and during "Hox hysteria" in the late 1990s many evolutionary developmental biologists linked these duplications to morphological novelties and greater developmental and morphological complexity. Genome duplications can free up genes to take on new roles or allow genes with multiple functions to subdivide them as the new copies specialize in different functions, as described in chapter 2.[30]

Whole-genome duplications have been associated with the origin of seed plants 390 million years ago, with conifers, and along the lineage leading to flowering plants about 310 million years ago. Asparagus, corn, and grasses like rice have each experienced whole-genome duplication. Although there may be cases in which whole-genome duplications have driven evolutionary innovations, few have been rigorously established. Several recent studies have recognized these lags and the problems they pose for connecting whole-genome duplications with evolutionary novelties. These lags are among the reasons why whole-genome duplication may be better considered a potentiating factor, a means of expanding developmental capacity, rather than a driver of novelty.[31]

I have focused on just a few of the many significant novelties among vascular plants. I have neglected the invention of lignin, the plant vascular system, and the remarkable ability of plants to synthesize a dazzling array of defensive chemicals. But as green algae moved onto land and generated bryophytes and vascular plants, the challenges of the inhospitable environment led these clades to acquire evolutionary novelties for support, reproduction, photosynthesis, and the transport of water and nutrients. But the acquisition of novelties exhibits a more sequential pattern than we found among animals, with the much later appearance of flowering plants. To be sure, the architecture of trees exhibits something of the early pattern of novelty and morphological disparity identified among animals. Plants also exhibit the now-familiar pattern of key novelties appearing well before the clades became ecologically significant. Molecular clock studies document a lag between the origin of the clade and its first appearance as fossils, and new studies of the dynamics of first appearance of clades support earlier suggestions of a further macroevolutionary lag between the appearance of taxonomically and morphologically diverse assemblages of flowering plants in the Early Cretaceous and the increased ecological dominance of the clade 40 to 50 million years later, in the Paleocene. We will return to innovations of flowering plants in chapter 8, with a discussion of their interactions with insects and other animals, whether their ecological spread was a response to the effects of the end-Cretaceous mass extinction, and an evaluation of the claim that plant evolution was fundamentally driven by novelties rather than changes in the external environment.[32]

Wings, Horns, and Helmets

Wings allow insects access to food and a means of escaping predators, help attract mates, and permit them to adapt and thrive almost everywhere on land. By any measure, insects represent a substantial fraction of the total species number on Earth, and their coevolution with flowering plants has been one of the greatest evolutionary success stories of the past hundred million years.

Wings and flight have appeared only four times, with insect wings the only one whose origin remains contentious. The origin of insect wings is an intriguing problem that illustrates the challenge in distinguishing between the three types of novelty I have identified. Despite the tremendous advances in comparative evolutionary developmental biology and in the reconstruction of gene regulatory networks, inferring homology and the origin of new characters remains tricky, as we will see.

Insects and their allies such as springtails separated from other arthropods some 450 million years ago. Insects have three thoracic segments behind the head segments, each with a pair of legs, with their wings on the second and third thoracic segments. The wings are connected to the body via a complex hinge structure. The first insects lacked wings, but the oldest insect fossils (about 325 million years ago) have fully formed wings capable of active flight. So, the fossil record is largely mute on the origin of wings, but there are some larval nymphs with hinged, lateral flaps along the abdominal segments. These may provide clues to the origin of wings, but studies of modern insect development have been more useful, even if they have not yet fully resolved the problem. Insects also have other outgrowths, among them horns on beetles and the spiky helmet of treehoppers. Some of the same developmental patterning modules involved in generating insect wings are also used in these later productions, which we will examine after considering the origin of insect wings.[33]

By the late nineteenth century, two competing theories for the origin of the insect wing had been developed based on modern insects, their developmental patterns, and the limited available fossils. The wings of pterosaurs, birds, and bats evolved from the front legs of their ancestors, and one of these hypotheses proposes that the insect wing derives from a transformed appendage, likely a gill. The alternative hypothesis traces insect wings to a new outgrowth of the body wall. These two hypotheses invoke very different views of homology between the wing and earlier appendages. The first alternative involves a very deep transformation but with continuity between the appendages of earlier arthropods and insect wings, while in the second scenario wings originated de novo via co-option of gene regulatory networks. Fortunately, there are specific predictions from each hypothesis about patterns of gene expression. The first test of these alternatives in 1997 compared two genes involved in insect wings with their homologs in crustaceans, concluding that the wing arose by repurposing structures originally involved in the formation of crustacean appendages.[34]

Crustaceans (crabs, shrimp, and their allies) generally have about eight segments in their appendages. During crustacean evolution some of these segments

shifted back into the body to form part of the body wall, a process that is particularly evident among the crustacean lineage that led to insects. Unresolved is
whether the genes that produce the segments closest to the body were later
co-opted for wing development, whether wing-formation genes were originally
active in the body wall, or whether both hypotheses are correct, with wings
generated by some genes originally expressed in the body wall and others expressed in proximal appendage segments (the dual-origin hypothesis). Some
of the ongoing disagreements over the origin of wings appear to me less about the
experimental data and more a reflection of philosophical expectations about
continuity and novelty, and even the evidence required to substantiate claims
for the novel origin of a structure. There is evidence for early origin of a gene
regulatory network involved in wing formation. This might be evidence of co-
option, but it is also possible, as I suggested for Cambrian bilaterians, that insect
wings share only a developmental patterning module.[35]

A brief note about the origins of the fine scales covering the wings of butterflies and moths: Wing scales are highly modified setae, or hairs, that serve
a variety of roles but are best known for the often-amazing color patterns
found on butterflies. Scales have evolved multiple times across distantly related insect groups, but a recent study of Cretaceous fossils has shown that
scales first appeared in the common ancestor of butterflies, caddisflies, and an
extinct clade, well before their extensive utilization by butterflies and moths.[36]

Beetle collectors and entomologists have long been attracted to the diversity of horns among dung and rhinoceros beetles, not only for their exotic
structures but also because they represent stunning examples of a biological
arms race. The horns range from subtle knobs to protuberances that double
the length of the beetle. Some have a single long horn, others have two antler-
shaped horns, all designed to help males battle over access to females. Over
the past couple of decades developmental biologist Armin Moczek and his
students have developed the horns of dung beetles into a remarkably accessible system for the study of evolutionary novelty, examining everything from
the genetics of the horns to the role of nutrition. Moczek's insights into novelty
are based on these long-term studies of the genus *Onthophagus*.

The horns of a dung beetles erupt from the wingless first thoracic segment,
immediately behind the head, whereas insect wings are attached to the next
two thoracic segments. This prothoracic segment also hosts the helmet of
treehoppers. One hypothesis is that the absence of wings on the prothoracic
segment has freed up the genes responsible for wing development for new
tasks, allowing their co-option into novelties such as horns. In this case the

horns would be serially homologous (as described earlier) with wings and should express similar genes.

Moczek's group undertook an extensive comparison of wing development and the underlying gene regulatory networks in flies and in beetles. From this work they have concluded that the horns of *Onthophagus* beetles are partial serial homologs of wings. The development of the horns in the first thoracic segment recruited elements of the gene regulatory network deployed for wing development in the second and third thoracic segments (providing the serial homology), after which horn development passes to other genes. Moreover, changes in gene expression patterns unique to *Onthophagus* established the potential for the subsequent co-option of the wing gene regulatory network to initiate the growth of horns.[37]

Treehoppers, or thorn bugs, have an elaborate shield with spines or subtle camouflage. While the wonderful variety of shapes taken by this helmet demonstrates the power of natural selection, the origin of the helmet has been more controversial. During development it arises from the shield on the back of the first segment of the thorax (the pronotum), and three alternative scenarios have been offered: the shield is an elaboration of the pronotum itself (and thus an adaptation rather than a novelty), it arose via co-option of the gene regulatory network responsible for producing legs to produce the long axis of the spines on the helmet (as with beetle horns), or via co-option of the gene regulatory network responsible for wing formation. The two co-option scenarios view the helmet as a serial homolog of either legs or wings. In the event, developmental studies have recently shown that formation of the helmet involves genes that are also active in forming the insect wing. This does not mean that the helmet represents a modified wing. Rather, as in other examples of co-option, a gene regulatory network has been repurposed to generate a novel structure, which was subsequently adapted to the elaborate structures found among the several thousand species of treehoppers.[38]

There are at least three different ways of evaluating these results for treehoppers and insect wings, with important differences for how we view novelty. The most common interpretation is that the generation of appendages constructed a gene regulatory network that has been reused in the development of wings and other outgrowths from the tergal body wall. In segments with wings this network is active during development, but in other segments it is available to be repurposed for other novelties. In *Onthophagus* the network has been recruited to initiate the magnificent horns and other structures, with other genes involved in their elaboration, fine-tuning them. The helmet of treehoppers also involved recruitment of the same set of genes. Thus, the horns and helmets represent serial homologs in which the wing gene regulatory network is used to generate different structures.

But recall my earlier observation that the gene regulatory network involved in wing formation is also found in crustaceans and thus was formed *before* insects appeared, much less insect wings. An alternative to the co-option story is that insect wings and other outgrowths share a common "morphogenetic field" (cells dedicated to a specific morphological outcome). From this perspective insect outgrowths are not novelties, but merely interesting cryptic persistence and reuse of a preexisting developmental potential. The third alternative appears superficially similar but is significantly different from both the co-option and cryptic persistence models. In this scenario the conserved gene regulatory network generates a developmental patterning module for bilayered, cuticularized body-wall outgrowths. The module can be deployed in a variety of circumstances, from insect wings to treehopper helmets. Unlike the morphogenetic field scenario, the module does not specify a particular morphological outcome but is more of a developmental subroutine available for use as required by evolution. From this perspective the module is homologous, but the structures are morphological novelties. My view of the developmental novelties of the Cambrian bilaterians is largely consistent with this proposal of conserved developmental patterning modules.[39]

Theory is easy. Reality is hard, as illustrated by these examples of novelty. In 2008 Moczek wrote, "Novelty begins where homology ends." But the studies of beetle horns and treehopper helmets illuminate the challenges of separating how much of a novel structure represents modulation of existing developmental pathways and how much represents genuine reuse of gene networks for a new function. These difficulties led Moczek and colleagues to revisit his earlier claims:

> Exactly where homologous relationships cease has become increasingly difficult to delineate, as findings in evolutionary developmental biology have forced a revision of homology away from a binary designation. . . . Here we show that a textbook example of evolutionary novelty, the prothoracic horns of beetles, derives partly from wing serial homologs, whose existence predates the origin of insects. On one side, this may cause us to question whether prothoracic horns should still be considered an evolutionary novelty. Alternatively, our results may serve to illustrate how substantial morphological innovation, rather than somehow emerging in the absence of homology, may instead be initiated through it.[40]

Insect wings, beetle horns, and treehopper helmets are novel structures, and through the interactions of insects and flowering plants, the first was a major contributor to the construction of the modern world. But the ease of repatterning existing developmental systems complicates clearly delineating novelty from refinement and adaptation. The construction of gene regulatory networks may generate one novelty, while simultaneously potentiating the generation of subsequent novelties by co-option of the networks.

Other Examples of Novel Characters

New cell types continued to evolve as plants and animals diversified, from the elaboration of photoreceptors in eyes to a menagerie of neurons in mammals. Although some new tissues and organs have appeared since the Cambrian, most animal tissues and organs originated with the initial explosion of animal architectures. New cell types arose as lineages split. New photoreceptors arose in arthropods and vertebrates after the clades diverged from the last common bilaterian ancestor, or between reptiles and mammals. The steps necessary for establishment of a new cell type may be elaborate, but there are examples of relatively simple regulatory switches between one cell type and another. *Nematostella* is a small sea anemone widely used to study cnidarian development. A characteristic feature of corals, jellyfish, and other cnidarians is the stinging cells known as cnidocytes. A single transcription factor can switch cell fates between piercing, venomous, harpoon-like cnidocytes and a second type that produces an ensnaring web. So, shifts between closely related cell types may not require pervasive repatterning of the regulatory state of a cell, but only a subtle change, even if the shift triggers a considerable change in morphology.[41]

PREGNANT MAMMALS AND STUNNING FISH

Two particularly well-studied examples of new cell types are decidual cells, which allow a fertilized egg to implant in the wall of the uterus in eutherian mammals, and the cells (electrocytes) that generate electric currents in fish. In each case the new cell type is associated with a new organ—decidual cells with the placenta, and electrocytes with electric organs in fish—and thus a new function. Placenta-like structures and electric organs have arisen multiple times, which provides more insight into the mechanisms associated with these novelties.[42]

Decidual cells characterize eutherian mammals, which include all living mammals other than marsupials like kangaroos, and the duck-billed platypus and the echidnas, which are monotremes. Decidual cells modulate interactions between the mother and the developing embryo. One might naively associate decidual cells and the placenta, but while this cell type is unique to eutherian mammals, over 130 vertebrate lineages from fish and frogs to mammals evolved some means of live birth requiring a placenta-like structure. So, placenta-like structures are a repeated morphological novelty. Most cases of live birth among snakes and lizards with decidual cell types are just one means of achieving this outcome.[43]

The evolution of live birth (viviparity) requires a loss of the eggshell; a means to supply nutrients, including calcium, to the developing embryo; methods of removing waste; and respiratory gas exchange. A placenta mediates these exchanges and provides the mother with greater control over the environment of the developing embryo than is possible with a shelled egg. This process involved recruitment of hundreds of genes, but a key component was the origin of the decidual cells. Progesterone and other cues at the origin of pregnancy cause endometrial stromal cells to differentiate into decidual cells, permitting the prolonged pregnancy typical of eutherian mammals. The underlying changes in gene regulatory networks arose by insertion of transposable elements. Transposable elements, or transposons, are short DNA sequences that contain their own regulatory elements and can insert into genes and modify their expression. They were first recognized in corn (maize) by geneticist Barbara McClintock in the late 1940s. The gene that codes for progesterone is one of those in which new transposable-element-derived binding sites have been identified, and these new regulatory sites allow the gene to operate in a new setting, while not disturbing the previous functions of the gene. These changes were followed by recruitment or co-option of other genes into a new gene regulatory network. It would be misleading to view the insertion of transposons as "the" event that generated a novelty. Rather, it was one of a series of changes found in the newly individuated decidual cell that contributed to formation of the eutherian placenta.[44]

Stunning your prey with an electric pulse is the most well-known use of bioelectricity, but most of the six different clades of "fish" (two clades of skates and four of bony fish) generate electric pulses for communication or navigation rather than defense or predation. Darwin cited electric organs as an important example of convergence in the *Origin*, but only recently has the extent of that convergence been appreciated. Electrocytes, the novel cell type underlying the generation of electricity, are derived either from muscle cells or directly from muscle-cell precursors, but in each case similar sets of genes are recruited. To generate sufficient voltage these cells are polarized and packed together end to end, like batteries in a flashlight. These packages of electrocytes and related structures together form electric organs. A recent study identified a configuration of genes whose activity was either reduced (lessening the similarity to muscle cells) or increased (to enhance cell size or electrical properties) to generate an electrocyte cell type. In terms of the framework developed in chapter 4, the repeated convergence and utilization of a similar suite of genes suggests that the genomic changes needed to generate an electric organ are relatively close to the normal regulatory state for fish muscle cells. So, while the formation of electrocytes and electric organs represents a true novelty, the novelty is easily accessible. Similarly, in chapter 4 I briefly discussed animal venoms, which represent repeated novel convergences upon a relatively small suite of accessible solutions constrained by available proteins for co-option and the physiological aspects of the prey.[45]

The placenta of eutherian animals and electric organs in fish each required the evolution of new cell types and novel developmental processes, and each arose multiple times. The repeated novelties associated with electric organs and placenta-like structures provide lessons in how similar suites of genes can be recruited, even across hundreds of millions of years. The mammalian placenta arose through cross talk between different tissues, which also features in the origin of feathers and may be a general feature of novelty.[46]

FEATHERS AND FLIGHT

Feathers range from the iridescent throat gorget of hummingbirds to the long indigo and black spatulate tail feathers of the turquoise-browed motmot in Central America. All feathers share a common suite of characters that distinguish them from the scales of reptiles, from which they are derived but have diverged into an incredible range of different character states. Flight and tail feathers have a strong, central rachis with barbs coming off to either side, but while down—the fluffy, plumulose feathers that provide the insulating undercoat for birds—still has a central rachis with barbs, the barbs lack the hooks that connect adjacent barbs. With the addition of stunning fossils of feathered dinosaurs from northeast China and elsewhere, the pathway leading from scales to feathers provides a wonderful case study of evolutionary novelty and subsequent innovation.

Patterning of structures in the skin is a critical novelty shared among all jawed vertebrates and is generated by interactions between the epidermis, the outer layer of skin, and the underlying dermis. Teeth, the denticles embedded in the skin of sharks and rays, the scales of fish, feathers in dinosaurs and birds, and the complex of hair and glands characterizing mammalian skin are all produced by the same developmental gene network, although the specific interactions may differ. This network probably arose with the origin of jawed fish in the Silurian as a simple interaction between two cell types. Hair and feathers appeared before the clades with which we normally associate them today: hair in Mesozoic mammals (on the Jurassic "beaver" mentioned earlier), and feathers are found in pterosaurs as well as dinosaurs, so must have originated in their ancestors, long before birds evolved from among the theropod dinosaurs. In the case of feathers, the conserved genetic interactions are responsible for the initial patterning of the bud leading to a feather, followed by more specific instructions that generate the actual feather. Additional novelties were required, principally gene duplications among the fibrous, structural proteins that comprise feathers.[47]

Feathers once seemed diagnostic of birds and likely closely associated with the origins of flight. Dinosaurs have each of the seven different types of feathers found among living birds, as well as others unknown in living birds. Flight is a behavioral novelty and required additional structural adaptations beyond the evolution of characteristic flight feathers, including the evolution of the wishbone, development of flight muscles, and a reduction in skeletal mass. The discovery of *Archaeopteryx* just a year after Darwin published *The Origin of Species* seemed to provide a missing link to the origin of birds, but may not be a bird at all, rather a closely related maniraptoran dinosaur without the skeletal changes required for flight. As is true of virtually all distinctive body plans, that of birds was assembled through a series of novelties and related adaptive changes, including a prolonged trend toward a reduction in size. But crucially, genetic changes needed to produce the feathers of birds appeared long before the morphological novelty of feathers arose in early dinosaurs. What has long seemed such a remarkable novelty for the first bird is underpinned by a long interval of scaffolding the potential for feathers, followed by the skeletal and other adaptations required to capitalize on this potential.[48]

Deep Transformations

There is little doubt about the novelty of the turtle shell or most new cell types. Other cases pose more challenges for individuation of new characters as a metric for novelty. Sometimes one could debate whether the genetic, developmental, or morphological change has been sufficient to warrant recognizing a new character rather than a deeply transformed character state. Systematists can tie themselves in knots debating the fine points of character homology and whether alternative character states are "sufficiently" different, or sufficiently individuated, to constitute a new character. To avoid debates over characters versus character states I recognize that some character transformations may be sufficiently deep to constitute an evolutionary novelty even if, strictly speaking, they do not represent the origin of a new character. Although Moczek was an early advocate for homology as a metric of novelty, as described earlier, he has come to appreciate the challenges to such an approach and has moved closer to the position I take here.[49]

Examples of novelties involving deep transformation include individuation of serially repeated limbs among arthropods, which has been particularly significant among crustaceans, and the fusion of the mantle tissues to form a tubular siphon in bivalve mollusks. Vertebrate limbs have undergone deep transformation in the transition from fins to limbs as a lineage of fish evolved into amphibians. Skates and rays are flattened fish with a skeleton composed of cartilage and are closely related to sharks. The broad wings of rays represent an elongation and extension of the fins, sweeping forward around the head. The origin of this novelty has been examined closely and provides another useful example of deep transformation.

Vertebrate Limbs

Four animal clades successfully managed the challenges of moving from the oceans or rivers onto land: arthropods, vertebrates, annelid worms, and mollusks. This transition required developing ways to support the body, ensure reproduction, and prevent dehydration and the loss of fluids. For snails the shell provided support, but only the fleshy fins of lobe-finned fish contained the combination of bones, muscles, and developmental patterning systems needed to put this lineage on a successful path toward limbs. Fossils of Devonian-aged vertebrates from 400 million years ago document many steps in this transition: the appearance of a neck as the skull separated from the shoulder girdle, transformation of the gill arches into the jaw and the ear, and the emergence of digits. To some surprise, paleontologists discovered that the earliest tetrapods had seven or eight fingers on each foot before later settling on pentadactyly. In addition, the swim bladder of fish (for buoyancy control) became the foundation for lungs. Among the many morphological novelties that arose during this transition, the best studied has been the origin of the limb from the paired fins of fish. Not all the events in the transition from fish to amphibian represent evolutionary novelties, but vertebrate limbs certainly represent a deep transformation of fins.[50]

The structure and positions of fins depend on the habitat of the clade, but they share a basic structure. Our bones share a common origin, as do the radial bones close to the body of a fish. The paired pectoral fins became the forelimbs of terrestrial vertebrates, while the paired pelvic fins developed into the hindlimbs, with the developmental mechanisms that generated the fins providing the scaffold for the changes required for limb development. In contrast, the rays of the fin have a different developmental origin, arising from ossifications in the skin. The fins of lobe-finned fish have a fleshy, scaly stalk with just a single bone attaching the fin to the body, unlike the multiple bones of other fish.

Developing limbs have been investigated in lungfish and primitive bony fish like gar and bowfin, revealing that the fin-to-limb transition is another case in which the critical developmental changes necessary for limb development were present in the last common ancestor, in this case that of all bony fish, before lobe-finned fish evolved and long before the transition to land. Establishing the proximo-distal axis along the limb involves several Hox genes, and one study identified changes in the level of Hox gene expression sufficient to produce skeletal changes in developing zebra fish embryos, like those that occurred during the origin of limbs. Despite the differing developmental origins of fin rays and digits, they share a common patterning system that predates the origin of paired fins. This patterning system was likely involved in controlling cell proliferation. These are not the only changes required for successfully moving onto land. Studies have also confirmed that lungs and swim bladders are homologous, and the genes needed for lung development were already present in early fish. Taken together, the regulatory machinery needed for limb and digit development, lung formation, and changes in the heart have deep roots in the ancestor of bony fish.[51]

Fins of Skates and Rays

One of my favorite memories from diving is rolling over 60 feet down off the Hawaiian island of Maui and watching three huge manta rays glide over me. The extraordinarily broad fin beautifully adapts many species of skates and rays to life on or close to the sea bottom. Rays and skates are cartilaginous fish, close cousins of sharks, with flattened bodies and greatly expanded pectoral

fins sweeping forward to meet at the head, transforming them into a swimming wing. But their enveloping pectoral fins are unlike anything in other fish.
Their distinctive glide is a swimming style impossible except for fish flattened
into a wing. Many evolutionary novelties have appeared among both bony fish
(Osteichthyes) and the cartilaginous fishes since they arose almost half a billion years ago, but the novel pectoral fins of skates and rays are particularly
striking. A study of development in the little skate illuminates the role of gene
regulatory co-option in the deep transformations of fins into wings.

The highly conserved gene regulatory module for vertebrate appendages,
including pectoral fins in fish and then limbs in amphibians, reptiles, and
mammals described above, has also been found in the posterior portion of the
fins of skates. As the fin develops a new patterning module appears toward
the head. This new module generates gene expression patterns partially mimicking normal limb development. Just as the fin of a fish has a growth gradient
from close to the body to the edge of the fin (proximo-distal), so skates and
rays have evolved a second proximo-distal growth gradient for the anterior
extension of the fin. The gene expression patterns are not identical between
the anterior and posterior growth zones. Rather, early in the evolution of
skates a portion of the gene network involved in normal fin development was
co-opted and became expressed in the anterior portion of the fin, generating
a second growth axis. This new gene network is not identical to the original
network, but the genes play similar roles and together they generate the novel,
expanded fin that gives manta rays their ghostly quality.[52]

Combinatoric Novelty

Earlier I described how the increase in available ingredients generated a
combinatoric expansion in recipes. Part of the power of combinatorics is assembling pieces that already work, whether genes and proteins, through
combining or rearranging already-functional sequences, or the evolution of
technology. The oxygen-transport protein hemoglobin in humans and other
jawed vertebrates is constructed of two α-globin proteins and two β-globin
proteins. Hemoglobins evolved from a single component, myoglobin, through
gene duplications and subsequent adjustments to enable the functionality of
the fourfold structure of hemoglobins (gene duplications have generated a
family of hemoglobin genes, with some used during early life and others by
adults). The bacterium *Escherichia coli*, which lives in the intestines of humans
and most other warm-blooded animals, provides a more powerful case.

Although most strains do not impact human health, others are associated with food-borne disease and other illness. Sequencing the genome of any individual *E. coli* produces about 5000 or so genes. But keep sequencing individuals, and one uncovers evidence that some genes swapped out for others. Only a small fraction of the 5000 genes are found in all individuals. A vast repertoire of upwards of 90,000 genes (called auxiliary genes) can be swapped in and out depending on environmental conditions or for antibiotic resistance. Here, the novelty is the pangenome itself, the capacity of many microbes to balance the preservation of a large library of potential genes with small genomes within individuals (thus facilitating rapid growth and reproduction). Combinatorics is also found in the recombination of regulatory elements, the fusion of proteins, and the horizontal transfer of genes from one species (often one very distant on the evolutionary tree) to another.[53]

One of the most profound examples of combinatoric novelty is the eukaryotic cell.

Origin of the Eukaryotic Cell and Early Symbionts

The eukaryotic cell is a chimera, a mélange of parts from here and there, and the riddle of their origin has absorbed generations of cell biologists. At least part of this interest was sparked by a 1967 paper where biologist Lynn Margulis championed the view that mitochondria and chloroplasts in eukaryotic cells originated by symbiosis of once-free-living cells. In this Margulis was absolutely right, but she pushed her argument to other components of eukaryotic cells, views that have not found subsequent support. Nonetheless, Margulis was a lifelong champion of the power of symbiosis to generate novel combinations.

In eukaryotic cells the DNA is packaged into a membrane-bound nucleus, and the cell also contains other specialized, membrane-bound structures, or organelles ("little organs"). The most important organelles are the mitochondria, which power cellular activities by converting carbohydrates and oxygen to energy. Eukaryotes also have an array of proteins that form the cytoskeleton that structures the cell and allows mobility. Animals, plants, fungi, various algae, and a host of microbes from slime molds to the intestinal parasite *Giardia* are all eukaryotes, with studies of DNA showing deep divisions dating to an early eukaryotic diversification. The origin of eukaryotes provided the scaffold upon which multicellular complexity has been built.

Collectively, the nucleus, mitochondria, and other organelles are housed in cells considerably more complicated than cells of the two other primary clades, the Eubacteria and Archaea. Eubacteria and Archaea are each single celled, lack a nucleus, and largely lack organelles, but have generated an incredible diversity of metabolisms. Archaea were only recognized in the 1990s when the late microbiologist Carl Woese found that a group of microbes had ribosomal RNA that was unlike that seen in either bacteria or eukaryotes. The first Archaea were found in inhospitable environments such as acidic mine waste, boiling vents in the deep ocean, and highly saline environments, but they have since been discovered in environments across the globe. The deep genetic differences between Eubacteria and Archaea are mirrored by fundamental differences in the structure of the cell walls and membranes, a division that dates to the last common ancestor of all life.

Recognizing the Archaea as distinct from other bacteria (Eubacteria, or "true bacteria") was critical to unraveling the origin of the eukaryotic cell. A group of Archaea known as the Lokiarchaea contain core cellular processes that are homologous to those in eukaryotes. Among these are the cytoskeleton, aspects of metabolism, membrane synthesis, and gene regulation. After the Archaea were described by Woese they were viewed as a primary division on the tree of life, along with the Eubacteria and eukaryotes. The discovery that the host cell of eukaryotes was essentially a Lokiarchaea means that this threefold division is no longer tenable, so today we view eukaryotes as nestled among the Archaea (figure 6.5).[54]

The structure and genome of mitochondria establishes their origin from among the Alphaproteobacteria. Beyond this simple statement are a host of unresolved questions, from why this event was successful (many theories view the mitochondria as "like the leftovers of indigestion"), whether the host was anaerobic or aerobic, and whether the symbiosis was a critical step in the origin of eukaryotes or a later addition. Once the mitochondrion was acquired, many of its genes were transferred to the host genome, others were lost, and a few host genes moved to the mitochondrial genome. In some eukaryotic lineages mitochondrial genomes shrank to only a few dozen genes. Several eukaryotes lack mitochondria entirely (such as the intestinal parasite *Giardia*), but their nuclear DNA contains remnant mitochondrial genes, indicating that they have lost their original mitochondria. The absence (thus far) of any eukaryote that has never had mitochondria supports arguments that the mitochondrial symbiosis was an early step in the process.[55]

Ingesting an alphaproteobacterium did not immediately generate a mitochondrion any more than the subsequent inclusion of a cyanobacterium instantly generated a chloroplast. The ur-eukaryote was a relatively complex

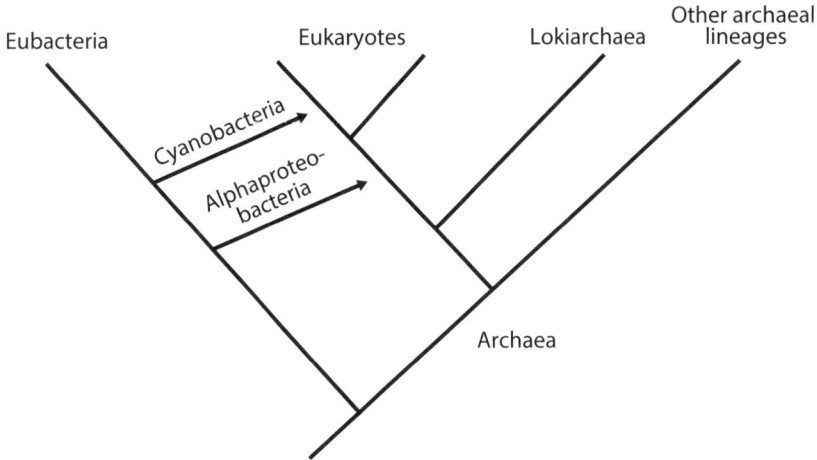

FIGURE 6.5. Basal tree of life with the relationships among Eubacteria, Archaea, and eukaryotes, including the position of the Lokiarchaea. Also shown are the symbiotic events leading to mitochondria and chloroplasts.

cell, containing many proteins once thought restricted to eukaryotes, as well as complex mechanisms for gene expression, protein regulation, and cellular activities, plus features permissive of acquiring stable symbionts. One issue that remains unresolved is whether the incorporation of Alphaproteobacteria coincides with the origin of eukaryotes or happened later. There are hints in molecular clock analyses that eukaryotes may have been assembled over hundreds of millions of years between the separation of the lineage from the Lokiarchaea and the last ancestor of all living eukaryotic cells (known as LECA, for the last eukaryotic common ancestor). Plants, algae, and some other groups also contain chloroplasts, photosynthetic organelles that arose through symbiosis of a cyanobacterium. As with mitochondria, the chloroplasts became domesticated: although chloroplasts retain a few hundred genes, over 90 percent of their genes have been transferred to the host genome. The host benefits from the carbon products of photosynthesis and controls the reproduction of the chloroplasts. Quite when this happened remains unclear, but well-preserved eukaryotes, including some multicellular filaments like living green algae, date to about 1.6 billion years ago.[56]

The origin of eukaryotes is often described as a unique or singular event in the history of life, and as a sudden evolutionary discontinuity: on Tuesday there was a eukaryote! Distinguishing among potentiation, novelty, and subsequent adaptation is to move away from narratives of discontinuity. For the origin of eukaryotes, the Lokiarchaea acquired features (I hesitate to describe them as adaptations) such as the elements of the cytoskeleton, gene regulation, and other proteins that established the potential for the successful acquisition of an alphaproteobacterial symbiont. But this new combination was only one step in the transformation of a Lokiarchaea into a eukaryote, as demonstrated by the lag between the first eukaryotic common ancestor, FECA, and LECA, during which this lineage acquired other novelties.

In a recent critique of claims of the uniqueness of the eukaryotic lineage, microbiologists Austin Booth and Ford Doolittle argued that the growing connections to Archaea showed that eukaryotes were less "special," and claims of uniqueness were nothing but storytelling. One of the strongest arguments in favor of a unique status has been that mitochondria provide four to five orders of magnitude more energy per gene than bacteria, permitting greater gene expression. If energy is this important the larger genomes of eukaryotes may have potentiated opportunities for inclusion of genes that selection would have removed from the smaller genomes of Eubacteria and Archaea. This is not the place for a detailed exegesis of this debate, but distinguishing among radical, generative, and consequential evolutionary events, a distinction often elided, recognizes that "uniqueness" has many flavors. As discussed later in this chapter, claims of uniqueness often rely on the view that a transition occurred rapidly (whatever that means for an event that occurred over two billion years ago). But novelty or innovation can be unique in cumulative effect without being instantaneous, and radical events carry no necessary implication of being generative nor consequential (and vice versa). The mitochondrion and chloroplast appear to be singular but incredibly consequential novelties in the history of life. Other research suggests that the earliest eukaryotes likely did not require mitochondria, which might have been a later addition. Booth and Doolittle fairly observe that claims for the unique nature of any evolutionary event require an objective metric, so I have great sympathy for their efforts to deflate the "majorness" of eukaryogenesis.[57]

The eukaryotic cell represents a classic "general-purpose technology," which facilitated new possibilities. Many eukaryotic clades contain multicellular lineages. Beyond the sophisticated developmental systems of plants, animals, and fungi, red and brown algae each have relatively sophisticated development, and some form of multicellularity has been found in at least 21 different eukaryotic clades. What does the distribution of multicellular forms across eukaryotes tell us about the genetic capacity for multicellularity? One possibility is that the various clades with complex development and those with simpler forms of multicellularity independently acquired such capacity. If this were true, we would expect that the developmental tools might be much different between vascular plants and animals, for example. But from the discussion of developmental novelty among bilaterian animals, one might wonder whether instead of being independent, co-option of common features might be reflected in the considerable overlap among their developmental tool kits.

During the COVID pandemic I began exploring these questions with developmental biologists Roberto Feuda and Arnau Sebe-Pedros (neither of whom should be blamed for the following). We hoped that the wealth of recent sequence data as well as new bioinformatic tools would shed light on the evolution of the eukaryotic regulatory genome. To my surprise, our studies combined with other recent reports showed that the last ancestor of all living eukaryotic cells had a sophisticated regulatory genome, containing a wealth of tools for controlling gene expression, such as transcription factors and small RNAs with regulatory functions. Other comparative studies reached similar conclusions. Between LECA and the origin of plants and animals there were many changes to the eukaryotic regulatory genome, including the expansion of transcription-factor gene families, which increased the combinations of transcription factors available to regulate gene activity. We easily eliminated the hypothesis that plants, animals, and other eukaryotic clades required independent genomic novelties to achieve multicellularity. Many of the similarities in the regulatory genome between plants and animals, and plausibly some other multicellular clades, likely arose because of what has been called "the parallel discovery of easily discoverable systems."[58]

The mitochondrion and chloroplast represent the primary eukaryotic symbioses, but the power of new combinations is further illustrated by the other symbiotic events that pepper eukaryotic history. Many reef-building corals have microbial dinoflagellates embedded in their tissues, and the carbon products generated by chloroplasts inside the dinoflagellates allow corals to thrive in nutrient-poor environments. Dinoflagellates are also symbionts within giant clams and sponges. Sea slugs (shell-less gastropods) graze on algae and then sequester their plastids in a portion of their gut, a process known as kleptoplastidy (something remembered by students of invertebrate zoology, just to be able to drop *kleptoplastidy* into casual conversation). Unlike chloroplasts and mitochondria, which have exchanged genes with their hosts to form more integrated complexes, dinoflagellates and other secondary endosymbionts can be expelled by their host and return to a free-living life. Other symbiotic relationships are also widespread: lichen represent a symbiosis between a fungus and an alga, many ants cultivate fungi or aphids, and there are mycorrhizal associations with fungi in the roots of some land plants.[59]

Symbiotic events represent horizontal transfer of entire cells, but horizontal gene transfer is common among Eubacteria and Archaea and occurs in eukaryotes as well. While this can scramble phylogenies, turning evolutionary trees into more complex networks, it is a potent source of novelty. Among Eubacteria and Archaea horizontal gene transfer is often associated with the acquisition of adaptations to highly saline, high-temperature, or similarly extreme environments. Multicellular organisms have exploited this pathway too: herring seem to have acquired their antifreeze proteins in this way, and then passed the genes forward to smelt.[60]

Earlier in this chapter I introduced transposable elements, or transposons, small pieces of DNA implicated in the origin of the mammalian placenta. More broadly, transposons are an important source of horizontal gene transfer. Eukaryotic genomes are commonly much larger than those of bacteria or archaea, and commonly the domains of a protein are separated by sequences that are not translated into protein (introns). That the modular structure of the protein is mirrored in the DNA structure provides an opportunity for rearrangement of the order of the modules, or domains, within a gene, as well as for insertion of new domains, often via transposons. Transposons may make up over half the total genome in some species and are an important source of novelty, particularly by adding transcription-factor binding sites to new genes. Among tetrapods there are at least 94 cases where different domains have been fused into new proteins, and in the livers of primates most new promoters and enhancers originated through transposition. I am not suggesting that all transposition events generate evolutionary novelties. Rather, transposition is an effective mechanism of recombination that can generate new regulatory patterns that may lead to a novelty.[61]

Fusion through symbiosis, exon shuffling, horizontal gene transfer, and the construction of pangenomes document the creative potential of combinatorics. Combinatorics rapidly expands evolutionary possibilities at lesser risk than attempting to generate entirely new features. New combinations are a route to individuation, the critical process that generates novelty. The *Pax* gene family of transcription factors arose by combining a *paired* domain with a *hox* domain. The origin of the family was a novelty, but not its subsequent duplication and divergence. An important question for future research will be to identify circumstances where novelty is likely to arise.

Novelty: Tempo and Mode

George Gaylord Simpson's *Tempo and Mode in Evolution* begins: "How fast, as a matter of fact, do animals evolve in nature? That is the fundamental observational problem of tempo in evolution." In chapter 2 I proposed that distinguishing novelty from other evolutionary processes reflected perceptions that some evolutionary changes were too rapid and too pervasive to be explained by natural selection. Huxley acknowledged the discontinuities evident in the fossil record, Goldschmidt invoked genetic repatterning as a mechanism for novelty, and other evolutionary biologists entertained ideas even further from consensus views of evolutionary mechanisms. The counterarguments to this position have viewed apparent discontinuities as failures of fossil preservation (as Darwin did) and rejected claims that evolutionary innovations require novel evolutionary mechanisms.[62]

Arguments about tempo and mode are often linked: apparently rapid transitions are seen as requiring alternative mechanisms. But rapid to a paleontologist can be tens to hundreds of thousands of years, which strikes a biologist as more than sufficient time to accommodate necessary evolutionary changes. Both the Ediacaran–Cambrian radiation and the origin of eukaryotes have been described as evolutionary discontinuities, with either implicit or explicit claims about rapidity. Yet molecular clock evidence suggests that hundreds of millions of years could have elapsed between the first eukaryotic ancestor and the last common ancestor of all living eukaryotes, an interval more than sufficient to encompass any conceivable evolutionary dynamic. The Ediacaran–Cambrian appearance of animals in the fossil record was relatively sudden (to a paleontologist, if no one else), but played out over millions to tens of millions of years and was built on a foundation of genomic and developmental novelties that had accumulated over several hundred million years.

This chapter links biological novelties to new patterns of genetic or developmental regulation. Not necessarily to specific motifs among gene regulatory networks, but to a regulatory state and often to the construction of new networks of interaction. Focusing on networks provides a perspective missing from purely sequence-based approaches to novelty. I have emphasized these interactions among regulatory genes for two reasons. They provide a critical mechanistic underpinning to our knowledge of the origins of evolutionary novelties, an underpinning that would not have been possible two decades ago. Indeed, advances in this area are being produced so rapidly that the details provided here are often limited to the most well-substantiated conclusions. Of equal importance, however, is that these regulatory networks illustrate the crucial role of networks in generating novelty while preserving the stability of basic architectures. Just as important as understanding the mechanisms generating novel morphologies is understanding how these novelties can be sustained over long intervals, and the networks of interactions underpinning character identity mechanisms address this challenge. Animal body plans have been a flashpoint for controversy among evolutionary biologists for decades because they capture this duality of change and conservation: the apparently discontinuous origin of architectures followed by their preservation for hundreds of millions of years.

Several of the cases described here highlight the origins of genetic or developmental novelties well before their full potential has been expressed in developmental patterning, or in the phenotype of the clade. This is true for many developmental novelties associated with the origin of animals. The developmental mechanisms that established limb buds, essential for the transition of fish onto land, preceded the appearance of limbs by about five million years. These developmental changes expand developmental capacity long before this capacity is utilized morphologically. Developmental capacity is the potential to generate a range of morphological and behavioral outcomes from a given set of genomic and developmental interactions, and it generates the opportunity space available to evolution. Much of that space may never be realized, or regions of that space may be occupied for hundreds of millions of years before vanishing (think of trilobites and dinosaurs). Many of the novelties discussed in this chapter have expanded developmental capacity for the clades that acquired them, sometimes incrementally, other times more radically. Distal enhancers greatly expanded the repertoire of regulatory interactions available to animals. With distal enhancers the number of transcription factors controlling the activity of a gene increased, allowing greater spatial and temporal control. Despite the incredible potential of distal enhancers, recall that they have not been widely deployed by sponges, cnidarians, or placozoans, becoming widespread only among bilaterian animals. Increased developmental capacity represents the tail of potentiation. Whereas potentiation is a necessary but not sufficient change for a specific subsequent novelty, increased developmental capacity opens a range of possible evolutionary novelties. Thus, the partial co-option of an appendage gene regulatory network to the anterior edge in the ancestor of skates and rays generated a capacity for new morphologies. To foreshadow later discussions, there are cultural and technological analogues to developmental capacity.

With the next two chapters our focus turns to the fate of evolutionary novelties, addressing the nature and dynamics of evolutionary innovation. Chapter 7 begins by exploring the factors controlling whether, and when, evolutionary novelties generate innovations. Here I contrast three scenarios: those where novelties directly drive innovation, those where environmental and ecological opportunity pulls novelty forward to generate innovation, and those that involve a process of construction. This leads to a consideration of two contrasting metrics for innovation: the standard view of counting the number of species and higher clades traceable to novelties, and an alternative view that considers consequences in terms of the ecological or network effects. Although in earlier work I have argued strongly for the latter, there are strengths and weaknesses to either a taxic or a network approach, and the most appropriate metric may often be dependent on the specific questions being addressed. The bulk of the chapter discusses modes of evolutionary diversification and the ecological drivers of innovation, before ending with a discussion of the contrasts between incremental innovations and those that are disruptive or radical. Chapter 8 employs the framework developed in chapter 7 to examine specific cases of innovation, continuing the story of many of the evolutionary novelties discussed here, particularly the origin of eukaryotes, the Ediacaran–Cambrian radiation, and the diversification of flowering plants, with the attendant impacts on mammals and insects.

7

Building Communities

THE DYNAMICS OF EVOLUTIONARY INNOVATION

PRIAPULIDS WERE off to a great start in the early Cambrian but were soon surpassed by annelids. The riotous early diversification of mollusks over tens of millions of years finally filtered down to a suite of architectures efficiently adapted to the vicissitudes of life. And of course, arthropods rapidly eclipsed their various cousins among the panarthropods. The wings of skates and rays allowed them to feed on crustaceans, clams, and other animals on the sea bottom (some species of rays later began filter feeding on marine plankton), and while there are more than 600 living species, skates and rays have not had the transformative impact on their ecosystems of the eukaryotic cell, arthropods, or the decidual cells that facilitated the evolution of pregnancy in eutherian mammals.

Novelties need not lead to innovations. Some novelties are associated with major evolutionary innovations, some contributed to the ecological and evolutionary success of the clades that possess them, while others persisted with little immediate impact. Macroevolutionary lags demonstrate that there is no necessary correlation between novelty and innovation. Novelties may be unsuccessful and disappear, others may persist for millions or tens of millions of years, successful within their clade but making little contribution to ecological communities. These persistent novelties represent latent innovations, banking increased capacity against a change in ecological or evolutionary circumstances.

Recognizing a decoupling between novelties and innovations is no more an argument that evolution has foresight than would be the suggestion that "key innovations" correlate with ecological opportunities. As with any form of variation, novelties represent character states that may be evolutionarily advantageous and persist because of natural selection or drift. Geneticist Michael Lynch has long argued that low population sizes in many plants and animals means that features may persist and become fixed without any selective advantage. He proposed that some features fixed during the early evolution of animals, even such characteristic features as the expansion of the number of Hox genes, might reflect chance (genetic drift) in small populations rather than selection, however useful such features may have proved subsequently.[1]

Historically, the novelties that have received the most attention are those linked to diversifications of clades through Simpson's model of adaptive radiation and those associated with the establishment of a new higher taxon (wings and birds, for example) or with ecological transformations such as oxygenic photosynthesis or grasslands. But Simpson's view of adaptive radiations conflated more species (or more inclusive clades) with ecological impact. He did not distinguish between the generative and the consequential components of evolutionary diversification. In part this may have been because when Simpson was writing in the 1940s to 1960s, we lacked the tools to interrogate these different dynamics more rigorously. Phylogenetic methods provide a ready means of clarifying patterns of taxonomic diversification (even if understanding causal triggers remains challenging), while advances in ecology, food webs, and network dynamics provide tools for understanding how novelties may contribute to new ecological networks. Clarifying the nature of evolutionary novelty now allows us to investigate how such novelties contribute to both generative (increased biodiversity) and consequential (ecological impact) aspects of innovation.

Why focus on innovation as a change in the structure of ecological communities rather than on the generation of new taxa? There are two primary reasons: First, adaptive radiations are just one of a range of modes of evolutionary diversification. Second and perhaps more importantly, adaptive radiations are not commonly associated with evolutionary novelty as defined here. This claim will doubtless strike many as either heresy or addled, but "key innovations" often do not meet the criteria for evolutionary novelty and in any case form only one mode of evolutionary diversification. In my view, key innovations all too frequently seem to be whatever someone happens to be studying. If I study eyes, then eyes become key innovations that explain many problems. In part the challenge arises because key-innovation hypotheses are fiendishly difficult to test.

Focusing on the transformative nature of novelties in generating new ecological structures, the consequences of novelty, is not an argument for ignoring the generative capacity of evolutionary novelties. Novelties have established new clades and triggered the diversification of clades containing novelties, as well as those with which they interact (think, for example, of the codiversification of insects with flowering plants). But whether novelties are tightly coupled with taxonomic diversifications is a more challenging issue. Where evolutionary novelties have generated new species and clades it has been because of the ecological roles the novelties have facilitated. So here I propose inverting Simpson's views to focus on how novelties contribute to ecological transformations, with taxonomic diversification as a consequence.

This chapter is a prelude to the case studies of evolutionary innovation discussed in chapter 8, beginning with a consideration of the factors that control the fate of evolutionary novelties. One focus of this section is the role of external perturbations such as rapid climate change and mass extinctions in establishing the conditions for successful innovations. The following section assesses the role of adaptive radiations, places them within the larger variety of evolutionary diversifications, and discusses how a taxic approach can contribute to understanding innovations. The bulk of the chapter discusses the construction of new ecological structures, in particular the role that some species play in modifying ecological relationships. One issue that will feature prominently in this discussion is the biological equivalent of the public goods discussed in chapter 3. Even with a focus on ecological dynamics, novelties can vary greatly in the impact of their associated innovations. The final section draws upon insights from economics into different types of innovations.

Fate of Novelties

Three issues influence our perceptions of the fate of evolutionary novelties. The first is the long-standing tensions in evolutionary biology between what philosopher of biology Peter Godfrey-Smith characterized as origins explanations versus distributional explanations; second, the challenge paleontologists face in identifying the rate at which novelties are introduced; and finally, uncertainty over whether successful innovation requires biotic crises or other disruptions.

Evolutionary biologists have often framed the tension between novelties versus innovations as differences between whether new variation drives evolutionary change, or whether preexisting variation is sufficient and evolutionary dynamics are driven by changing distributions of clades and their characters once they arise. Many have assumed that ecological and environmental opportunity pull novelties along. In other words, the generation of novelties, like all sources of evolutionary variation, is sufficiently frequent that it required no special attention. I noted in chapter 3 that many economic historians have debated the parallel question of the drivers of change in technology. Alternatively, the supply of evolutionary novelties may vary through time, if it is influenced by circumstance, with genomics and development pushing out novelties for vetting by natural selection and drift.[2]

The second issue is that the fossil record records only those novelties that are sufficiently successful to persist and lucky enough to be preserved. Many novelties, perhaps most, fail. How many novelties were generated but disappeared without leaving a trace? It is possible that the rate of novelty generation is constant, but what varies over time is how frequently these novelties contribute to innovations. Establishing the frequency of novelties and how the rate of generation of novelties differs between major clades or over time are fascinating questions, but ones it may be impossible to fully resolve. Novelties that persist but do not immediately contribute to innovation represent latent potential that increases the capacity of a clade to respond to future opportunities. By recognizing the contribution of potentiation to facilitating innovation we acknowledge that innovation is contingent.[3]

Even though some novelties disappear quickly, other novelties persist and may become important features in the life of a clade. Some fraction of novelties led to innovations through the generation of biodiversity or by contributing to the construction of new ecological relationships. The secondary symbiosis of chloroplasts described in the preceding chapter illustrates how novelties diffuse across clades, sometimes forming new combinations and generating subsequent innovations. In the past many students of cultural and technological innovation had dismissed comparisons with biological innovation, arguing that while an engineer in Kansas can easily borrow an agricultural invention from a Swede, combine it with technology from Germany to invent something new, and sell it on to a farmer in Peru, biological evolution lacks a similar capacity. But this is just what happens among many microbial clades, where borrowing is routine. Even clades with complex development such as plants and animals transfer information between species through transposons and similar genetic elements.

Finally, to what extent does innovation require or exploit disruption? a question often posed in cultural and economic contexts. Mass extinctions and other biotic crises have occurred repeatedly through the past 550 million years. There were doubtless similar crises before the advent of animals, but mass extinctions among microbes are unlikely, and in any event we lack a sufficiently well-documented fossil record to be able to identify mass extinctions before about 550 million years ago. The five canonical mass extinctions as recognized from a synoptic compilation of the marine fossil record occurred at the end of the Ordovician (444 million years ago), in the Late Devonian, at the end of the Permian (251 million years ago), at the end of the Triassic (199 million years ago), and at the end of the Cretaceous (66 million years ago), although the Late Devonian seems to have been an interval of repeated crises rather than a single, discrete episode, and some have questioned whether the end-Triassic event was of a similar magnitude to the other episodes. Beyond these canonical events, however, at least 18 other significant biotic crises have been recognized through the Ediacaran and Phanerozoic.[4]

Mass extinctions have often been invoked as triggers for evolutionary novelties. Increased diversity and morphological novelties in flowering plants, insects, and placental mammals have been tied to the aftermath of the end-Cretaceous event. Equally impressive were widespread diversifications among vertebrates in the aftermath of the end-Permian mass extinction, when the first dinosaurs, mammals, and turtles as well as numerous other now-vanished vertebrate lineages appeared. Similarly, some paleontologists have connected the magnitude of the Cambrian diversification to a possible late Ediacaran mass extinction. Here I simply note the frequent claims that innovation follows mass extinctions and might also follow lesser biotic crises. Longitudinal studies through long-lived clades identifying evolutionary novelties and intervals of innovation have emerged in the past few years, particularly for vertebrates, yet the relationship between mass extinctions and subsequent innovations still needs a fuller exploration.[5]

The canonical mass extinction episodes were associated with rapid and pervasive changes in global climate, but there were other episodes of rapid climate and environmental changes as well. If climate change is correlated with innovation, deglaciation after widespread continental glaciation in the early Permian and a sharp, 5°C–8°C increase in global temperature about 55 million years ago (the Paleocene-Eocene Thermal Maximum, or PETM) are among many episodes that may have been implicated. It has been suggested that the crown groups of many marsupial and placental mammals radiated after the PETM, with consequent impacts on terrestrial vertebrate communities. Birds may have diversified at the same time, although it is far from clear that the PETM was associated with substantial ecological innovation. During the Paleozoic and Mesozoic numerous episodes of widespread low-oxygen conditions developed in the world's oceans, including during the early Cambrian and the Cambro-Ordovician transition and in the wake of the end-Permian mass extinction. Rachel Wood and I proposed a model with novelties developing at times of low oxygen in the oceans, and subsequently spreading as innovations once higher oxygen levels returned. The fossil record provides many opportunities to test the role of different types and magnitudes of environmental disruption in generating innovation.[6]

Perhaps the most famous example of increased diversity in the wake of a mass extinction is the burst of modern birds in the 20 million years from the Late Cretaceous to the Paleocene. During this span all the major clades of birds were established, from flamingoes to owls and songbirds (modern birds, or Neoaves, includes all birds other than ostriches, waterfowl, chickens, and their allies). Placental mammals and flowering plants underwent diversification as well, with many new major clades and large numbers of significant novelties, as discussed in the next chapter. Innovations are also associated with other biodiversity crises. In the aftermath of the great end-Permian mass extinction, innovation was widespread among invertebrate clades. As two examples, the modern scleractinian corals appear in the Triassic (although they must have a largely missing history in the Paleozoic), and marine gastropods shifted from feeding on detritus, plants, or filtering plankton to largely predatory lifestyles. But the importance of biodiversity crises as a means of "opening up ecospace" requires rigorous testing. Devonian to Early Triassic ammonoids (a now-extinct clade of coiled cephalopod mollusks) provided just such an opportunity to evaluate bursts associated with two of the five major mass extinctions as well as several other biotic crises. The shape of their shells is highly informative about their ecology. An analysis of the morphological disparity of the clade at very high temporal resolution revealed no bursts following either of the biotic crises, despite the rapid taxonomic increases.[7]

Mass extinctions, rapid climatic shifts, and other large-scale biodiversity crises are the most extreme examples of environmental perturbations that might foster evolutionary novelties. But perhaps innovations need not require the widespread destruction that accompanies mass extinctions. Aside from the odd hurricane, tropical rain forests and adjacent shallow-marine ecosystems experience less environmental variability than temperate and polar regions. Tropical forests, coral reefs, and adjacent habitats are the most diverse on the planet, and fossil evidence suggests a long history of high tropical diversity. It might be the case that these species reflect the ability to finely divide resources, multiplying essentially similar species but with little novelty or innovation.

But studies across clades on land and in the oceans have identified the tropics as a cradle of innovation, not just diversity. For example, there is evidence for preferential origin of durably skeletonized orders of marine animals in tropical ecosystems, and others have suggested that the increased productivity and energy availability in the tropics enhances the success of those novelties that are energy intensive. But claims for the tropics as a cradle of diversity must still be treated with caution. Diversity is high in the tropics, but the generation of novelty corrected for diversity may not be any higher than in other regions. More rigorous treatment of generation of novelty in a phylogenetic framework will be needed to evaluate the role of the tropics.[8]

Successful innovation runs from nearshore habitats into the deep sea. Over the past 250 million years new orders of sea urchins, corals, and other marine invertebrates have most commonly appeared in nearshore habitats. Although studies identifying specific novelties and resulting innovations are needed, several aspects of these studies strongly implicate novelty and later innovation: These new clades appear at low diversity in nearshore areas and only later expand across the shelf, often with a lag of millions of years. Moreover, there is no preferential nearshore origin of genera or species, which appear across the depth gradient.[9]

Other geographic factors may also play a role in the fate of novelties and their success as innovations. Giant kelp (a type of brown algae) contains a novelty that allows it to withstand intense grazing pressure from marine herbivores, and this novelty evidently arose and spread from the North Pacific. This region was also the home to a clade of now-extinct mid-Cenozoic, four-legged mammals known as desmostylians. Powered by their massive limbs, this wonderfully bizarre clade of marine herbivores ripped vegetation from shallow seas. The North Pacific was also the ancestral home of a variety of walruses and sea lions that fed by crushing clams, snails, and other hard-shelled invertebrates, a lifestyle that greatly increased the rate of disturbance in these communities. More generally, a comparison of the number of innovations in the temperate North Pacific with the Northern Atlantic and temperate southern oceans indicates that ecologically important novelties were more common and arose earlier in the North Pacific and were more likely to spread to other temperate oceans. An explanation for this pattern that is consistent with studies of the tropics, described above, is that the North Pacific is the largest and most productive of these ecosystems (and likely the oldest), with heightened competition and greater available energy contributing to the origin of these novelties and their eventual success.[10]

A caveat to these studies is that they reflect how novelties become success-ful as innovations, not the necessarily the circumstances under which novel-ties first arise. But a common thread through these examples is a restructuring of ecological communities. One of the longest-standing debates among ecolo-gists is whether assemblages of species are organized into discrete, replicable ecological communities, which might therefore be more resistant to insertion of a new species, or whether the composition of ecological communities is relatively open. The apparent link between innovation and the aftermath of biotic crises has fueled the assumption that the latter create opportunities, just as Mayr and Simpson argued. If communities are relatively open, as suggested by most ecological studies, distributions reflect individual adaptations of a species and their interactions with competitors, predators, and other species with which they interact. Simple co-occurrence of a suite of species provides little evidence about whether their communities are open or closed. For ex-ample, plants may have similar requirements for soil type, rainfall, and tem-perature but little ecological interaction. Their co-occurrence does not imply the structuring and interdependence necessary of closed ecological communi-ties. Barnacles and muscles are commonly found in the high intertidal of rocky shores along the Pacific coast of North America, where they can withstand periodic exposure and extreme temperature fluctuations. Crabs and a variety of clams and snails are found further down, with kelp beds and tracts of sand dollars in the subtidal. Predation and competition are certainly important in structuring these communities—the lower limit for the mussels is often set by the range of predatory starfish—but species primarily respond to specific en-vironmental tolerances rather than biotic integration. If most communities are relatively open, the insertion of species with novel traits should not be restricted to the aftermath of biotic crises. At least in theory, open communi-ties should be relatively more accommodating to the introduction of new species.[11]

Generating Innovation

A fundamental challenge in many sciences is that what one wants to measure is often different from what one can measure, or at least measure easily. Econo-mist Simon Kuznets introduced gross domestic product as a measure of eco-nomic activity in 1934, and despite its well-known flaws (several of which Kuznets pointed out), the quarterly release of GDP estimates is eagerly awaited by journalists, investors, and politicians (most of whom should know better).

Evolutionary biologists face similar challenges in analyzing large-scale evolutionary patterns, and this is particularly true in trying to understand the nature and impact of evolutionary novelties and innovations. In the project described earlier, using short-lived families of marine animals (higher taxa) as a metric for ecological opportunity, Valentine, Sepkoski, and I were a long way from what we really wanted to measure, but it seemed like a good idea at the time. Our study was hardly unique in relying on higher taxa, and even with the far more sophisticated tools available today, studies of evolutionary diversifications persist in conflating distinct issues: increases in the number of taxa, normally species or genera within a phylogenetic framework; diversity of form, or morphological disparity; increases in ecological variety; and the impact of evolutionary novelties. Often one assumes that what one is interested in, such as the impact of novelties, covaries with something one can measure, like the number of taxa, and then measures the latter. Specific cases may exist where species numbers provide a useful proxy for other dimensions of biodiversity, but that is something to be demonstrated, rather than assumed.[12]

Here I explore the limitations of relying on adaptive radiations as a general scenario for innovation, despite expansions of models of evolutionary diversification over the past decade or so. This sets up the next section, which examines alternative metrics and why a new generation of tools is required to provide greater insights into innovation.

The Poverty of Adaptive Radiations

Events as diverse as the diversification of Darwin's finches in the Galapagos Islands, cichlid fish in the Great Lakes of Africa, and placental mammals following the disappearance of dinosaurs 65 million years ago have been described as adaptive radiations. Across South America, birds, plants, and monkeys all exhibited rapid diversification beginning 65 million years ago. Several thousand different studies of both fossil and recent adaptive radiations have been published over the past few decades, revealing a greater diversity of patterns and underlying processes than can be comfortably encompassed under a single conceptual metaphor. As one example, in Australia there have been diversifications of geckos and other lizards, eucalyptus trees, and spiny trapdoor spiders during continent-wide drying over the past 10–15 million years. To some, continent-wide diversifications stretch the bounds of an adaptive radiation, as does application of adaptive-radiation models to events that played out over millions to tens of millions of years in the fossil record.

Recall that adaptive radiations represent a response by closely related and thus ecologically similar species to new opportunities and competition for resources. These opportunities may arise following a mass extinction or other biotic crisis; invasion of an island, an archipelago such as the Galapagos Islands, or the continent of Australia; or from an evolutionary novelty. Most adaptive radiations are bounded in the sense that however great the initial ecological opportunities, the multiplying species and their attendant specializations eventually exploit all available possibilities. As resources become limited the rate of diversification slows, and the adaptive radiation transitions to a more stable dynamic. If the adaptive radiation includes an increase in the number of species, the rates of generation of new species and of change in morphology are expected to decline during this transition.

Botanist Tom Givnish has emphasized that key early workers, from vertebrate paleontologists Osborn and Simpson to ornithologists David Lack and Mayr and botanists Stebbins and Sherwin Carlquist, viewed an increase in ecological roles as a key feature of adaptive radiation, independent of any changes in the number of species. They were equally accepting of an adaptive radiation involving just a few species spread across a range of different ecological roles, exemplified by Darwin's finches of the Galapagos Islands (the focus of Lack's work), or dozens of new species, but the generation of new species was secondary to the expansion of ecological roles. It follows from this focus that adaptive radiations could arise within a species, with humans as the obvious example. But this is also true of microbes with a pangenome, such as *Escherichia coli*, or the minute photosynthetic bacterium *Prochlorococcus* described later in this chapter.[13]

Despite the initial emphasis on ecological roles, the spread of phylogenetic methods and an increasing emphasis on quantification led some evolutionary biologists to conflate rapid, explosive increase of the number of species with an adaptive radiation, with later studies linking adaptive radiations to initial bursts of evolutionary activity generating increases in the number of species, in body size, in ecological functions, or in morphological disparity. Significantly for understanding novelty and innovation, this later work has recognized a link between morphology and an increase in the number of species, most obviously during the Cambrian radiation of animals, introduced in the previous chapter. A characteristic of the Cambrian diversification in many clades is rapid increases in morphological disparity, unmatched by a similar increase in the number of species. Other cases of decoupling have since been identified from the fossil record but have also been criticized by biologists who see little evidence for such "early-burst" patterns among modern adaptive radiations.[14]

The earliest history of echinoderms provides a beautiful example of decoupling between taxonomic diversity, morphological novelty, and ecological function. Sea urchins, starfish, brittle stars, crinoids (feather stars), and the sluglike holothuroideans make up modern echinoderms. But these five modern classes are a pale remnant of the incredible diversity of early Paleozoic forms, with a range of interesting, sometimes bizarre morphologies. At least 20 classes and some 50 orders have been described from the lower Cambrian through the end of the Ordovician. Based on a detailed phylogenetic analysis, a recent study examined patterns of morphological change and functional diversity. This study showed that morphological evolution outpaced ecological change, a pattern inconsistent with adaptive radiation scenarios. Unlike earlier work on morphological disparity in other clades, the earliest echinoderms had generalized morphologies and feeding habits, staking out the basic boundaries of the morphospaces and ecospaces before they were filled in by subsequent evolution. As with other studies of disparification, this is a much different evolutionary dynamic than expected under an adaptive radiation scenario, or indeed most other scenarios of evolutionary diversification.[15]

While building phylogenies emphasizes the importance of taxa and their lineages, the resulting phylogenies provide a foundation for the interrogation of changes in morphology, size, ecological function, and biogeography. Rather than relying on ad hoc stories about how these patterns are related, a variety of explicit models have been introduced against which empirical studies can be compared. This allows rigorous tests against patterns expected from Brownian motion or other processes. If the phylogenies can be carefully calibrated in time through well-dated fossils and molecular clocks, then reliable estimates of evolutionary rates can be established, bringing to fruition the dream Simpson had in writing *Tempo and Mode in Evolution*. Here I will skip the analytical complexities, but such studies have recognized pulses in many evolutionary diversifications. Such pulses of diversification have been identified in sea urchins through the past 250 million years, in canids (dogs), and in the Carboniferous to Triassic diversification of reptiles.[16]

Early bursts are not unique to biology but have been associated with cultural and technological events too. Patterns seemingly mimicking an adaptive radiation occurred with the onset of personal computers (my first was the luggable and loveable Osborne), in the 1990s with the first stirrings of the internet, and in the late 2000s after the introduction of the first iPhone. Despite

the many pitfalls to embracing biological analogies for cultural and technological change, it can also be instructive to consider the analytical and conceptual
challenges in different domains.

Another route to testing the relationship between novelty and adaptive
radiation is to examine studies of the latter and identify the occurrence of
novelties, which might then lead to evolutionary innovations. I discovered a
compilation of 1030 adaptive radiations gleaned from 730 papers published
between 2003 and 2013. Of these, 70 recorded key innovations (48 different
papers). I examined each of these 70 cases and identified 10 examples of
evolutionary novelty (5 with temporal lags, and 5 without). These novelties
included antifreeze in Antarctic fish, cacti, ice plants, and the evolution of
electric fish. Although I knew of other studies that had not been included but
documented novelties associated with evolutionary diversifications, my objective was to evaluate the number of cases in this unbiased compilation. From
this analysis I concluded that it is difficult to argue that evolutionary novelties
are linked to adaptive radiations or to evolutionary innovations. I am certainly
not denying the existence of adaptive radiations (or novelty, for that matter).
Simpson's hypothesis can be saved only by adopting his view that "every
marked expansion of a group, whether it be a genus or a phylum or the whole
animal kingdom, is an adaptive radiation."[17]

A variety of modes of diversification have been recognized by evolutionary
biologists, often placing diversifications within a phylogenetic context, quantitatively assessing changes in morphological disparity, dating divergences
with molecular clocks, fossil evidence, and other data, among other improvements. The *Anolis* lizards of the Caribbean are a well-studied example of replicate or convergent radiations, in which similar morphologies repeat within
the same clade on adjacent islands. Many evolutionary biologists find such
radiations compelling evidence for the long-term persistence of adaptive
peaks within an evolutionary landscape. Although islands provide numerous
examples of adaptive radiations, on continents both adaptive and nonadaptive radiations have been well documented. Nonadaptive radiations reflect
multiplication of lineages within a clade with limited expansion of adaptive
variety, with one example being the diversification of neotropical mice and
rats after they migrated from North America. This may seem an oxymoron,
but nonadaptive radiations illustrate why focusing exclusively on taxic diversity is a poor metric for adaptive radiations, much less evolutionary innovations. Finally, there are cases where rapid phenotypic divergence is accompanied by little increase in taxonomic diversity, a pattern that bears the hideous

neologism of *disparification*. A recent example comes from the diversification of frogs in Madagascar: adaptive variety increased with no proliferation of new lineages.[18]

This brief synopsis suffices for three observations regarding their relationship to evolutionary novelty and innovation: First, the classic adaptive radiation scenario promulgated by Osborn and enthusiastically adopted by Simpson, Mayr, and other participants in the Modern Synthesis captures only a fraction of the range of evolutionary diversifications examined with more robust methods of phylogenetic analysis and modeling. This leads to the second key finding, which is that there is no general correlation between bouts of adaptive radiation and evolutionary innovation. Finally, as emphasized by studies of morphological disparity, and by the evaluation of phylogenetic, morphological, and ecological evolutionary patterns in Cambrian and Ordovician echinoderms described above, morphological novelty is often decoupled from taxonomic diversity.

In my own work I have been interested in how feedback and interactions among species can influence diversifications, leading to the progressive exploitation of new ecological opportunities—opportunities that were not present when the radiation began, which I described as constructive radiation. In the next section we turn to their role in evolutionary innovations.[19]

Constructing Innovation

When Spanish explorers first sailed up the Chesapeake Bay in 1570, surfaces above nine meters water depth were covered by the American oyster, *Crassostrea virginica*. The Powhatan and other Native American groups had harvested oysters for thousands of years, building extensive shell middens along the coast (oyster middens are found from Florida north into Maine). After the British explorer John Smith founded the colony of Jamestown on Powhatan territory in 1607, oyster reefs on the Chesapeake were considered a threat to navigation, so the Royal Navy later used them for target practice. Well into the nineteenth century oysters filtered the entire volume of the bay in two or three days, removing bacteria and generating clean water in which an amazingly diverse ecosystem thrived. But by the 1970s, overfishing of oysters had caused the population's collapse. So few oysters remained that one complete filtration of the volume of the bay required a year, perhaps longer. In consequence, the bay changed from a healthy ecosystem with blue crabs and menhaden to one with excess nutrients and jellyfish.

A liter of seawater from the open ocean in the tropics and subtropics contains 100 million cells of *Prochlorococcus*, a minute, micron-sized (0.001 mm) cyanobacterium. They are both the smallest and the most abundant photosynthesizing organism on Earth, generating much of the organic carbon feeding into the base of marine food webs, particularly in nutrient-poor, open-ocean settings. Despite the considerable genetic and metabolic diversity across individuals of *Prochlorococcus*, all belong to a single species. Together, they generate about as much organic carbon as all the Earth's agricultural crops.[20]

There are dozens of oyster species spread across more than a dozen genera, but the ecological impact of oysters is a collective function of individuals, not their taxonomic diversity. *Prochlorococcus* has a vast pangenome but constitutes a single species. An individual *Prochlorococcus* may be the least significant thing in the oceans. Fecal pellets of most animals are vastly larger. But collectively, *Prochlorococcus* underpins a significant fraction of global primary productivity. The impact of oysters and *Prochlorococcus*, like that of humans, is collective, not individual. Individually, humans are capable of remarkable feats of creativity (Shakespeare, Ada Lovelace, Beethoven, Admiral Zheng He, Darwin, Maya Lin), while groups of humans have perpetrated almost unfathomable cruelty. Human accomplishments from the origin of language and the domestication of plants and animals to the construction of mathematics and the growth of technology are fundamentally collective achievements, however much popular Western history fetishizes individual genius. The creativity of individual geniuses is deeply embedded in social, cultural, and economic systems. Collectively we have had an impact on the Earth perhaps only paralleled by *Prochlorococcus* (although *Prochlorococcus* is almost certain to long outlive us).

But counting species tells us nothing about the collective impact of individuals. Evaluating the generation of new taxa is certainly a component of evolutionary innovation, but this section will sketch out approaches involving the construction of new ecological relationships, including those with the potential to expand biodiversity. My goal here is to expand the boundaries of research programs and encourage the development of new tools.

Growing Ecological Networks

Having taken the decision to define evolutionary innovations by some measure of ecological interactions, a variety of different approaches could be adopted. One might compare species composition before and after an innovation, calculating some measure of the dissimilarity between assemblages. This approach has been used to chronicle the appearance of new communities over the past hundred years because of biodiversity loss and human disturbance. Because a simple use of dissimilarities results in a continuous range, this basic approach has been extended by comparing the dissimilarities to a baseline developed by examining community change through time. In this way new ecological assemblages that stand out from the more routine dynamics of community change can be readily identified. One might extend this approach by comparing ecological assemblages before and after an ecological crisis. After taking into consideration differences in preservation and other artifacts, this would provide a relatively straightforward means of identifying innovations.[21]

The notion of an ecospace provides an alternative method of characterizing innovations, one focused on function rather than species. In chapter 4 ecospaces were defined by attributes such as mobility, habitat type, and feeding habitat (e.g., filter feeding). One version for marine invertebrates has 27 states and 60 total characters. While this is sufficient to generate a complex, multidimensional space, it is too coarse to be affected by any but the most substantial evolutionary innovations. For specific suites of novelty it might be possible to modify this approach, increasing the specificity of the characters and examining the impact of novelties on an ecospace by tracking a clade through time, much as we can track changes to the morphology of a clade through time on a phylogeny.[22]

Food webs have been a useful tool of ecologists since the early 1900s. These networks depict the relationships between consumers and predators in a community. Depending on the data available to an ecologist, a food web can span from primary producers through herbivores to various levels of predators. In these networks each species is represented as a node, with the connections between nodes representing trophic relationships between species. Ecologists understand species differ in abundance, often by orders of magnitude, and that some species prey more heavily on one species than another, differences that can be reflected as changes in the strength of connections between species. Ideally, this would be reflected in networks with nodes and connections of different sizes. Until recently, however, the math to analyze such network dynamics did not exist. It would be nice to think that recent improvements in the analysis of food webs reflect a growing appreciation for their importance, but the truth is that new tools were needed to understand the operation of power grids and financial networks. Mathematically these problems are very similar.[23]

Over the past decade or so the study of ecological networks has been expanded to include host-parasite interactions and mutualist networks, such as those between flowers and their pollinators. But these are just components of broader, multidimensional ecological networks. The structure of these networks influences their stability during times of crisis and how readily new species can invade. Collaborations between ecologists, archaeologists, and others have generated food webs encompassing human cultures within their natural settings.

In my view the future of studies of ecological networks, including how innovations change their structure, lies in exploring the full range of interactions. Consider beavers. They feed on the soft tissues underneath the bark of trees and grub around for shoots of new vegetation. In a food web, beavers are connected to willow, aspen, or whatever trees grow in their habitat. But like the oysters of the Chesapeake Bay, they are premier ecosystem engineers, gnawing down trees, digging up mud, and damming rivers and streams as they refashion their environment. As the beaver activities wax and wane, the water levels of their ponds rise and fall, with consequences for adjacent plant communities and fluctuations in populations of ducks, herons, and geese, and woodpeckers hammering away on the dead snags surrounding a pond. Beaver engineering enhances the success of beavers, while creating new habitats for dozens of other species through ecological spillovers. By forming wetlands, beavers change the hydrology of rivers, the cycling of nutrients, and the quality of waters downstream. Indeed, beavers are so beneficial that states in the Rocky Mountains are seeking to encourage their return to areas they inhabited before nineteenth-century demand for beaver hats led to their extirpation. Perhaps most importantly, ecosystem engineering activities generate an ecological inheritance that can persist for far longer than the 10- to 12-year life-span of a beaver. Salmon and brown bears are rarely considered part of the same ecological community, but along rivers in the Pacific Northwest and Canada, as the bears feed during salmon runs, they transfer nitrogen and other nutrients from the salmon to habitats along the rivers. This effect requires both the bears and the salmon; neither alone is sufficient.[24]

Charles Darwin's final book, published in 1881 just six months before his death, was *The Formation of Vegetable Mould, through the Action of Worms, with Observations on Their Habits.* In this Darwin summarized decades of observation and experiment on the earthworms in his garden at Down House, along with accounts from across Europe, to illustrate how earthworms have constructed the soil. Earthworms, beavers, oysters, and coral reefs are examples of animals whose activities can modify their own physical or chemical environments and those of other species. Darwin was just the first of many biologists to document how many plants, animals, fungi, and even some microbial groups actively modify and construct the environments in which they live. Simpson describes such activities in *Major Features of Evolution*.[25]

These activities have been described as ecosystem engineering or niche construction. These concepts are similar and generate ecological inheritance (such as beaver dams) that modifies the selection active on the species itself. Beavers illustrate the difficulty in distinguishing niche construction from ecosystem engineering, because their lodges and dams are an ecological inheritance that benefits later generations of beavers as well as having pervasive and long-lasting effects on other species. Beaver dams may last for generations of beavers, but their effects are still measured more in ecological than evolutionary time. Other types of ecosystem engineering can have a cumulative effect that is very long-lasting, even on an evolutionary timescale. Reefs can persist for thousands of years (depending on how rapidly sea level changes), and reef-building activities have been almost continuous since the Cambrian, with fossil reefs built by microbes known from rocks about two billion years old. The accumulation of shell beds in shallow seas has modulated the sea bottom, as has active burrowing and bioturbation. Because the two ideas are so similar, I will use *niche construction* and *ecosystem engineering* interchangeably in the remainder of the book, somewhat favoring ecosystem engineering, because establishing fitness differences, a key component of niche construction, in the deep past rarely lies within the purview of the paleontologist.[26]

These nontrophic interactions illustrate the need for a broader approach to ecological networks. Traditional food webs are just one example of a broader class of species interaction networks. Ecological networks encompass multiple types of ecological interactions: feeding, competition, ecosystem engineering, and others. Representing such a variety of interactions requires a more sophisticated class of graphs known as multigraphs. In a multigraph the nodes still represent species (more precisely, trophic species, which are collections of species that have identical ecological interactions), but nodes may be connected by multiple edges. Two species that compete but also prey upon the same species have two trophic interactions and thus two links connecting them in a multigraph. In binary graphs and multigraphs interactions are between species, but the interactions can themselves interact. To return to the Chesapeake Bay, oysters are critically important to the bay, but so are menhaden, small fish that filter plankton and are food for many predators. A more accurate depiction of the filtering of the bay is that it is a consequence of oysters and menhaden, which jointly deliver cleaner water to blue crabs and other species. When ecological interactions affect not just the species but other interactions, we need to move up another level in network complexity to hypergraphs. Study of a suite of arthropods associated with a shrub in the Brazilian tropics revealed a hidden network of such higher-order interactions, which impact the arthropod species in ways that would not have been apparent through a standard food web approach. While study of multigraphs and hypergraphs is an active area of research, fewer tools are available to investigate ecological or evolutionary problems than the bipartite food networks described earlier. But as these ideas develop, applying them to evolutionary novelty and innovation is a great opportunity for those with skills in biology and physics or for collaborations between the two fields. Current limitations constrain the case studies of innovation in the next two chapters to qualitative rather than quantitative discussions.[27]

An objection to using ecological networks as a foundation for understanding evolutionary innovations is the challenges of building such networks in deep time. If modern food webs required a detailed knowledge of the feeding habits of each species, based on observation, gut contents, or similar information, then this would be a valid criticism. In fact, food webs for living species are often constructed based on functional and phylogenetic information. This has allowed the reconstruction of food webs from the recent past as far back as the Cambrian. A more compelling concern is that fossilization generally preserves only durably skeletonized taxa and misses soft-bodied forms. When some colleagues and I reconstructed food webs of the Cambrian Burgess Shale communities we had the advantage of the exquisite preservation of these soft-bodied animals, sometimes including gut contents. A recent analysis has shown that ecological and preservational signals can be carefully disentangled, suggesting that a reliable reconstruction of ecological food webs to understand the impact of innovations is not impossible. For events with little fossil record, including innovations associated with the earliest evolution of animals or the origin of eukaryotes, the challenges are much greater. The magnitude of the transitions associated with these innovations may be sufficiently great that we can still learn a great deal about them absent a detailed ecological network.[28]

There is a further complexity in understanding now novelties change the dynamics of ecological networks. In chapter 3 I introduced the idea of public goods, economic goods like the calculus, or knowledge more generally, that are non-excludable and non-rivalrous. Economists such as Paul Romer have argued that public goods play a disproportionate role in fostering economic growth, and I have extended this argument by suggesting an analogy between various economic goods and biological goods. As with economic growth, biological novelties that generate public goods have made significant contributions to the impact of evolutionary innovations. Some niche-constructing and ecosystem-engineering activities not only benefit the species that constructed them (as long-lived beaver dams may benefit generations of beavers) but may also spill over to impact other species, even entire ecosystems. Oysters in the Chesapeake Bay generate at least two public goods that benefit other species: their reefs provide a hard substrate for attachment of species that would not survive on muddy or sandy bottoms, and by filtering the bay they produce cleaner water that allows more diverse and productive ecosystems. Their filtration is a trophic interaction between the oysters and their food, but the effects of that activity percolate through the entire bay. In a network sense, public goods have widespread direct and indirect effects and thus require a more complex network approach.[29]

Impacts of Innovation

Joseph Schumpeter distinguished between incremental and radical innovations, in which the former contribute to an ongoing process of change in technology, and the latter are disruptive, triggering widespread changes in manufacturing processes and the replacement of one class of goods by another. Schumpeter's insight has been adopted, challenged, refined, and renamed by subsequent generations of economists, and many economists have proposed their own distinctions among inventions and innovations. But the general utility of recognizing that technological innovations differ in their impact remains. Extreme cases of radical innovations may result in general-purpose technologies that impact most preexisting spaces and open or construct vast new spaces in which evolution can operate. Classic examples of these are the steam engine, electric motors and the electrification of society, the internal combustion engine, and semiconductors and the spread of information technology. There are human cultural analogues to general-purpose technologies: language, writing, mathematics, and agriculture. An intriguing question is whether the idea is also applicable in the biological realm, an issue to which we will return in chapter 10.[30]

Different types of innovation can be placed in the context of the evolutionary spaces described in chapter 4: incremental innovations represent the exploitation of new opportunities within existing spaces, along the lines explored by Andreas Wagner and his colleagues, while radical or disruptive innovations are those that generate new evolutionary spaces and are likely to be less common. Viewed as evolutionary spaces, the distinction between incremental and transformative innovations can be applied across biological, cultural, and technological domains. In principle, one could distinguish between incremental and radical novelties, but our focus on novelties is character based, while the generative or consequential ecological and evolutionary effects of novelty are a dimension of innovation. Several examples of radical innovations will be discussed in the next chapter: the singular origin of oxygenic photosynthesis (photosystem II) in cyanobacteria, later transferred via symbiosis into a variety of eukaryotes; the eukaryotic cell itself with mitochondria, which has enabled far more complex ecosystems than appears to have been possible with Eubacteria and Archaea alone; the spread of mycorrhizal associations of fungi with various plants and other fungal innovations, which have been essential contributors to nutrient distributions within terrestrial communities. There are far more radical innovations than can be discussed in this book, but those that generate public goods are more likely to radically transform ecosystems, because the spillover effects have greater reach through the ecological networks.

Radical innovations are not limited to public goods. The origin of flowering plants (angiosperms) involved important but not immediately obviously transformative novelties. Recall that the key novelties associated with flowering plants apparently originated well before the spread of the clade. But flowers and their associated changes with the origin of flowering plants opened a vast opportunity space for the clade. The spillover effects from this innovation expanded and restructured terrestrial ecosystems as insects diversified through intimate coevolution with many plant clades. As in technology, it is the spillover effects from the innovation that may have the greatest impact. The initial flowering plants revealed vast developmental and ecological spaces for evolutionary exploration.

Just as Darwin and Wallace independently discovered the principle of natural selection, and Newton and Leibniz calculus, biological innovations can arise independently in different clades. Some of these have been more incremental than radical, such as the repeated acquisition of similar toxins across widely disparate animal clades (presumably because of the limited range of potential starting proteins and the biochemical similarities in targets), but other convergent innovations have been transformative. A through-gut in animals allowed animals to package fecal matter into pellets, which in turn improves the sequestration of the carbon-rich matter in sediment. Some evidence suggests that the cumulative impact of fecal pellets may have transformed global oxygen and carbon cycles. Grasses provide a second example. They differ in the pathways they use to capture carbon dioxide. All plants share the ancestral C_3 pathway, but C_4 grasses have arisen independently in many different tropical and subtropical clades, because this pathway is more efficient in hot climates. C_4 grasses have transformed grassland environments over the past 15 to 20 million years. That C_4 grasses have appeared repeatedly indicates that this novelty is readily accessible for C_3 grasses as well as being of significant value.[31]

Although there was likely microbial life on land long before 500 million years ago, plants, animals, and fungi began to invade the land only during the Ordovician and Silurian. This gave the marine realm a three-billion-year advantage in generating novelties and innovations, from metabolic pathways to multicellularity. Paleontologist Gary Vermeij compared the origins of high-performance and metabolically expensive innovations since the Ordovician in the oceans and on land. Vermeij used three criteria to identify innovations: First, they must provide a new capacity and competitive superiority, which implies that they are energetically expensive. Second, it must be possible to identify or reliably infer the presence of the innovation in fossils, thus excluding everything from C_4 photosynthesis to culture. Finally, the innovation must have arisen more than once, and in evolutionarily disparate clades. These criteria reflected the questions in which Vermeij was interested, but they effectively eliminated a host of post-Ordovician innovations. It is unclear how significantly these criteria bias the conclusions.

Vermeij concluded that the opportunities for innovation in the oceans are more limited than on land, for 11 of the 13 innovation he considers either appeared first on land or are found only on land. From this he concludes that this represents an "irreversible transition" from land as a recipient of innovations from the ocean to a provider of them to marine ecosystems, increasing overall productivity and performance. There are two challenges to this conclusion. First, Vermeij's conclusion is at least partly predicated on the interval he considers. His analysis begins with the Ordovician, so it misses the long history of earlier novelties and innovations. By the end of the Great Ordovician Biodiversification Event, fundamental marine innovations had appeared and there were greater opportunities on land. Extending the time span under consideration back to 600 million years ago would swamp the terrestrial results with the innovations associated with the initial rise of animals. Second, by only considering innovations in one or more clades the analysis emphasizes accessible novelties (chapter 4). Why unique high-performance and metabolically expensive novelties are excluded from the analysis is murky.[32]

The persistence of novelties that have not contributed to innovations represents a bank of latent or potential innovation, which may be engaged when future conditions are appropriate. There are emerging opportunities to integrate comparative developmental studies with phylogenetics and the fossil record to explore the acquisition of latent novelties within clades and their expression as innovations. Promising areas for such studies are recoveries after both the end-Permian (251 million years ago) and end-Cretaceous (66 million years ago) mass extinctions. Dinosaurs, mammals, turtles, and a variety of other vertebrate clades appear soon after the end-Permian mass extinction, and some of the novelties underlying these clades were likely generated before the extinction. Tracing these novelties and how the new clades influenced the construction of new, post-extinction communities is already underway, but there is much more to learn about the dynamics of this process.

The varying fates of evolutionary novelties reflect the contingent circumstances in which they arise. If novelties often require potentiating factors to facilitate their success, then focusing exclusively on ecological opportunity as the controlling factor provides a blinkered view of the dynamics of evolutionary innovation. Innovations may generate new dimensions of biodiversity. The recognition of the range of modes of evolutionary diversification has spurred diverse research programs and increasingly sophisticated quantitative techniques for assessing the dynamics underlying phylogenetic patterns. But in this chapter, I have argued that focusing exclusively on the generative dimensions of evolutionary innovation blinds biologists to the network effects of important novelties and their spillover effects on other species, and on ecological networks more broadly. Importing a goods approach from economics to biology provides critical insights into the consequences of innovations, and particularly into why some innovations are radical or disruptive and others are more incremental. Oysters, *Prochlorococcus*, and humans each represent species with transformative powers through their impacts on broader ecological networks. Although network dynamics involves a powerful set of techniques, which have been applied to questions far beyond their home in statistical physics, applying them to biological problems of evolutionary innovation has been poorly explored. Despite the efforts of decades of ecologists in studying the processes of community assembly, there is still considerable work to be done in understanding and quantifying how innovations might restructure communities, whether ecological disruption or climate change is an essential step, and whether innovation is more probable in communities with different structure. There are great opportunities for collaborative projects for evolutionary biologists, ecologists, paleobiologists, and others to employ and expand network tools.

I am acutely aware of the rudimentary nature of suggestions in this chapter that innovation can be recognized by changes in network structure of ecological assemblages. Much more needs to be done in this direction. Distinguishing between incremental and radical or transformative innovations is an initial step toward understanding whether there are different classes of innovation. Despite the evident contingencies in the evolutionary success of innovations, can we identify factors that may enhance the likelihood of their success or failure? Taken together, studies on the fates of evolutionary novelties may help identify the factors controlling the generation of evolutionary innovations.

With this foundation, the next chapter examines a range of evolutionary innovations. Together, these examples illustrate how contingency may structure the fate of evolutionary novelties, the role of potentiation in the generation of innovation or novelty, and the importance of networks of ecological interaction.

8

Ode to *Opabinia*

I HAVE always been happiest with the interval from the earliest glimmerings of animal life through the Ediacaran-to-Cambrian Explosion of animals, the exquisite reefs of the middle Permian, and the convulsive end-Permian mass extinction, to the origins of the modern world in the Triassic Period. The rocks are fascinating, outcropping in wonderful places to do fieldwork, from South Africa and Namibia to China and back to the American Southwest. But avoiding the past 200 million years has largely freed me from having to confront the issue of progress. Few paleontologists (and no one else) seem overly concerned with whether the brachiopods and sponges of the mid-Permian reefs in the Guadalupe Mountains of West Texas are more "advanced" than their Ordovician ancestors, 300 million years earlier. But with the Triassic came the earliest mammals, dinosaurs, turtles, more modern marine ecosystems, and a raft of questions about progress and complexity: Has the history of life been trending toward plants, animals, and even microbes that are more complex, more sophisticated, better adapted, more powerful? Somehow just *better*? From a meager beginning with minute cells, life spent several billion years exploring the metabolic opportunities of microbial life and constructing reasonably sophisticated microbial communities, before the evolution of the eukaryotic cell and eventually multiple avenues of multicellularity allowed even greater opportunities.

Life began minute, with single cells. This allowed far more room to grow larger than smaller (although viruses arguably plumbed the smaller opportunities). Plants, brown algae, fungi, and animals are all large, complex multicellular clades, and their appearance over the past billion years suggests a trend toward increasing complexity. But the increased size and complexity of plants and animals could have arisen passively, through diffusion toward the larger end rather than having been driven toward greater size or complexity. Size we can measure, but distinguishing diffusion from a driven trend requires rigorous analysis. Complexity is more difficult, but equating a trend with progress requires some agreement on what progress means. Darwin concluded the *Origin* with his view that natural selection would inevitably drive organisms toward increased adaptation. Prolonged trends driven by natural selection should produce organisms with higher fitness than their ancestors. Such logic has led some to equate such trends with a sort of evolutionary progress. For example, Gary Vermeij has forcefully advocated the view that organisms have progressively acquired more power through time. Historically, notions of progress have often been tied to visions of the evolutionary Chain of Being described in chapter 2, leading successively to humans, European societies, and white males. A crusty residue of such thinking persists, of course, even if we recognize that it has zero scientific validity, and recent advocates for progressivist evolutionary views have worked to cleanse it of this historical baggage. Innovation has often been interpreted as a sign of progress, whether in culture, technology, or biology. The success of novelties as innovation is often contingent on the ecological and environmental settings in which organisms exist. If, as historian Will Durant once opined, "Civilization exists subject to geological consent," the same is true of evolutionary innovations.

Here we focus on the dynamics of innovation, building off the considerations in chapter 7 to examine a suite of case studies. These examples include the Ediacaran-Cambrian origin of animals and the origin and early history of eukaryotes; the rise of flowering plants and mammals, which transformed terrestrial ecosystems from the Late Cretaceous into the early Cenozoic, interrupted of course by the bolide impact that triggered the end-Cretaceous mass extinction; the spread of grasslands, beginning about 25 million years ago; and some smaller-scale innovations associated with the origin of snakes, electric organs in fish, and dung beetles, and the role hippos play in transforming their environment. The chapter closes with a consideration of the nature of social and cultural innovations. Social and cultural novelties are widespread, stretching back to microbial assemblages, but they lead us into one of the most transformative suites of novelty and innovation in the history of life: human culture, language, and technology.

The Ediacaran–Cambrian Explosion of Animal Life

As nineteenth-century geologists assembled the geological time scale, defining intervals based on characteristic assemblages of fossils in Europe, they confronted the striking absence of fossils prior to the appearance of trilobites in

Cambrian rocks. The implications of the absence were unclear, and they had no idea of the age of these rocks. That animal fossils appeared abruptly in the fossil record, representing arthropods, mollusks, and other animal clades with eyes, limbs, and guts, was troubling to Darwin and to many other paleontologists. The easiest explanation was a long gap in the fossil record sufficient to account for what must have been the prolonged evolution of early animals.

When I first visited Cambrian boundary sections in Siberia, many paleontologists believed that the Cambrian Explosion was just that—the explosive evolution of animal clades within a few million years. As we explored outcrops along the Lena and Aldan Rivers it seemed that we could almost put our fingers on the Explosion. Some paleontologists disagreed, of course, but often their intuition was based on assumptions about how long evolution "should" take, rather than a defined understanding of the temporal dynamics. Two years later I made my first visit to China, and already the evolutionary picture was becoming more complicated. As we examined the rocks along the Yangtze River and then mines and road cuts in Yunnan Province, events confined to a single bed in Siberia stretched across dozens of meters of rock.

Our far more detailed understanding of the Ediacaran and Cambrian fossil record reflects studies of the relationships between rocks in different regions and the plausibility of missing time; development of a rigorous temporal framework with which to evaluate evolutionary rates; a reasonable understanding of animal phylogeny, which informs numerous molecular clock studies of the timing of the origin of major clades; and finally, geochemical proxies for environmental conditions and how they may have influenced biological evolution. By no means are all our questions resolved, and there is still much to do. But our understanding of these events now is far more sophisticated than it was in 1990.

In chapter 6 I examined genetic, developmental, and morphological novelties and how they are associated with the origin of animals and the appearance of morphologically sophisticated bilaterians. Recall two key points: First, molecular clock studies of DNA from living organisms indicate that major animal clades arose tens of millions of years before any of their representatives appear as fossils, with sponges and cnidarians likely arising perhaps 100 million years earlier. The oldest fossil records of red and green algae date to about one billion years ago, and rocks dating to 800 million years ago contain fossils representing a variety of eukaryotes, reflecting the diversification of crown-group eukaryotic clades. A reasonable assumption is that the lineage leading to metazoans and their single-celled allies (the Holozoa, see chapter 6) originated about this time. Second, many of the novelties underpinning the developmental processes required to build animal architectures appeared earlier than they were utilized. In other words, there was widespread potentiation of developmental processes, which were later co-opted for morphological architectures, particularly those of bilaterians like arthropods, mollusks, and vertebrates. Here we consider factors driving these innovations, changes in the physical environment, the construction of new ecological networks, and the appearances of morphological and behavioral novelties across every major animal clade. To make matters interesting, feedback between these components increases the degree of difficulty. I will consider the pattern of innovations in four phases: an early, missing phase suggested by molecular clock data; the soft-bodied Ediacaran fauna; the Ediacaran–Cambrian transition, beginning with the final, Nama phase of the Ediacaran and stretching into the earliest part of the Cambrian; and the full burst of bilaterian clades with the appearance of widespread predation and animal-dominated complex ecological networks.[1]

The interplay between the physical and biotic aspects of the environment underlies critical transitions in the history of life. It is impossible (in my view and that of many of my colleagues) to understand the novelties and innovations absent an understanding of climate, oxygen levels, and related features of the physical environment. This major transition in the history of life occurred in a world utterly unlike today or the recent past. The Cryogenian Period is defined by the Sturtian glaciation from 717 to 658 million years ago, and the shorter-lived Marinoan glaciation from about 650 to 635 million years ago. Glacial ice near sea level at what were then equatorial latitudes was early evidence that these were events unlike more recent glacial epochs, where continental glaciers emanated from polar regions. During the Sturtian and Marinoan, ice extended to the equator, a condition known as a Snowball Earth. This seemingly bizarre situation has been controversial, but the geological evidence almost guarantees that even if the Snowball Earth hypothesis turns out to be incorrect, any replacement will be just as strange. Each glacial phase ended abruptly amid rapidly rising temperatures, a huge pulse of sediment coming off the continents, high alkalinity, and a widespread lid of freshwater on the surface of the oceans. Each of these would have created intensely challenging environments for the survival of any early animals—together the biological impact must have been as great as the largest Phanerozoic mass extinctions.

The slow drift of the continents led to a period of continental emergence during the late Neoproterozoic. As mountains slowly erode, many key nutrients for life, particularly phosphorus, are delivered by rivers to the oceans.

Some evidence suggests that widespread erosion during this emergence increased nutrient supply. The oceans of the early Ediacaran (~630 million years ago) likely had trivial amounts of oxygen, perhaps only 0.1 to 1 percent of current levels, and consequently had high levels of iron, low levels of phosphorus, and possibly low levels of critical trace elements such as molybdenum. Such an ocean would have limited the primary productivity upon which ecosystems depend, and thus total biomass would likely have been much lower than in modern oceans. Some evidence suggests that such conditions persisted in some regions even as increased nutrients fueled higher productivity in shallow, nearshore settings.[2]

Oxygen has been invoked as a driving factor in the evolution of early animals since the 1960s, albeit commonly without specifying just how changing oxygen levels facilitated novelty or innovation. An increase in oxygen levels after about 800 million years ago has been invoked by the "oxygen control hypothesis" as an ecological opportunity to which early animals responded with increasing body size, a zoo of new clades, and construction of new ecological communities. Early advocates of this scenario did not appreciate that animals must have previously acquired the developmental capacity to respond to these new opportunities. While possessing the variation needed to respond to a new island habitat is plausible for some adaptive radiations, having lineages poised to generate new cell types, tissues, organs, and morphologies is an entirely different dynamic. Moreover, relating increased oxygen levels to evolutionary innovation is challenging, since oxygen availability could influence the origins of animal multicellularity and of eumetazoans or bilaterians, the acquisition of large body size, and the formation of skeletons during the earliest Cambrian. These distinctions are important, as resolving them will help discriminate between increases in oxygen levels or stability that potentiate developmental or morphological novelties versus increases that potentiate innovation.[3]

As I write this, multiple geochemical proxies support the broad conclusion that available oxygen remained at about 1 percent of present levels until after 800 million years ago, and perhaps into the Ediacaran, when the levels began increasing, albeit in an unstable, pulsed, and highly variable pattern, with low oxygen levels persisting through the Cambrian and into the early Paleozoic. Some evidence from South China, for example, suggests that a wedge of oxygenated shallow waters overlay more anoxic deep waters. Increased oxygen levels may have been a general potentiating factor for the early evolution of animals, but work I have been doing with Rachel Wood of the University of

Edinburgh and her group has established that most novelties among Ediacaran organisms appeared before pulses of oxygenation, although diversification appears to have been later. Thus, at present it is not possible to unambiguously relate increased oxygen levels to the origin of animals or to the acquisition of critical developmental novelties. As discussed below, a better case can be made that the acquisition of large body size across many metazoan clades and skeletonization among some are linked to an increase of and greater stability of oxygen levels.[4]

If the recent molecular clock results are roughly correct, there is a missing history of the basal metazoan clades from their origin and initial diversification until the mid-Ediacaran (about 576 million years ago). Such a divergence pattern implies a lag between the origin of the major bilaterian lineages, such as lophotrochozoans or panarthropods, and the acquisition of the characteristic architectures by which we know these clades today. These lags between developmental and morphological novelties, and between the novelties and ecosystem innovation, are a characteristic feature of the Ediacaran-Cambrian diversification. Some geologists have suggested that early sponges and burrowing bilaterians may have helped mediate oxygen levels by helping to sequester carbon, which frees oxygen for the atmosphere. Under the right conditions sponges can filter remarkable volumes of water, and in doing so significantly engineer their environment. The greatest puzzle for me has long been cnidarians, for their greatest novelty is the cnidocyte, the explosive stinging cell characteristic of the clade. If you have been stung by a jellyfish, you have experienced the collective effects of cnidocytes and may understand why they are so effective for predation and defense. Cnidarians split into the anthozoan (corals and sea anemones) and hydrozoan (*Hydra* and relatives) clades very early in animal history. Both clades possess cnidocytes, but why did such a novel and energetically expensive cell type evolve long before other animals? Although today some cnidarians feed on fish, others have more manageable prey such as animal larvae. But a cnidocyte is far too expensive to waste on harpooning a bacterium. This conundrum remains unresolved, but nonetheless the presence of cnidocytes before 635 million years ago reveals something of the nature of the earliest animal ecosystems.[5]

This first phase continues after the end of the Marinoan glaciation at 635 million years ago. The early Ediacaran rocks document a diversity of ornamented single-celled eukaryotes, many types of algae, and several enigmatic forms that some have described as animal fossils, possibly related to cnidarians. But the phylogenetic placement of these fossils is too poorly established and too little is known of their biology to make any claims about either novelties or innovation.[6]

Finally, this phase overlaps with the known fossil record of the Ediacaran biota (phase 2), for molecular clock evidence suggests several cnidarian lineages, the rapacious, predatory ctenophores, and a host of bilaterian lineages were established by about 550 million years ago, tens of millions of years before any of their representatives first appear in the fossil record. The difficulty is that because of the pattern of independent co-option of developmental genes described in chapter 6, we do not yet know when most of these lineages acquired their characteristic morphological novelties, nor when they began to have an ecological impact. At present, there is no evidence that any of these ghost lineages significantly influenced Ediacaran ecosystems during phase 2, but these ecosystems may have had greater complexity than is apparent from fossil evidence.

The Ediacaran fauna dating from 576 to 539 million years ago is the next phase of animal evolution, with a variety of fronds, disks, and other remains of soft-bodied organisms, which have been described in Australia, Canada, Namibia, China, Ukraine, Russia, and elsewhere. Paleontologists are fortunate that at some localities extensive surfaces are exposed that essentially preserve single ecological communities, allowing the assemblage to be censused, spatial relationships examined, and possible ecological interactions identified. This is possible in Newfoundland and China, but particularly at Nilpena National Park in South Australia, where paleontologists Mary Droser and Jim Gehling have been excavating since the early 2000s with their students and colleagues. The geology of the site allows them to peel the beds back one by one, progressively revealing the rich complexity of these ecosystems. This is one of the most diverse Ediacaran fossil sites with more than 40 described species, some of which are known from hundreds of specimens.[7]

Ediacaran fossil assemblages at Nilpena, Charnwood Forest in England, the Shibantan Member high above the Yangtze River in China, and along the coastline of Newfoundland are packed with fossils. The abundance of these organisms yields important insights into ecological interactions. Most Ediacaran organisms either lay on the sea bottom or were attached to it by stalks and holdfasts, extending into the water to produce a tiered structure like the levels in a forest. Tiering allows different species to preferentially feed from different levels and creates eddies that can either aid or hinder the feeding of organisms further downstream.

Mobility was a significant feature of life at Nilpena and assemblages of a similar age. Many animals leave tracks, trails, and burrows, which may become fossilized to produce trace fossils, an important line of evidence for early animals. Among these are the ghostly "footprints" of successive feeding spots on microbial mats by a large form known as *Dickinsonia* and the feeding traces of *Kimberella*, a probable bilaterian that has been found commonly in South Australia and the White Sea coast of Russia. But the diversity of Ediacaran trace fossils remains low when compared with the Cambrian, with all the traces horizontal. While some of the animals that generated the traces seem to have been burrowing underneath widespread microbial mats, none show any evidence of vertical burrows. Mobility requires that these animals had muscles, along with the required nervous and sensory systems.[8]

We can apply what we know of the processes underlying living animals to infer the developmental processes (but not specific genes) responsible for features such as patterning of the body from front to back. Scott Evans, a former student of Mary Droser, led this research while he was a postdoctoral fellow with me. We focused on four well-studied Ediacaran organisms: the circular, triradiate *Tribrachidium*; *Dickinsonia*; *Ikaria*, a small wormlike form that Droser and Evans had described from Nilpena; and *Kimberella*. *Tribrachidium* was multicellular with several cell types and tissues and could produce axial polarity. To these features *Dickinsonia* added left–right symmetry, and its mobility required muscles and a nervous system, while *Ikaria* likely had a through-gut. Finally, *Kimberella* possessed a discrete head, a central nervous system, and more sophisticated developmental patterning, including an extrudable mouth and segmentation. Since we know when the same processes appear on the animal tree, we were able to suggest phylogenetic positions for these Ediacaran forms. We concluded that *Tribrachidium* exhibited patterning consistent with a cnidarian-grade eumetazoan; *Dickinsonia* an early bilaterian, but without the full developmental capacity of extant, crown-group bilaterians; *Ikaria* a bilaterian; and we concurred with several previous studies that *Kimberella* was likely a lophotrochozoan. Other Ediacaran macrofossils likely shared some of these features as well. So, despite earlier bilaterian divergence 550 million years ago, demonstrable lophotrochozoans and other clades do not appear as fossils until after about 538 million years ago, and arthropods after 522 million years ago (although for more on the latter, see the next section).[9]

To summarize, this phase encompassed the first ecologically and morphologically diverse assemblages of animals with significant novelties in mobility, the ability to sense and respond to the environment, and simple skeletons.

Together these generated innovations by constructing increasingly sophisticated ecological structures that foreshadowed Phanerozoic marine communities, such as vertical tiering. To be sure, microbial mats remained the foundation of food chains, and predation was almost entirely absent, so the ecological networks remained far flatter than those that developed later.

Although the base of the Cambrian long seemed to represent a fundamental discontinuity, work over the past few decades has questioned this view. In recognition, I combine the late Ediacaran assemblage and the earliest Cambrian into a single phase 3. To be sure, the Cambrian is different, with an explosion of new behaviors reflected as new types of trace fossils, reflecting large body size and new means of feeding, and the appearance of morphological novelties among bilaterian clades, but this phase is best seen as transitional between the low-dimensionality ecosystems of phase 2 and the full complexity of ecological networks in phase 4, which represent the full flowing of Phanerozoic marine ecosystems. Many Ediacaran macrofossils characteristic of phase 2 disappeared, perhaps due to a mass extinction, leaving more depauperate communities in their wake. This phase began with an increase in organisms dwelling in tubes, including several early skeletal forms, apparently reflecting an increase in gregarious feeding behavior. Reefs begin to proliferate, and there is a steady increase in the architectural intricacy of trace fossils.

Two significant novelties are first evident from trace fossils and document critical ecological innovations: vertical burrowing and the many-legged trails of the first arthropods. Trackways produced by multiple paired appendages, a significant morphological novelty, first appear in the Shibantan Member in China, dating to between 551 and 541 million years ago. Paired sets of dimples, some straight, others curved, are a centimeter or so in width, and some disappear into burrows, but they are more irregular than later tracks from arthropods and annelids. It would be easy to claim these are the earliest traces of something like an onychophoran, tardigrade, or arthropod, but the paleontologists who described them were appropriately cautious, concluding that they were made by an unknown bilaterian. Burrowing into sediment rather than gliding or bulldozing along the surface requires muscles to pull against something. Vertebrates have bones, arthropods an exoskeleton, and mollusks their shell, but unskeletonized forms use a fluid-filled body cavity known as a coelom. Thus, the advent of vertical burrowing has been interpreted as reflecting the presence of coeloms, probably in several different clades of bilaterians. Since the early 1990s the base of the Cambrian has been taken to be the first appearance of vertical burrowing in a characteristic trace fossil. This trace fossil was likely a priapulid worm, a clade of carnivorous worms common in Cambrian deposits but (regrettably) less well known today. Recent studies in Namibia have shown that complex burrowing behavior is a feature of late Ediacaran ecosystems, further smearing out innovations.[10]

These traces document the expansion of ecosystem engineering. A full understanding of ecosystem engineering will emerge through future work, but enough is known to suggest that burrowing was likely responsible for the disappearance of widespread microbial mats, as well as increased bioturbation of the sediment. Burrowing irrigates the sediment, and as oxygen levels increased this generated oxygen-rich sediments, leading to other changes in geochemistry. Based on studies of modern oceans this seems likely to have increased primary productivity, providing more food for expanding animal ecosystems. But as I pointed out previously, the increases in oxygen seem to have followed increases in developmental capacity and morphological novelty. Overall, ecosystem engineering helps communities bootstrap continuing generation of new ecological roles and species diversification.[11]

One of the most significant ecological innovations was the expansion of predation in the early Cambrian, which is well illustrated by the arthropods. Molecular clock estimates date the origin of ecdysozoans (the broader clade of which arthropods are a part) to the early Ediacaran between 636 and 578 million years ago, Panarthropoda to 616–562 million years ago, and the arthropod crown group to the middle to late Ediacaran (589 to 540 million years ago). The multiple paired legs of arthropods produce characteristic trace fossils as they skitter across the seafloor. Taking the paired trackway of the Shibantan deposits in China as the oldest evidence for Panarthropoda yields 12 million years between the origin of the clade and these fossils, while the distinctive scratch marks and tracks of arthropods dating to about 531 to 525 million years ago yield a gap of 31 million years. In either case, these trace fossils appear several million years before arthropod exoskeletons were sufficiently mineralized to preserve as fossils. This illustrates the critical information trace fossils can provide on the abundance and behavior of both readily fossilized clades and clades largely absent from the fossil record. This lag reflects the progressive hardening of the arthropod exoskeleton described in chapter 6, which is related to changes in ocean chemistry as well as expanding ecological complexity.[12]

Not until phase 3 do we find the characteristic innovation of the Cambrian, the appearance of skeletonized animals from sponges to arthropods. Although many of these clades first appear in the fossil record in the next phase, this phase features many small, skeletonized shells, tubes, and plates, collectively known as the small shelly fauna. Both fossil and molecular clock evidence has established that these clades arose during earlier phases, and in the case of sponges and cnidarians perhaps during the Cryogenian. Many different clades of animals developed the ability to biomineralize during the early Cambrian, although they accomplished this task in very different ways, from the spicules of sponges and the shells of mollusks to the mineralized carapaces of arthropods and the bones of the earliest fish. More broadly, the skeletonization marks considerable expansion of the complexity of ecological networks, driven by predation, through-guts (a complete digestive system with separate mouth and anus), and image-forming eyes. Particularly noteworthy is the growing importance of ecosystem engineering, as described previously. Here environmental potentiation and ecological innovation fed back to clades to generate further evolutionary novelty, a pattern that likely also featured prominently in the events of the final phase.[13]

The first trilobites appeared 521 million years ago, marking the transition to phase 4: the appearance of many large-bodied bilaterian clades and complex marine ecosystems. By this time the Cambrian Explosion is essentially complete. All major clades of marine animals have debuted, from sponges and mollusks to vertebrates, although some are represented by now-extinct lineages rather than members of the crown group. For example, although there are a variety of mollusk-like shells, it is unclear that crown-group gastropods or crown-group bivalves were present until later in the Cambrian. These marine ecosystems exhibit the layered complexity of modern ecosystems, with specialist filter feeders, detritivores, and predators. An important factor in our understanding of these events comes from the assemblages preserving soft-bodied organisms in China, Australia, and Greenland (the best known of these is the Burgess Shale fauna from British Columbia, Canada, but it is about 15 million years younger). These faunas preserve details of the soft-part anatomy that is rarely preserved in fossils, including the anatomy of the gut, fine structures on appendages, and even details of brain anatomy. Consequently, we have a far better understanding of the extent of this evolutionary explosion than would otherwise be the case. Arthropods are the most diverse animals among Cambrian ecosystems and provide a useful basis for examining novelties and innovations during this episode.[14]

The panarthropod clade includes modern tardigrades (water bears), onychophorans, and arthropods. So many extinct representatives have been recovered from the Chengjiang and similar faunas that paleontologists have been able to reconstruct their early history in remarkable detail, including the novelties in the transformation of the stubby-limbed lobopodians into the jointed-limbed, skeletonized arthropods with a multipart head and various specialized appendages. The major panarthropod clades had diversified into many subclades by the time they first appeared in the fossil record 521 million years ago.[15]

Reconciling estimates of divergence times from molecular clocks and the first appearances of clades in the fossil record is less of a problem than it might appear. There is no reason to expect the appearance of morphological attributes of a new clade to coincide with the divergence of sister clades. Indeed, quite the opposite. The generation of individuated, novel morphological characters requires time, even more so if novel morphologies require new cell types or tissues with attendant genetic or developmental novelties. To restate one of the central theses of this chapter, novelties provide the capacity for further novelties and innovations, but that capacity need not necessarily translate into further evolutionary change.[16]

Starfish, sea urchins, brittle stars, and crinoids each have fivefold, or pentaradial, symmetry, leading many of us to assume that this feature is characteristic of all echinoderms (holothuroideans, or sea cucumbers, are the sole modern clade of echinoderms without pentaradial symmetry). But most of the dozens of Cambrian and Ordovician clades of echinoderms lacked pentaradial symmetry. Echinoderms as a clade are defined by the presence of a mesh-like, calcitic internal skeleton, or stereom, and all but the earliest echinoderm clades share a remarkable suite of novelties in the water-vascular system (of which the tube feet of starfish are a component). Among the 21 formally recognized echinoderm classes are clades with spiral and pentaradial symmetries as well as some bizarre asymmetric forms. Overall, early Cambrian echinoderms displayed very high morphological disparity and low taxonomic diversity, and recent estimates suggest there were more than 30 individuated clades in the early Paleozoic, with at least 10 known from the Cambrian. Despite phylogenetic uncertainty about the relationships among these clades, each is associated with one or more morphological novelties, and they occupy different regions of the echinoderm morphological space. As with arthropods, the high morphological disparity was apparent when echinoderm clades first appeared in the fossil record about 521 million years ago. Recall from the previous chapter that a comparison of their morphological and ecological diversification found that morphological novelties preceded ecological success. Moreover, morphological change in general evolves more rapidly than does ecological function, at least in the early history of this clade.[17]

Integrating the details about novelty described in chapter 6 with the perspective provided here suggests lags after the origin of developmental novelties and likely also after the appearance of at least some morphological novelties. The patterns of acquisition of morphological novelty and ecological innovation have been best studied among arthropods and echinoderms, but the patterns in these clades seem likely to hold among other bilaterian groups. The major increase in morphological disparity and the establishment of major clades largely occurred between about 540 and 520 million years ago, with morphological disparity outpacing taxonomic diversity or ecological function. The trace fossils produced by the jointed appendages of arthropods and the diagnostic stereom of echinoderms provide high confidence that the characteristic architectures of these groups evolved close to when they first appear as fossils. Yet molecular clock data suggest clade divergences predated the appearance of these groups. These patterns are best explained by the early potentiation of developmental capacity, likely through gene co-option.

Origin of Eukaryotes and Early Forms

The rise in oxygen between 800 and 500 million years ago was the second transformation of planetary oxygen levels. Earlier, the Great Oxygenation Event about 2.4 billion years ago occurred as the spread of oxygenic photosynthesis allowed oxygen to build up in the oceans and atmosphere. The Great Oxygenation Event may have been a prerequisite for the success of eukaryotes, as eukaryotic cells are generally larger and thus more energetically expensive than bacterial or archaeal cells. Oxygen is essential for modern mitochondria, and thus it seemed plausible that their origin was linked to increased levels of atmospheric oxygen. When the late Lynn Margulis invoked symbioses as critical steps in the origin of eukaryotes in 1967, she linked these events to increased oxygen levels and the arguments by geologists such as Preston Cloud. While not all eukaryotes require oxygen (what are known as obligate aerobes), it is necessary for most. Chapter 6 described the foundational novelties associated with the origin of eukaryotes, including their origin out of the Lokiarchaea, the combinatoric exchange of genes between bacterial lineages and archaea, the eventual symbiosis to generate mitochondria, the later origin of chloroplasts, and the progressive growth in cellular complexity between the first eukaryotic common ancestor (FECA) and the last eukaryotic common ancestor (LECA).[18]

The resulting eukaryotic cell was one of the major evolutionary transitions in the history of life, but despite its undoubted potential there appears to have been a considerable lag between the origin of eukaryotes and their impact on Proterozoic ecosystems. Recognizing this lag is dependent upon timing critical events in the early history of eukaryotes. In chapter 6 I hinted at molecular clock evidence suggesting that assembling the eukaryotic cell may have required several hundred million years (in other words, between FECA and LECA), with an origin of eukaryotes between 1.9 and 1.6 billion years ago (although possibly much earlier). Here we examine molecular and fossil evidence for the early history of eukaryotes, the relationship of these events to major changes in the global environment, the divergence of crown-group eukaryotes, and the question of whether limitations on Proterozoic primary productivity retarded eukaryotic evolutionary dynamics. No surprise, I interpret this evidence as supporting a model for the progressive assembly of the eukaryotic cell, followed by a lag of hundreds of millions of years until environmental changes (potentiation) permitted the growth of eukaryotic-dominated ecosystems. Key events in this timeline are illustrated in figure 8.1.[19]

Paleontologists have relatively little to offer by way of constraints on early eukaryotic history, as it is challenging to identify definitively eukaryotic features among single-celled, organic-walled microfossils. Rocks dating to 1600 million years ago in China preserve relatively large microfossils with complex cellular walls and other features consistent with eukaryotes, and slightly younger rocks from northern Canada contain a more diverse assemblage of likely eukaryotic remains, some with surface ornamentation. Although we can be relatively confident that eukaryotes had evolved by 1600 million years ago, none of these forms have unequivocal evidence of being post-LECA eukaryotes. Not until rocks from about 1000 million years ago are remains of red and green algae found, making them the oldest definitive fossil evidence for crown-group eukaryotic clades. Vase-shaped tests (a test is a shell for a single cell) and scales dating to about 800 million years ago are evidence of predatory eukaryotes and defensive structures.[20]

New analytical techniques, broader coverage of different clades, and additional sequences now provide molecular clock estimates for the divergence between Archaea and eukaryote lineages at ~2700 million years ago and LECA at 1870 to 1680 million years ago, with a mid-Proterozoic divergence of crown-group lineages (the living eukaryotic clades), which is consistent with other recent studies.[21]

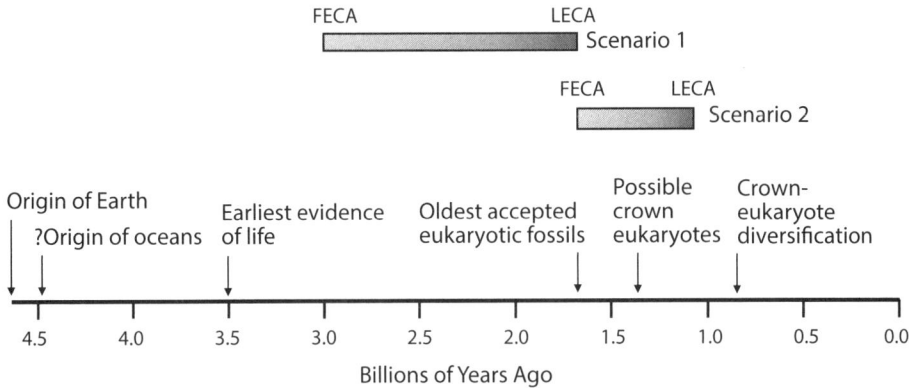

FIGURE 8.1. Timeline of key events in the origin and early history of eukaryotes. Scenario 1 involves the first eukaryotic common ancestor (FECA) older than 3.0 billion years ago, with the last eukaryotic common ancestor (LECA) about 1.6 billion years ago, as discussed in the text. Scenario 2 posits a much younger FECA about 1.6 billion years ago, with LECA about 1.1 billion years ago.

The idea that since existing mitochondria operate by aerobic respiration (using oxygen to metabolize nutrients and produce energy), the symbiotic origin of mitochondria necessarily occurred in an oxygen-rich environment has a long history. Such a reliable energy supply has allowed eukaryotic cells to be larger, with greater internal complexity. But there are both biochemical and geochemical reasons for questioning this scenario. Several alternative scenarios place the origin of eukaryotes in largely anoxic environments, and a currently leading model posits that eukaryotes emerged from close ecological consortia of bacteria and archaea in which each partner depends upon products cast off by another partner, known to microbiologists as a syntrophy. In such a setting mitochondrial metabolism relies on hydrogen rather than oxygen to transfer electrons. There are several such models, which differ in their details about the nature of the consortia, the details of which are not important here. But the key point is that the organisms generating hydrogen become tightly linked to those consuming the hydrogen to the point that evolution favors the generation of a new combination, which became the eukaryotic cell. In this case oxygen may have played little role in the origin of eukaryotes. Although this will be obvious to many readers, I need to emphasize that this last scenario effectively decouples the origin of mitochondria via symbiosis from their conversion to operating with oxygen (aerobically). All known living eukaryotes postdate the

great divergence of eukaryotic clades, and the distribution of aerobic mitochondria across the tree of eukaryotes is consistent with a single origin of
mitochondria, but multiple conversions to an aerobic means of operation.
Consequently, even the mode of operation of mitochondria in LECA is uncertain: low levels of oxygen were certainly required for mitochondria, but the
mitochondria may have used oxygen when available but not required it. The
lengthy gap between FECA and LECA and the accumulation of novelties between them raises the possibility that some record of this transition might be
preserved among Proterozoic fossils.[22]

Organisms need to be abundant and in the right environments to be preserved as fossils. Many of those estimating eukaryote divergences may downplay
the possibility of lags before new clades become ecologically significant. But
eukaryotes may not have become ecologically important until hundreds of millions of years after the divergence of eukaryotic crown clades (a scenario reminiscent of the evolutionary pattern associated with animals). Support for this
scenario is found in fossil lipids, preserved remnants of the fatty acids that make
up cellular membranes. Each major clade of eukaryotes produces characteristic
lipids known as sterols; cholesterol and testosterone are perhaps the most widely
known. After burial, chemical processes convert sterols to steranes: more chemi-
cally stable forms which can be preserved for hundreds of millions to a billion
years. For example, dinoflagellates produce a structure known as dinosterol.
Such chemical fossils, or biomarkers, provide critical insights into Proterozoic
life. (The rise of biomarker studies has, however, forced paleontologists to con-
sort with those who enjoyed organic chemistry in college.) While the oldest
likely sterane biomarkers date to about 1650 million years ago, steranes become
far more abundant in rocks younger than about 800 million years ago. One interpretation of this pattern, given the conclusive evidence for earlier eukaryotic
fossils, is a delayed increase in the abundance of eukaryotes.[23]

The world in which eukaryotes originated and diversified was alien to our
modern Earth. The oceans and atmosphere were largely free of oxygen before
the onset of the Great Oxygenation Event about 2400 million years ago, which
boosted oxygen levels to perhaps 1 percent of present levels (today's atmosphere is about 21 percent oxygen at sea level). During the Snowball Earth
event about 2200 million years ago, pervasive glaciation covered much of the
globe. Although the Great Oxygenation Event was once viewed as a trigger for
the origin of eukaryotes, the molecular clock evidence establishes that it preceded the origin of eukaryotes by hundreds of millions of years. If so, then
while increased oxygen levels may have permitted the early evolution of eukaryotes, the last common eukaryotic ancestor possessed mitochondria and

generated products such as sterols and thus plausibly required some level of oxygen, however slight.[24]

So, what do we know of Proterozoic oxygen levels? Recall from the earlier discussion that we lack a means of directly assessing past oxygen levels. Rather, geochemists rely on geochemical proxies that provide indirect evidence of past oxygen levels, with iron, chromium, and molybdenum featuring in recent studies. These proxies each have their idiosyncrasies, and they vary in reliability. Most geochemists working on this problem take the view that from 1800 to 800 million years ago oxygen levels were low, somewhere between 1 and 10 percent of present atmospheric levels, and oceans likely had relatively high levels of iron, although localized oxygen oases may have been present.[25]

Before turning to the implications of these low oxygen levels for Proterozoic eukaryotes, I should mention some opposition to this. Paleontologist Susannah Porter has been a staunch advocate of a relatively young LECA, around 1100 to 1300 million years ago. Molecular clock studies suggest the major eukaryotic superclades diverged within about 300 million years of LECA. The first appearance of fossil representatives of these clades between 1050 and 750 million years ago is consistent with a late date for LECA. There are several dozen localities that preserve eukaryotic fossils between 1700 and 800 million years ago, so the late appearance of eukaryotic crown groups is a relatively robust result. Porter's argument for a relatively late LECA would make the pre-1100-million-year-old microfossils representatives of now-extinct, stem-group eukaryotic clades. In particular, she suggests that aerobic respiration and possibly mitochondria were acquired relatively late in eukaryotic history, closer to LECA.[26]

Low Proterozoic oxygen levels may have left the oceans sufficiently starved of nutrients to retard eukaryotic evolution and the formation of higher productivity ecosystems. As in modern economies, the diversity and productivity of an ecosystem are controlled by the amount of readily available energy. In this sense (and in this sense only), low-productivity Proterozoic ocean ecosystems were like early modern Europe, where the wood fires and the horse as the prime means of transportation constrained economic opportunities. Phosphorus is a critical nutrient for life, and today most phosphorus is recycled by a variety of eukaryotes, but such recycling requires oxygen. At lower oxygen levels, such as during much of the Proterozoic, recycling would have been much less effective. If this were the case, life may have been limited to areas near rivers delivering phosphorus into the sea.[27]

To summarize, although most living eukaryotes require our richly oxygenated world, connecting the origin of eukaryotes to the endosymbiosis of mitochondria and the Great Oxygenation Event has not been substantiated. Rather, the intimate syntrophies between archaeal and bacterial cells that led to FECA likely formed in largely oxygen-free environments, and it was in such settings that the transition from these syntrophies to the remarkable cellular complexity of LECA occurred, which did include mitochondria. The eukaryotic cell is a unique transition in the history of life, but until the recent explorations of the Lokiarchaea many had viewed this as a relatively rapid evolutionary transformation driven by the advent of oxygen-rich environments. Instead, it now appears that the transition to LECA occurred over hundreds of millions of years, and the formation of the mitochondrion was not an early step in this process but may have been relatively late. Finally, the crown-group eukaryotic clades probably diverged sometime between 1600 and 1000 million years ago when the deep oceans were free of oxygen, but shallow marine settings had dynamic oxygen levels of 1–10 percent of present atmospheric levels. By 1000 million years ago some more complex eukaryotic lineages were diversifying. Yet the full ecological impact of eukaryotes was not felt until closer to 800 million years ago, as increased oxygen levels in the deep ocean facilitated more complete recycling of nutrients, particularly phosphorus, and expansion of eukaryote-dominated ecosystems.

Three important conclusions emerge from this scenario: First, the evolutionary novelties predate their utilization as ecological innovations. Second, both biological adaptations and changes in the physical environment (both oxygen and nutrient levels) potentiated evolutionary innovations. Finally, the novelties in LECA provided a pathway to thriving in higher-oxygen environments and new evolutionary opportunities for complex, differentiated, multicellular algae, fungi, plants, and animals. Remarkably, LECA had the capacity to manage a transition from a low-oxygen world to one more like today. As Lynn Margulis and Preston Cloud recognized in the 1960s and 1970s, the dynamics of eukaryotic novelty and innovation cannot be understood apart from a complex dance of environmental potentiation and biological response as these lineages co-constructed evolutionary innovations.

Late Cretaceous Mammals and Flowers, and Cenozoic Consequences

I will skip dinosaurs. Not because dinosaurs lacked fascinating novelties or even because they get too much press, although they do. Dinosaurs dominated Mesozoic terrestrial ecosystems from their initial diversification near the end of the Triassic until the demise of the non-avian dinosaurs during the end-Cretaceous mass extinction. But the more telling story of innovation and novelty is found among mammals of the past 200 million years and their interactions with flowering plants and insects.

When I was a student, Mesozoic mammals were envisioned as rat-sized nocturnal creatures cowering in fear of lumbering dinosaurs. But discoveries over the past few decades have revealed a vibrant world of Mesozoic mammals, particularly during the Cretaceous when new clades arose in concert with the diversification of flowering plants and insects. Several novelties associated with flowering plants and mammals were discussed in chapter 6, and here we take this story forward to consider the transformation of terrestrial ecosystems. These changes are inextricably tied to the mass extinction at the end of the Cretaceous and the subsequent radiation of birds and placental mammals, illustrating the complex interactions involved in the success of novelties. We end this section with briefer discussions of several Cenozoic innovations, including the transformative spread of grasslands after the Miocene and the rise of plants with C_4 photosynthesis, concluding with an innovation driven by combinatorics: the association of nitrogen-producing root nodules with legumes and other plants, the foundation for pervasive increases in terrestrial productivity.

These cases reveal some general points about innovation: The developmental capacity of a clade is defined by the suite of novelties possessed by or accessible to the clade. If that space is relatively limited, as it appears to have been for Mesozoic mammals, then future novelties are limited as well. The study of mammalian evolution has been caricatured as "the history of teeth through time," but teeth not only preserve well as fossils but also provide important insights into dietary preferences, ecology, and other aspects of extinct clades. Studies of teeth in living mammals have shown that relatively small genetic changes can produce significant changes in tooth morphology. Collectively these illustrate the contingent pathways of innovation out of which our modern world has been constructed.

Mammals

Mammals evolved in the wake of the end-Permian mass extinction, as did dinosaurs, turtles, ichthyosaurs, plesiosaurs, and several other vertebrate clades. In Late Triassic times one would have been hard-pressed to predict that dinosaurs would dominate the following 150 million years, much less give rise to birds, or that mammals would achieve remarkable diversity in the Jurassic and Cretaceous. Mesozoic mammals include the Jurassic beaver-like animal I mentioned earlier, something like a flying (actually, gliding) squirrel, diggers, carnivores, and others. Far from being rat-sized insectivores, placental mammals emerged from the devastation of the end-Cretaceous to produce elephants, hippos, and, well, us. Turtles have lumbered through it all.[28]

But the Mesozoic diversification was limited, at least in comparison to the explosive diversification of placental mammal clades after the end-Cretaceous mass extinction. The diversification of placental mammals encompassed many developmental, morphological, and behavioral novelties. Some see this as an explosive response to the mass extinction, while others see signs of a much earlier, Cretaceous origin for the clade. No fossil placentals are known from the Cretaceous, although molecular clocks project their origin back into the Triassic. This disjunction between fossil and molecular data has persisted for more than two decades, despite new fossil discoveries and great improvements in molecular clock methods. In fact, both the molecular and fossil data could be largely correct, reflecting a decoupling between lineage divergence and the later acquisition of characteristic morphologies. But there is more to the limited morphological disparity of Mesozoic mammals. A recent study of morphological space through mammalian history has shown that Mesozoic mammals were more constrained than those of the Cenozoic. Most Paleocene fossils represent stem placentals, while crown-group clades appeared during the Eocene (after 56 million years ago), including the first bats, primates, and whales. In the southern hemisphere marsupial clades also diversified during the Eocene. The appearance of placental crown groups in the Eocene was accompanied by an expansion in brain size across multiple clades, particularly in sensory regions. Although these changes in encephalization were not themselves an evolutionary novelty, greater relative brain size is correlated with improved cognitive ability and complex behavior. Thus, these changes were a prerequisite for behavioral innovations.[29]

An essential contribution to the restructuring of Cretaceous terrestrial communities was the evolution of the three-cusped, tribosphenic molar discussed in chapter 6. By comparing the ecospace occupation of modern small-mammal communities to exceptionally preserved fossil mammal communities from the Jurassic, Cretaceous, and Eocene, paleontologists have shown how mammals with this novel molar interacted with diversifying flowering plants.[30]

Chen and colleagues constructed an ecospace from body size, dietary preference, and mode of locomotion, similar to ecospaces described in chapter 4 but modified for small mammals. Some 240 possible combinations each approximate a different ecological role. Applying this ecospace to 98 modern mammal communities illustrated the importance of climate, habitat type, and vegetation. Five fossil assemblages were also examined. Despite the disparity of Jurassic and Early Cretaceous mammals, this analysis shows that the introduction of the tribosphenic (and pseudotribosphenic) molars coupled with the spread of angiosperms allowed Mesozoic mammals to greatly expand the variety of food types. For example, species feeding at multiple trophic levels increased, as did the number of herbivores after the Cretaceous. Earlier types of mammalian dentition do not appear to have been highly evolvable. The tribosphenic molar has more cutting ridges and a combination of a cone-like cusp on the upper teeth with a basin in the opposing teeth on the lower jaw, which allowed animals equipped with this dentition to shear and crush food. Coupled with introduction of a back-and-forth grinding motion of the jaw, such mammals had access to a greater variety of food types. Recall from chapter 6 that the formation of a tribosphenic molar is not developmentally challenging; only simple genetic and developmental changes are involved, so this type of tooth arose independently in several different clades of eutherian mammals. Although the earliest tribosphenic teeth appeared in the Jurassic, this innovation became important from the Late Cretaceous into the early Cenozoic as mammals took advantage of the new opportunities presented by flowering plants.[31]

Darwin's Abominable Mystery

Charles Darwin christened the sudden appearance of flowering plants in the fossil record as an "abominable mystery," in many ways more troubling than the explosive appearance of animals at the base of the Cambrian. Despite the obvious success of angiosperms, with several hundred thousand living species, their origins and early history have been challenging to untangle. Most living species belong to clades that first emerged during the Cretaceous, but the success of angiosperms seems far from obvious in their early history, with a macroevolutionary lag of 40 to 50 million years between the early novelties and divergence of major angiosperm clades in the Early Cretaceous and their ecological dominance in the Paleocene. The carpel is one of the critical novelties associated with angiosperm origins, and as discussed in chapter 6 this female reproductive structure enclosing the ovules allows a range of modes of fertilization. The key genes associated with carpel formation were assembled sequentially before being co-opted to form this structure. This is only one of the important novelties associated with the origin of angiosperms. The specialized structures of the flower and double fertilization were also significant, but the timing of these structures is not well resolved. Structural changes improved rates of nutrient flow and photosynthesis. Flowering plants also produce chemicals that increased defenses against herbivory. Within angiosperms macroevolutionary lags are found with grasses, as discussed later in this chapter.[32]

The evolution of flowering plants is a clear case of decouplings between the origin of a clade based on molecular clock estimates, the acquisition of key evolutionary novelties, diversification into multiple clades, and their later expansion to ecological significance (figure 8.2). No unambiguous fossil crown angiosperms have been recovered from rocks older than about 135 million years ago. Tiny fossilized flowers, pollen, and larger plant fragments recovered from both the northern and southern hemispheres date to about 130 million years ago and represent most major plant clades, from basal angiosperms (*Amborella* and water lilies) to the more complex monocots and basal dicots. This pattern is striking. Angiosperms appear in the fossil record having achieved most of their phylogenetic diversity and morphological disparity, and with considerable geographic coverage. By 30 million years later these groups were joined by the major eudicot clades. The origin and early diversification of flowering plants are missing, but how long is this history? Is this pattern purely a reflection of low abundance and poor fossilization, and restricted distribution, or are other factors involved?[33]

A recent molecular clock study covering about 85 percent of all living angiosperm families provides some perspective. This study suggested that angiosperms diverged from gymnosperms (pine, gingkoes, and cycads) 350 million years ago, with the origin of the angiosperm crown group in the Late Triassic 209 million years ago. These results are roughly congruent with previous studies, albeit with greater precision. There is a lag of roughly 70 million years between the molecular clock estimates and the first unequivocal fossil angiosperms, a result consistent with major divergences among angiosperm clades during the Jurassic. So, while this study confirms the rapid generation of angiosperm subclades in the Cretaceous, we are left with the inference of a missing pre-Cretaceous history of crown angiosperms.[34]

Although I have emphasized molecular clock studies, it is possible to use the distribution of fossils early in the history of a clade to infer the age of origin of each clade independently of a molecular clock. When this approach was applied to the origin of angiosperms the results were similar to the molecular clock study just described, with Jurassic crown angiosperms and a rapid

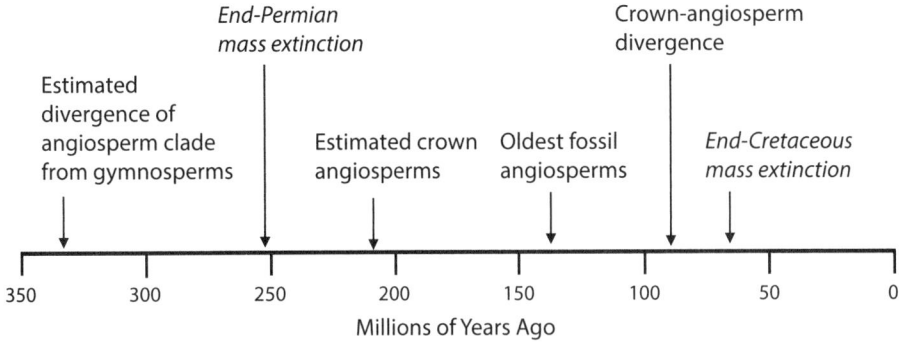

End-Permian
mass extinction

Crown-angiosperm
divergence

Estimated
divergence of
angiosperm clade
from gymnosperms

Estimated crown
angiosperms

Oldest fossil
angiosperms

End-Cretaceous
mass extinction

| 350 | | 300 | | 250 | | 200 | | 150 | | 100 | | 50 | | 0 |

Millions of Years Ago

FIGURE 8.2. Timeline of key events in the origin of flowering plants, with divergence estimates based on molecular clock analyses and fossil evidence.

accumulation of family-level diversity through the Cretaceous, followed by much lower Cenozoic rates. Thus, both fossil and molecular studies support a pre-Cretaceous origin of the angiosperm crown group, a conclusion that does not rest on the disputed claims of Jurassic angiosperm fossils.[35]

But if we were to visit a Late Cretaceous landscape would flowering plants be significant if not dominant parts of the flora? The next best thing to time travel is a fossil deposit preserving a snapshot of a living assemblage, rather than the more common accumulation of easily preserved species. Just such a deposit is exposed for more than four kilometers along Big Cedar Ridge in central Wyoming, where my colleague Scott Wing and his research group spent much of the 1990s censusing this 73-million-year-old flora. They measured the abundance of 122 different taxa at more than 100 sites, and their analysis showed an inverse relationship between diversity and abundance. Ferns and palms covered much of the site but had relatively few species. As expected, many angiosperm taxa were recovered, but their abundance was relatively low. One might dismiss this evidence for decoupling between the number of taxa and their abundance as a happenstance occurrence, but subsequent work has confirmed that while flowering plants had certainly diversified into the full range of major structural and phylogenetic groups by the Late Cretaceous, there were substantial macroevolutionary lags between the origin of families and diversification of each lineage into modern species (the crown of each family). The mean lag between the origin of families and their crown ages is 37 to 56 million years across the angiosperm tree. This result is consistent with the inference from the Big Cedar Ridge flora that the phylogenetic diversification of angiosperms was decoupled by tens of millions of years from their ecological dominance in the Paleocene and Eocene. The report of this study hypothesized that the lag involved "synergistic interactions between evolutionary innovations and ecological context."[36]

The vital role of insects in the pollination of flowering plants has led many to assume a causal relationship between innovation and diversity of the two clades, a connection first proposed by Darwin. These interactions are an excellent example of how new ecological relationships are constructed, from the diversity of different pollinators and insects with specialized mouthparts to "mine" leaves to the new possibilities presented by leaf litter. But flowering plants and insects are only a part of a deeper transformation sometimes described as the Cretaceous terrestrial revolution. The diversity of other land plants expanded during the Cretaceous, as did that of fungi, spiders, snakes, amphibians, and other vertebrates. One hypothesis for this diversification is that higher Cretaceous productivity allowed many plant and animal species to persist through increased dispersal despite relatively low densities.[37]

Insect pollination was not a unique innovation of flowering plants. It first appeared among now-extinct Mesozoic gymnosperms and other clades. Indeed, early flowering plants co-opted these preexisting processes. Nonetheless, insects provide possibly the best avenue for testing claims for a Cretaceous terrestrial revolution. One could test the coevolutionary hypotheses by examining correspondence between diversity of fossil insects and angiosperms. But the most detailed compilation shows no pronounced Cretaceous expansion, rather a more irregular increase into the Cenozoic with substantial diversifications of several insect clades during the Cretaceous and Paleogene. Relationships between insects and flowering plants have also been tested by examining estimates for the crown ages of key clades of insect pollinators. Almost all existing insect pollinators belong to one of five clades: the Coleoptera (beetles), Diptera (flies), Hymenoptera (bees and wasps), Lepidoptera (butterflies and moths), and Thysanoptera (thrips). Molecular clock estimates place the crown ages of all five clades in the Permian (albeit with huge uncertainties spanning the Carboniferous to Triassic), well before any credible evidence for angiosperms, much less sufficient abundance to drive a coevolutionary response among these insect pollinators. Evidence that these clades diversified in the latest Cretaceous and early Cenozoic is consistent with the view that the angiosperm-driven terrestrial innovation largely occurred between 65 and 50 million years ago.[38]

The ecological impact of flowering plants seems to have begun during the Paleogene with the rise of modern forests, and by the Eocene angiosperms fully dominated most environments, and terrestrial ecosystems reached their modern biodiversity. In the next chapter I suggest that one of the most significant human novelties of the past few centuries was the invention of innovation itself through scientific research and technological change. Similarly, one of the most significant novelties of angiosperms was their ability to iteratively enter and reenter adaptive zones. This evolutionary flexibility allowed them to rapidly accommodate climatic and other environmental changes. Flowering plants and their relationships with insects lie at the core of this ecological transformation, although the dynamics remain incompletely understood. It is a plausible and testable hypothesis that increased biodiversity after 110 million years ago reflects the greater evolvability of angiosperms relative to gymnosperms and other plant clades. Pollination by insects, birds, and eventually bats; herbivory; niche construction; and increased productivity expanded ecological networks. By one estimate more than 50 percent of all living species are either angiosperms or insects, fungi, and other organisms dependent upon flowering plants. But the mass extinction greatly complicates understanding this episode, as exemplified by the evolutionary history of the once-dinosaur clade that survived the extinction, birds.[39]

Flying Dinosaurs

Understanding novelty and innovation among animals during the Ediacaran to Cambrian, and particularly the exquisite fossil arthropods, required reconstructing their phylogenetic relationships. Phylogeny is equally critical to reconstructing the early history of birds, but unlike the Cambrian arthropods, they have one of the best-resolved early fossil histories of any major clade. This is a bit of a surprise, since birds fall apart more readily than most invertebrates. Reconstructing the early history of birds has been greatly aided by exquisite fossils such as the 150-million-year-old *Archaeopteryx* from the Solnhofen Limestone in southern Germany, and more importantly, a variety of feathered dinosaurs and early birds from the younger, Early Cretaceous Jehol biota from northeastern China.[40]

Although feathers and powered flight are the most distinctive features of birds, each of the seven distinct types of feathers found in modern birds has also been identified among other dinosaurs. Feathers may have become widespread among dinosaurs for display, with vaned feathers appearing in maniraptoran dinosaurs, for example. Maniraptors appeared during a prolonged reduction in size among clades ancestral to birds, a process of miniaturization that hollowed out bones, altered posture, and allowed access to other critical adaptations, such as flight and changes in skull structure. Only later were feathers co-opted to build wings for powered flight. Egg brooding, the wishbone, and likely flight are not unique to birds but shared with many related dinosaur clades. Other characters commonly associated with birds—including the beak, rapid growth, and the fused tail and keeled sternum to which flight muscles attach, which both aided flight capacity—evolved later in the Cretaceous. Changes in body size were critical to the evolution of birds, as they were for the evolution of whales and for bilaterian animals in the early Cambrian. In each case body-size changes permitted access to adaptations that would otherwise have been impossible. The size changes alone are not evolutionary novelties, as they do not represent individuated characters or combinations. They are best seen as necessary but not sufficient potentiating events that facilitated subsequent novelties.[41]

The earliest birds rapidly diversified into archaic clades occupying regions of morphospace distinct from dinosaurs, particularly as their limbs and pectoral girdle adapted to flight. Other changes included expansion of the forebrain, adaptations to rapid growth, and higher metabolic rates.[42]

The bolide impact 65 million years ago eliminated archaic birds, leaving a few lineages to found the explosive expansion of modern birds. An analysis of whole genomes from 48 species across the major clades of living birds revealed a rapid latest Cretaceous to Paleogene radiation (Neoaves: all living birds from pigeons to parrots, except ratites, waterfowl, and chickens). This burst of new species had established essentially all the modern orders of birds by 50 million years ago, an extraordinary episode of evolutionary creativity, but the ecological contributions are perhaps more profound. The earliest Late Cretaceous Neoaves were ecologically like modern shorebirds. Several of these lineages survived the mass extinction, and from them evolved grebes, doves, cuckoos, and hummingbirds, quickly followed by a rapid diversification of "core waterbirds" (including pelicans, herons, penguins, and loons, among others) and "core landbirds" (encompassing eagles, owls, woodpeckers, bee-eaters, parrots, and songbirds). The ecological explorations of the possibilities of flight were pervasive, generating almost all major clades of birds within a few million years.[43]

The explosive diversification of birds after the end-Cretaceous mass extinction coincided with a burst of placental mammals and the expansion of flowering plants and insects. While it is relatively easy to track increases in the number of taxa, each of these groups adopted new ecological habits and expanded geographically, and together they built the foundations of modern terrestrial ecosystems. We have the greatest control on the ecological expansion among flowering plants, which did not dominate plant assemblages until the late Paleocene or Eocene. Quantitative assessment of these innovations, of how these new ecological communities were structured, and how they differed from previous ecological communities is a project for the future. The expansion of birds, mammals, flowering plants, and insects continued in the early Cenozoic, modulated by considerable changes in climate and accompanied by evolutionary novelties in whales, bats, butterflies, various clades of social insects, and cacti, among others.

The spread of grasslands and C_4 grasses, beginning about eight million years ago, triggered many downstream effects. Focusing just on Africa, large hippos speciated as they adapted to different basins across Africa, as did gazelles, and other grazing animals and elephants spread. This transition was at least partly driven by widespread aridification across Africa, which allowed grasslands to replace woodlands, grazers to replace browsers, and restructuring of vertebrate communities. Here I will examine the interactions between elephants and grasses because of the transformative effects as grasslands spread, and because innovations in elephant behavior seemingly preceded the acquisition of morphological novelties by several million years.[44]

The African record of elephants and their relatives (proboscidians) illustrates the restructuring of terrestrial communities. This group first appeared about 55 million years ago and persisted at low diversities in Africa and adjacent Arabia until about 22 million years ago, when they dispersed through Europe and Asia, eventually migrating into the western hemisphere. The number of species, forms, and ecological roles expanded as new functional types appeared. As climate became warmer and dryer during the Miocene, increased C_4 grasslands spread at the expense of C_3 shrubs, heavily impacting many browsers. The African shovel-toothed gomphotheres were largely browsers, but as grasslands spread, selection on the rounded cusps of their molars favored the evolution of transverse lophs, which fused into the parallel structures now characteristic of elephant teeth. To resist the abrasive effects of the siliceous phytoliths found in grasses, their teeth became higher. In many animals geochemical studies can reveal their primary food types. With elephants, we can track the relative amounts of C_3 and C_4 plants in their feed by comparing the carbon isotopes of their teeth to the surrounding soil. Grasses began spreading about 10 million years, ago and by 8 million years ago there was a rapid shift among elephants from feeding on C_3 to C_4 plants. Importantly, however, the changes in tooth structure and height, as well as necessary accommodations in the skull, lag behavioral changes by about 3 million years. Elephants were the dominant large herbivores, but similar changes in tooth structure occurred among many grazers as grasslands spread. But there is a catch to this story. The C_4 grasses had lower productivity than the shrubs they replaced, and they grow seasonally. Consequently, the number of coexisting species declined, shrinking elephant ecospace.[45]

Big Things in Small Packages

Native Americans may not have known about nitrogen fixation, which allows certain bacteria to capture nitrogen from the air and convert it to a form more amenable to plants, but they knew beans, corn, and squash as the three sisters. The sisters are complementary: corn provides support for the beans, shade from squash leaves retains moisture and reduces weeds, and the nitrogen-fixing bacteria on the roots of the beans help fertilize all three plants, increasing their yield. Plants require nitrogen for growth, but despite its composing about 78 percent of the air, lack of nitrogen often limits plant growth. The nodules solve this problem, and farmers rotate other crops with plantings of clover or alfalfa to scale the effects. As with C_4 photosynthesis and venoms, the multiple appearance of these novelties in different clades is strong evidence for potentiation. Nitrogen fixation is an example of seemingly smaller-scale innovations with profound ecological impact.

The bacteria that fix nitrogen from the air are found in nodules along the roots of beans and a few other plants, such as roses, but these nodules are scattered across plant phylogeny, which has caused considerable uncertainty about their origin. Did nodules with nitrogen-fixing bacteria arise once somewhere near the base of flowering plants, later experiencing massive losses across most plant clades, or was the key to this combination of plants and bacteria the acquisition of the *capacity* to form nodules, which was only later adopted within legumes and a few other clades? Although gene losses are common, it seems unlikely that so many clades would have eliminated such a useful acquisition. On balance it seems more likely that the potential for nodularization is common across vascular plants, but only a few times has this been converted to a novelty. The factors that led to the independent establishment of root nodules in different Cenozoic clades remain to be explored, but this example illustrates the importance of combinatoric novelty (of nitrogen-fixing bacteria with plants) and potentiation (of the ability to form root nodules near the origin of angiosperms) in generating an evolutionary innovation that underpins the productivity of terrestrial ecosystems, as well as much of modern agriculture.[46]

Snakes also arose in the Mesozoic but experienced a pronounced diversification after the Cretaceous, as they adapted to a much wider array of foods. The loss of appendages and expansion of vertebrae are the most apparent features distinguishing snakes from lizards, but their diversity of feeding habits is associated with behavioral novelties (such as constricting,), plastic and dynamic skulls to facilitate swallowing larger prey, and a witches' brew of venoms and venom delivery systems. These features have made the early history of snakes a fascinating problem since the nineteenth century. When snakes originated in the mid-Jurassic, they had extremely high rates of phenotypic change, based on a combination of molecular and fossil data. Indeed, these rates were the highest of all known diapsid reptile lineages. The authors of this study rejected adaptive radiation as an explanation for these patterns. Rather, they identified a lag between the establishment of diapsid lineages after the end-Permian mass extinction and their later taxonomic diversification, and similarly the rapid morphological changes associated with the origin of snakes were decoupled from taxonomic diversification. The explosive early changes in snakes remain poorly understood, but the large numbers of transposable elements found in snake genomes is consistent with their role in repatterning gene regulatory networks. By the Early Cretaceous snakes had begun to feed on lizards, but it was not until the Eocene that their dietary preferences expanded to fish, reptile and bird eggs, mollusks, salamanders, birds, mammals, and much else.[47]

Ecosystem engineering and niche construction have been prominent components of many innovations. The destruction caused by elephants has often led to conflict with farmers in rural regions of India and Africa, while innovations in other clades have also spread novelties that generate pervasive habitat transformations. In the preceding chapter I discussed the partial co-option of the limb gene regulatory network to generate the anterior "wing" of skates and rays (batoids). Most skates and rays are messy eaters, bulldozing through the sea bottom in search of crustaceans, worms, clams, and other invertebrates, using flattened plates in their mouth to crush their prey. They excavate shallow pits on the seafloor, and collectively their activities cause substantial disturbance. The skates and rays exemplify the post-Paleozoic expansions of shell-crushing predators, and many of their prey acquired novel morphologies and new behaviors. Mantle tissue covers the inside of mollusk shells, leaving a free space open to the water for the gills. Post-Paleozoic bivalves developed the ability to fuse the fleshy parts of mantle tissue to form inhalant and exhalant tubes for feeding and ventilation of their gills. This freed bivalves from needing to live close to the seafloor and allowed them to adopt new types of feeding, particularly filter feeding by deeply buried bivalves. The new post-Paleozoic bivalve clades are all siphon feeders. The burrowing capacity of bivalves also increased greatly because of mantle fusion. Collectively, batoids and bivalves illustrate the increased bulldozing of the seafloor by many different clades over the past 250 million years. Most of the clades that dominated the shallow seas of the Paleozoic could not have survived in post-Paleozoic oceans. Many had been eliminated by the end-Permian mass extinction, but the survivors had to quickly adapt or suffer the same fate.[48]

Behavior

Rattlesnakes in the western United States have never bothered me. Not that I have not encountered many, sometimes more intimately than I would have preferred. Indeed, I greatly prefer them to copperheads in the southeastern United States or the mambas and boomslangs of southern Africa. Partly this reflects my own learning—I have been around rattlesnakes since I was 11 or 12 years old and have captured many—and partly this reflects the kindness of rattlesnakes in announcing their presence. The rattle represents an interesting example of a behavioral change becoming encoded and elaborated in morphology. Rattlesnakes belong to the viper clade (family Viperidae), which forms part of a vast clade of modern snakes, the colubrids. Many viperids rattle their rattle-less tails as a warning, and recent work found that the vigor of their vibrations increases in species that are phylogenetically close to rattlesnakes. Thus, the development of a rattle in the terminal segments of a rattlesnake took advantage of an already widespread behavior. The vibrating tail behavior represents a potentiating feature (not all vipers develop rattles), which morphological adaptations exploited to generate a morphological novelty, which in turn generates an impact on the ecological communities favored with rattlesnakes.[49]

Behavioral changes may facilitate morphological or other novelties. This base of the Cambrian System is defined by the first appearance of a burrow in rock outcrops along the coast of Newfoundland. This distinctive, corkscrewing burrow was probably constructed by a priapulid worm searching for food through shallow sediments and was just a preview of widespread behavioral

innovations across many early Cambrian animals. In this case the worm's burrowing behavior was possible only because it possessed the necessary body
architecture, in particular, a fluid-filled body cavity known as a coelom, against
which muscles can contract.

Climbing trees has allowed several different lineages of tree frogs access to
juicy ants, beetles, and other invertebrates. The various types of tree frogs have
independently acquired similar novel structures on their toes, as well as other
skeletal and muscle changes to facilitate climbing. And most tree frogs are
small, the better to maneuver on twigs and leaves. That these novelties have
arisen repeatedly suggests that they are relatively accessible to frogs. Elegant
experiments revealed that even a terrestrial South American toad possesses the
behavioral flexibility and manual dexterity to safely navigate a set of rods, despite climbing not being a part of the behavioral repertoire of the lineage.[50]

In chapter 3 I introduced Imo, the Japanese macaque who discovered the
advantages of washing sweet potatoes in water before eating them, and of separating sand from grains of wheat by scattering the mixture over the water and
scooping up the floating grains of wheat. No doubt Imo was a creative monkey,
but the long-term impact of these novel behaviors required not just a creative
individual but also persistence within her band, which was accomplished
through social learning. Imo's approach to food preparation reflected macaques' plastic response to a new environment, a plasticity that is common
across many different animals, from differing architectures in reef corals depending on wave energy to responses to predators among arthropods.

Mary Jane West-Eberhard and more recently David Pfennig and his colleagues have been at the forefront of studies integrating phenotypic responses
to environmental changes as a basis for evolutionary novelty, which is sometimes described as a behavioural-plasticity-first approach to novelty, as introduced in chapter 3. As discussed in chapter 7, work on niche construction is
an alternative path to addressing these questions, focusing on the relationship
between organismal activities and the environment. Earlier I described how
activities can modify the selective environment of a species through construction of ecological inheritance. But cultural niche construction is more complex,
as it can influence both natural and cultural selection. Cultural species, including humans, can generate ecological inheritance, from fertilized fields to free-
ways. Niche-constructing behavior does not require a species to have the ability to easily adjust to a different environment (whether one it confronts or one
it builds), but phenotypic plasticity can aid in the conversion of creative individual behaviors into persistent attributes of a species. When coupled with

social learning, phenotypic plasticity can generate cultural traditions, but phenotypic plasticity and social learning are costly strategies for a species. Genetically encoding favored behaviors reduces these costs. If the new behavior and social learning persist long enough, genetic variation may eventually arise that encodes, or partially encodes, the behavior as an instinct. This canalization of the new behavior generally results in a faster response to the environmental challenge at lower cost and preserves the response in future generations.[51]

If behavior can lead innovation, are some aspects of behavior more likely to generate novelty? And are clades with certain attributes more likely to acquire behavioral novelties than others? There are suggestions of circumstances that may promote behavioral novelties, including an attraction to novelty (neophilia) and exploration, individual learning, insight or creativity, and behavioral flexibility. In surveying which aspects are most likely to generate behavior novelties, ecologist Marlene Zuk and her colleagues identified predator avoidance, foraging and food preferences, mating and mate attraction, and movement (migration and dispersal) as attributes most commonly associated with novel behavior patterns and as avenues into novel morphologies.[52]

If organisms have the capacity to respond in different ways to changing circumstances (phenotypic plasticity), does this provide a basis for the eventual incorporation of a learned, social, or cultural activity? The Baldwin effect, as this is known—after James Mark Baldwin, who first described this process—is one component of the coevolution between genes and culture, not just in humans but in many other species as well.[53]

Social and Cultural Innovations

One can be reasonably confident that sulphur-crested cockatoos did not evolve in an environment rich with garbage bins. But cockatoos are intelligent and social birds, and the residents of Sydney provide a ready supply of food, thoughtfully marking the bins with red tops to signal the edibles inside. Sometime before 2018 a few birds had worked out the complex steps of prying open the top of a garbage bin and how to walk the cover back to the hinge before flipping it over. This novel behavior was initially noted in only 3 suburbs, but it did not take long for other birds to catch onto the possibilities. Social learning being what it is, by late 2019 cockatoos were flipping garbage lids in at least 44 suburbs of Sydney.[54]

Many animals change their behavior as experience or trial and error leads them to new behaviors. Advantageous changes in behavior may lead to more offspring and feed into future evolutionary success. Both Imo and her macaques and the sulphur-crested cockatoos of Sydney represent examples where new behaviors are transmitted through social learning, not just from parent to offspring but also horizontally across a single generation. Often unrelated adults may serve as teachers or role models, leading to oblique transmission. Such examples of behavioral innovation have attracted attention, as they are central to discussions about animal mind, social evolution, brain evolution, and social intelligence.

Sociality is common among some microbes. The social amoeba *Dictyostelium* lives individually under most conditions, but when resources wane individuals aggregate into a large multicellular body that extrudes a multicellular stalk and generates spores. This alternation between individual and social phases has long made *Dictyostelium* a model for studying cooperation and sociality. Although an amoeba such as *Dictyostelium* displays cooperation and social behavior, there is no evidence for social learning. If social learning between individuals persists and becomes widespread within a population, it eventually becomes a tradition, with traditions woven together to form a culture. A decade hence, all the sulphur-crested cockatoos in Sydney may be flipping red-lidded garbage bins, or perhaps the behavior will have died out. Some social learning is too transient to form a tradition. The whereabouts of tasty fruit may vary from week to week, whereas knowledge of accessible water sources may be longer lasting. Some might still object to the application of the term *culture* beyond humans, but by now it is beyond dispute that social dynamics and transmission of culture are widespread among animals. The transmission of bird and whale song, tool use by New Caledonian crows, and foraging behaviors among bees all involve social learning, the formation of traditions and establishment of cultures.[55]

The capacity for social learning requires cognitive ability, including the ability to store and recall social information (Lulu stole food from me last week, but Tabby helped me get the food back). These cognitive capacities also facilitate investigations of social learning through field programs, experimental work with colonies of animals, and simulation studies. Social learning and culture have been described not only in birds, various primates, and some cetaceans but also in some fish, as well as in insects such as bees.

Cultural inheritance, like niche construction, is an alternative means to genetic inheritance of intergenerational information transfer that effects fitness. The various approaches to human cultural evolution described in chapter 3 differ in the relative significance they place on genetic versus cultural inheritance. Evolutionary psychologists, for example, posit that Darwinian evolution has generated specific genetic adaptations for cultural traits and place less emphasis on cultural inheritance, while advocates of gene-culture coevolution examine feedback from cultural change to genetic inheritance. Other cultural evolutionists place greater emphasis on cultural evolution as involving a system of cultural inheritance. Chimpanzees (*Pan troglodytes*) in Africa display considerable cultural variation in the use of hammers as nut crackers. This behavior is widespread in far-west Africa but absent in East Africa, even where appropriate nuts and rocks are present. Archaeological evidence reveals that nut cracking with stones has persisted for at least 4300 years. Such cultural traditions are examples of cultural inheritance.[56]

Species with cultural inheritance can rapidly acquire and transmit new information about their environment not only from parent to offspring but also horizontally to other members of a single generation and obliquely to nonrelated members of different generations. Such patterns of cultural inheritance have been found in primates, birds, and bees, with suggestive evidence from field studies of whales. Although evidence for culture is widespread, less is known about the fitness effects of cultural inheritance. Social learning and cultural inheritance are general-purpose novelties that can enhance the fitness of species by improving reproductive success, predator avoidance, or resource acquisition. But most importantly for us, cultural inheritance allows novel behaviors to persist long enough for them to serve as the foundation for new developmental or morphological novelties via gene-culture coevolution.[57]

While culture is widespread, whether cumulative cultural evolution is also found among animals has been more contentious. Cumulative cultural evolution involves the accumulation of cultural changes over time, building upon one another as a "cultural ratchet." The next chapter explores such aspects of cumulative culture as language, math, and technology, characteristic of human cultural change. Although Harvard cultural evolutionist Joe Henrich, for example, has invoked cumulative cultural evolution as characteristic of humans, there is growing evidence that it occurs among some animals as well, with the best evidence from chimpanzees, the pattern of changes in bird song, the evolution of songs in humpback whales, and even in the behavior of bighorn sheep. The most interesting cases of human cumulative cultural evolution are of collective intelligence, where the knowledge is distributed across individuals rather than in a single individual. There is little evidence of collective intelligence in animals other than humans.[58]

The cultural ratchet is one of several positive feedbacks that have been proposed as components of cumulative culture. Animal behaviorist Kevin Lala has studied how social learning can enhance social tolerance, tool use, and foraging, leading to increased brain size in primates. Larger brains lead to more efficient and accurate social learning, improving a variety of cognitive skills such as theory of mind, leading to another round of improvements in the proficiency of social learning. This cultural drive hypothesis entrains novel behaviors (which Lala terms innovations) by individuals and small groups to enhance the abilities of the species through gene-culture coevolution.[59]

The most extreme social innovation is the transition to eusociality among some insects, shrimp, and mole rats, with more complex eusociality in ants, bees, and termites. Eusociality is marked by mother and adult offspring in overlapping generations, which permits division of labor and the formation of specialized castes separating reproductive and worker functions. Eusociality among bees, ants, and termites involves a single reproductive female, who may be served by hundreds to thousands of sterile workers, soldiers, and other castes, who ensure the functioning of the colony and reproductive success. That so many individuals apparently sacrifice their Darwinian fitness for the success of the colony had been a consuming problem for evolutionary biologists until William Hamilton and others clarified the nature of kin selection. Some cases of eusociality required significant evolutionary novelties but generated increased functional complexity and major changes in ecosystem structure. The repeated cases of eusociality within the clade of ants, bees, and termites (order Hymenoptera) has suggested that there are genetic features of the clade potentiating the transition, and other work has identified ecological characteristics that also predispose lineages.[60]

A recent analysis of formicoid ants supports the model of novelty and innovation proposed here. These are the most diverse clade of ants (over 12,000 species), with numerous subclades having a high degree of eusociality, with large colonies and extensive development of nonreproductive castes, as well as long-lived queens. Leaf-cutter ants, with their extensive agricultural mutualisms with fungi, belong to this clade, as do various swarming army ants. The genomic analysis found that the lineage leading to the formicoid clade experienced high selection in the Early Cretaceous, with three areas with the highest rates of change also important in establishing eusociality: gene silencing helps ensure workers are sterile, histone acetylation is involved in caste differentiation, and recycling of cellular debris (autophagy) aids longevity of the queen. Together these changes enhanced the ability of colonies to generate castes for division of labor. No evidence was found for subsequent rounds of rapid evolution in the lineages leading directly to eusocial groups, so the important genomic changes for eusociality arose during about 20 million years of the Early Cretaceous, prior to fossil evidence for eusociality. I interpret these results as strong evidence for early potentiation of the capacity for euso-ciality, followed by a lag before eusociality appeared within a subset of clades, just as we have seen in numerous other cases. The acquisition of the genomic potential was effectively decoupled from the later formation of new levels of social complexity among these ants, with the ecological impact of these changes dependent upon the size and abundance of the colonies.[61]

This chapter has surveyed evolutionary innovations ranging from the origin of eukaryotes and of animals to rattling rattlesnakes. These innovations span a range of scales to emphasize that while some innovations can transform the course of life on Earth (eukaryotes, animals, vascular and flowering plants), the majority are of lesser magnitude. Some cases suggest behavioral changes may establish the conditions for subsequent changes in development and mor-phology. The evolution of elephants is a situation where the fossil record is sufficient to support a behavior-first model, but in most cases the fossil record will not be sufficiently obliging to allow rigorous tests of alternative models. Most of the examples from earlier sections of this chapter provide evidence for macroevolutionary decoupling between novelties and innovations, a pat-tern inconsistent with behavior-first models. That innovations arise through a multiplicity of pathways is to be expected, for they can be contingent upon environmental conditions, ecological circumstances, and the availability of appropriate novelties. By decoupling novelty from innovation, the conceptual framework advanced in this book acknowledges the possibility, indeed the likelihood, that transformations of ecological networks have occurred without a significant contribution from novelties.

Given the focus on novelties, I have not detailed innovations absent evolu-tionary novelty, but in deference to curious readers the Great American Inter-change provides a ready example. The closing of the Isthmus of Panama 2.7 million years ago allowed migration of terrestrial and freshwater species be-tween North and South America through Central America. Parrots, hum-mingbirds, opossums, armadillos, and the now-extinct ground sloths moved north from South America, while a larger number of species headed south, from snakes and turtles to condors, deer, and rabbits. But the transformation of South American ecosystems came from the inward migration of placental

carnivores, including a variety of cats and canids (wolves). The complex ecosystem changes are not relevant here, but this is a clear example of innovation
(changes in network structure and taxonomic diversification of some clades
within South America) with little evolutionary novelty.

Social learning, tool use, and culture have appeared repeatedly among various animal groups, with the most informative cases among the relatively large-
brained whales and corvids, as well as primates. Social learning, culture, and
possibly cumulative cultural evolution are effective novelties that enable the
acquisition of other behaviors, from predator avoidance to new skills in
acquiring food, but debate continues about their extent. There is little debate,
however, over the essential role of cumulative cultural evolution in human
evolution. Human evolution encompasses a range of novelties, including language and social systems, and those surely rank among the most significant
evolutionary transformations in the history of life on Earth. Chapter 9 examines novelty in cultural and technological contexts and argues for a close similarity to the processes in biological evolution, while acknowledging the unique
dimensions of cultural and technological change.

9

On the Shoulders of Giants

THE DYNAMICS OF CULTURAL
AND TECHNOLOGICAL INNOVATION

HUMAN INTELLECT, language, and culture were once ascribed a miraculous origin, lacking antecedents among even our closest relatives and sustaining the comforting view of human uniqueness. Philosopher John Locke took this view, presumably derived from the first verses of the Book of Genesis. In the *Origin* and *The Descent of Man* Charles Darwin challenged the uniqueness of humans by examining similarities between us and primates. Since then, comparative studies of primates and humans have exploded and been joined by discoveries of a rich fossil record, detailed genetic comparisons, and theoretical models. Yet even today some scholars champion seemingly miraculous views of the acquisition of human novelties, with Noam Chomsky's views of the origin of human language being among the most well-known. Chomsky has argued that language appeared suddenly just before human migration out of Africa and generated both a means of interpersonal communication and "the stuff of thought," as Steven Pinker put it in his 2007 book. In later work Chomsky identified the critical acquisition as recursive combinations of elements into larger elements, whether of internal thoughts or sounds (phonemes) into meanings and words into phrases, sentences, ad infinitum, an ability he described as "Merge."[1]

Chomsky's view that human language was a sudden, instantaneous evolutionary novelty is of a piece with older views on the origin of higher taxa, Goldschmidt's "systemic mutations," and similar sudden transitions discussed earlier. Indeed, the debates over the origin of language capture in microcosm many of the issues raised in this book about rapidity, apparent discontinuity, and pervasive change. How sudden was this transformation? Was the success of language predicated on prior, enabling mutations? Did the capacity for language arise before we have evidence for widespread language? Throughout this book I have argued that apparently sudden transformations disappear as potentiating factors are identified, novelties distinguished from innovations, prior novelties investigated, and the context of innovations examined.

Human societies with language are the final major evolutionary transition proposed by Maynard Smith and Szathmáry, although this is the only such transition that occurs within a single lineage. And language is just one of the many novelties that have generated successive innovations among humans over the past several hundred thousand years. We have big brains, upright posture, opposable thumbs, mathematics, have tamed energy sources from fire to nuclear, developed art and music, and produced increasingly complex social organizations. Any account of novelty and innovation must necessarily address the biological and cultural processes that have produced human culture and technology. But those biological and cultural processes are deeply intertwined, with biology often potentiating cultural novelties, which have fed back to influence our biological evolution. Such gene-culture coevolution affected the gene involved with digestion of dairy products, and improved the ability to handle secondary compounds from plants and disease resistance, among others.[2]

Because of these complexities, in this chapter I have combined the treatment of biological and cultural novelty and innovation into a single account, applying the conceptual framework to the biological origins of cumulative cultural evolution, outlining the early potentiation of novelties such as language and mathematics, and then describing several cultural and technological cases of novelty and innovation. As in earlier chapters, I illustrate cases of individuation, deep transformations, and combinatorics, alternating between cultural and technological cases. *If* there is a common conceptual framework for investigating these novelties and innovations across biological, cultural, and technological domains, then greater understanding of human novelties and innovations may also provide useful insights into those deeper in the history of life.

Simplifying considerably, we can group ideas about the foundations of human culture into the big brain, modular brain, and cultural brain hypotheses. Brain volume expanded rapidly along the line leading to *Homo sapiens*. From this many have concluded that language, culture, and technology reflect our *individually* greater cognitive capacity—the big brain hypothesis. Each of us is simply smarter than ancestral humans because we have larger brains. One flavor of this argument views increased cognitive capacity as reflecting an intuitive understanding of causality in the physical world coupled with social cooperation. Others have argued that greater general intelligence was the driving factor for cultural expansion, linking the increased size of the neocortex to progressive increases in behavioral complexity, which allowed us to form larger social groups. While there is little doubt that increases in brain volume have created the capacity for cultural, social, and technological change, focusing exclusively on the volume of the brain or the neocortex neglects dimensions of human cultural evolution.[3]

Chomsky's identification of Merge as the critical innovation leading to human language is a flavor of the modular brain hypothesis, in which specific brain structures are associated with human capabilities such as language, tool use, and theory of mind. Brain imaging studies of people from different cultures, languages, and educational backgrounds have been used to test this hypothesis. Finally, the cultural brain hypothesis does not deny that an expansion of brain volume significantly increased cognitive capacity but places greater emphasis on the importance of social learning and cumulative cultural evolution. Field studies show that imitation of prominent individuals or a community consensus underlies many human behaviors. Such findings challenge claims that individual abilities to infer causality have been fundamental. In fact, individually, most of us are fairly poor at correctly identifying the causal reasoning underlying many behaviors and habits.

Mathematics illuminates these contested issues. Plato, Descartes, and some more recent philosophers have viewed numbers as a universal, formal system independent of humans, with mathematics existing in a realm of pure thought. Experimental work identifying the ability to count among pigeons, honeybees, and other animals has been taken as evidence that this facility may be widespread (although considerable investments of training time and equipment are necessary to achieve such results). Yet a study of 189 Australian aboriginal languages found 139 where counting was limited to three or four, and an additional 21 with an upper limit of five. Many human languages only discriminate among one, two, a few, and many, and lack capacity for symbolic manipulation of quantities. Cognitive scientist Rafael Nunez of the University of California usefully distinguishes *quantical* abilities, a biological capacity to discriminate between approximate differences, present in all humans and at least some other animals, from culturally developed *numerical* abilities to manipulate exact, symbolic representations. Numerical skills were critical for trade, money, and finance, with their influence spreading to every part of our lives, so that today even the concept of number is polysemous. The cultural foundations of quantitative abilities have been supported by studies showing that people from different cultures use different brain areas. As we acquire mathematical skills, different areas of the brain are recruited depending on our language and cultural background.[4]

No one imagines that *Homo erectus* or early *Homo sapiens* engaged in algebraic topology. No module for double-entry bookkeeping or linear algebra is hidden in our brains. In the conceptual framework developed in this book, simple quantical abilities, language, and the plasticity of our brains potentiated our ability to acquire numeric skills. But these nascent abilities in each individual are insufficient. Social learning is required. Quantification and mathematics, like language and most technologies, are cumulative activities. We do not need to individually work out the concepts of differential geometry. Rather, we have accumulated vast mathematical and quantitative knowledge stored in books (and now digitally), as well as institutions for teaching. Social learning passes on this understanding to serve as a foundation for further mathematical knowledge.

In principle we all have the potential as children to speak Urdu, Tewa, or French, and there is no doubt that properly taught any of us could master quantum physics, play the Brahms's Violin Concerto, or hit a four-seam fastball. At least in principle. Our achievements in mathematics, music, and baseball rely on general human intelligence and cumulative cultural evolution. The difference between our potential and realized lives depends upon the culture in which we are embedded and the opportunities we have for training and learning. Thus, contrasting genetic versus cultural evolution misses the point. Human cultural evolution employs our genetic and developmental substrate, with gene-culture coevolution feeding back to influence genetic evolution. Cultural evolution can adjust to changing circumstances far more rapidly than genetic evolution and provides an alternative means of transmitting information to subsequent generations.[5]

Cumulative Culture, Language, and Collective Intelligence

Human interactions, learning, and the preservation and transmission of knowledge from one generation to the next have been the foundation of human novelties and innovations. We act cooperatively to accomplish tasks beyond the capacity of a single individual or even a family. This is true of hunting, agriculture, and the functions of modern societies. Underlying these achievements are our capacity for cumulative cultural evolution and the acquisition of language. Each of these represents an evolutionary novelty, but they are also technologies that have facilitated many other novelties and innovations.

Cumulative Cultural Evolution

How do you solve a problem? Broadly, there are two alternatives: figure out a solution yourself, or find someone to teach you. The effectiveness of each alternative depends upon the task. For some problems trial and error may work, but many problems will prove easier to solve with advice from someone who faced a similar challenge. We have highly evolved capacities for learning and for learning how to learn. In childhood we learn to discriminate between situations where individual (asocial) learning or intuition is sufficient from those where learning from someone else (social learning) is preferable. We employ a variety of "cognitive gadgets," in the words of psychologist Cecilia Heyes, to enable social learning. Even infants integrate clues about a potential teacher's skill, competence, and standing to identify preferential role models. Early learning often involves vertical transmission of information from parents, and later from mentors. Other learning comes horizontally from our own cohort, or is sourced obliquely, as from unrelated adults. Social learning also greatly reduces the cost of acquiring information, which was particularly important as humans moved into new environments or adopted new means of gathering food, whether by hunting, gathering, fishing, or later by domesticating plants and animals.[6]

If cumulative cultural evolution remains a contentious subject for humpback whales and other species, the transmission and modification of cultural practices is a defining feature of humans. Our ancestors accumulated tools, traditions, beliefs, and practices, all of which could be tested and improved. These refinements might involve improving the shape and thus power of a bow, adjusting planting times, or understanding which foods to gather and how best to process them. This process of testing and refining produces a ratcheting effect, generating cultural practices that none of us could have accomplished acting individually, which can persist for generations. Groups with more advantageous cultural practices could expand and possibly displace other groups. Many cultural traditions and technologies are complex, hierarchically structured recipes involving multiple, sometimes counterintuitive, steps. For some, the primary driver of human evolution over the past million plus years has been cumulative cultural transmission, or, as anthropologist Joseph Henrich of Harvard put it, "Culture makes us smart."[7]

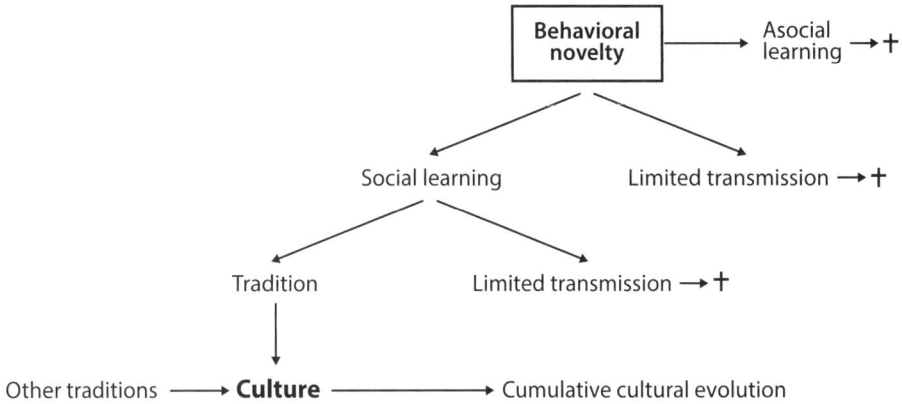

FIGURE 9.1. Schematic of the relationship between behavioral novelties and cumulative cultural evolution. A behavioral novelty may lead to asocial learning and possible niche construction but does not persist beyond an individual. Behavioral novelties leading to social learning may have limited transmission (for example, about an ephemeral food source) or may lead to more persistent social learning. Persistent social learning may develop into a cultural tradition, which if combined with other traditions forms a culture. Persistent social learning establishes the conditions for gene-culture coevolution and may potentiate genomic, developmental, or morphological novelties. With appropriate cognitive capacity and other conditions, culture may become cumulative, establishing cumulative cultural evolution. Crosses mark the loss of behavioral traits.

This dynamic of cumulative cultural evolution is captured in figure 9.1, with a new behavior arising through asocial learning being captured and transmitted socially. If the new behavior is advantageous, social transmission will continue, along with improvements. If a behavioral novelty is no longer useful, it may be dropped from the cultural repertoire (denoted by crosses on the figure). As with the combinatoric expansions of evolutionary spaces described in chapter 4, cultural ratchets enable the progressive expansion of cultural opportunities. Recall that Kauffman named this pattern the "adjacent possible," while others have described this as the "zone of latent solutions." Ultimately the open-ended nature of human cumulative cultural evolution can generate new cultural design spaces. Domestication of plants and animals and the origins of agriculture are widely discussed examples of such cultural niche construction, and others are discussed later in this chapter.[8]

Cumulative cultural evolution also involves feedback from cultural processes through natural selection to influence genetic evolution. Gene-culture coevolution has influenced our capacities for social learning, our abilities to copy the physical actions of others, and to intuit their motivations, thinking, and beliefs. A common example of feedback from culture to genetic evolution is the persistence of the lactase enzyme into adulthood with the spread of cattle domestication and dairy farming. Lactase allows mammals to digest the milk protein lactose and its production normally shuts down after weaning, but the spread of dairy farming allowed adults to continue to exploit the nutritional benefits of milk. Other examples include changes in tooth size, the robustness of our jaws, and changes to our digestive systems with the spread of food preparation and cooking, longer childhoods to facilitate social learning, and changes in the structure of the foot, particularly the arch and lower leg, allowing long-distance running to chase down prey.[9]

The interplay among individual learning, cultural transmission, and biology is complicated. How we learn as individuals is heavily influenced by biases that affect the "learnability" of different parts of culture. Among individuals these learning biases influence social learning and cultural evolution (as shown in figure 9.1). Since social learning can feed back and influence subsequent social learning, it produces cumulative cultural evolution, which in turn shapes the cultural evolutionary landscape. Over longer intervals of time, the cultural landscape can feed back to influence biological evolution. The complexity of these feedbacks emphasizes the nature of cumulative cultural evolution as an evolutionary novelty and its foundational role in human innovations.[10]

Language

Underpinning the social learning and teaching that generate human cumulative cultural evolution is language, and more recently writing. Language allows us to communicate with others who share our language (or closely related languages), while separating us from other language communities. With language we coordinate activities, facilitate cooperative hunting and foraging, plan battles, and spread religious ideas. Language enables teaching and the transfer of collective knowledge to new generations, preserving that knowledge into the future. Language does not stand apart from human cultural evolution. It might be seen as one of the primary technologies through which we achieve cumulative cultural evolution. When, how, and why language originated are pressing questions of cultural evolution to which we do not know the answers, but I agree with many who argue that culture did not "cause" language, nor did language "cause" culture. Rather, language and human culture coevolved, with gene-culture coevolution and cultural niche construction actively controlling the process.

Dozens of scenarios for the origins of human language have been proposed, although fewer seem to have been rigorously tested. As described earlier, two broadly different views on the origins of human language can be distinguished. The first is the nativist position of Chomsky and his allies that language is universal, has a biological foundation, is essentially about internal thought rather than being primarily social, and evolved suddenly at a discrete point (although these components could be separated to some extent). A more diffuse group argues for a cultural basis of language. To rephrase this controversy in terms of the conceptual framework of this book, a Chomskian views human language as a genetically encoded novelty, with some advocates identifying language-specific modules within the brain. The alternative is that the human brain has evolved cognitive structures that serve as the biological potentiation for a cultural novelty. I will not resolve this controversy, but I will argue for the latter position, emphasizing evidence for a lag between biological potentiation and cultural novelty. Language has been a critical component facilitating cumulative cultural evolution, which is the innovation phase in this scenario. We begin with a brief sketch of these alternative views.

Children rapidly acquire a rich ability to communicate in their native language. During the 1950s and 1960s studies suggested that native language competence seemed to outpace exposure to language, which was one line of evidence leading Chomsky to argue that humans share a biologically encoded universal grammar. If we think of language in terms of the evolutionary spaces described in chapter 4, Chomsky's claim is that all human languages share a common structure that defines and constrains the space of possible languages. Consider two phrases: *Birds fly* and *Sally kisses David*. These can be combined into *Birds fly when Sally kisses David*. This process can continue through multiple rounds, combining syntactic structures, which Chomsky termed Merge. Merge provides structure to the universal grammar. For him, this capacity is the critical novelty in the origin of human language, which he dates to about 80,000 years ago. For Chomsky, this is the only plausible explanation for how rapidly children acquire language (known technically as the poverty of stimulus problem) and for similarities in grammar and other properties across languages. Chomsky followed Darwin in believing that human thought is dependent upon language, a view consistent with Chomsky's views on the fundamentally internal role of language. It thus follows that human cumulative culture absent language would have been limited or nonexistent.[11]

Chomsky's views on universal grammar and Merge are diffuse and largely theoretical, making them challenging to test. Two sympathetic responses to these difficulties have emerged. The first is a more nuanced but still biologically centered view with a hard-wired, language-specific domain or module within the brain, but without necessarily accepting the strong requirements of the Merge scenario. Psychologist Steven Pinker is sympathetic to Chomsky's views but grounds ideas about a universal grammar in specific regions of the brain that provide the general structure of language during early childhood. The late Derek Bickerton sought a middle course, based on his studies of creoles. Despite the name, Hawaiian Pidgin is not a pidgin but a full language, or creole, that arose by combining Hawaiian and English, with admixtures from other languages. Bickerton was sympathetic to a universal grammar and a discontinuous origin of languages, like Chomsky, but he viewed language as enabling thought and proposed a model for the advent of language that was firmly rooted in a rapid evolutionary transformation of brain structures (in contrast to the more gradual transformation advocated by Pinker).[12]

Individual and social learning, including biases in how we learn, provides a culture-based alternative to nativist views and emphasizes the social nature of language, with a gradual transition from animal communication to human language. While Pinker and most constructivists acknowledge a role for gradual change and selection, they differ in the relative importance of culture. Some, such as psychologist Michael Tomasello, have primarily focused on the development of something called shared intentionality, our ability to cooperate, among early hominins as a necessary prelude to the development of language. The importance of gesture in communication has led some to suggest that it played a prominent role in the generation of protolanguage. Linguists Morton Christiansen and Nick Chater, for example, have made a strong argument for the importance of improvisation as a step toward language, while Ronald Planer and Kim Sterelny emphasize the importance of gesture. Whatever its origins, the dominant view is that language is an outgrowth of social activity and provides a foundation upon which human cultural evolution has been built. An interesting corollary of the view that language is adapted to the brain *rather than being initially shaped by the structure of the brain* and emerged from gesture and improvisation is that language could have had multiple origins in different populations rather than a single origin. Thus, there is no necessary reason to assume that a single, common human language ever existed.[13]

The feedback among biology, learning, and culture clarifies how language may have emerged as a cultural process. As Kirby put it: "Linguistic structure emerges as a natural outcome of cultural evolution once certain minimum biological requirements are in place." But historical linguists have shown that this increased complexity is balanced by a drive toward speed and efficiency. Simplicity makes a language easier to learn, but complexity ensures that language captures the range of meanings of the speaker. Since communication is challenging when meaning is ambiguous, this provides a drive for increasing the complexity of language. Moreover, as we spread into new environments languages would have diverged and diversified.[14]

The origin and expansion of human language accompanied cognitive changes as well as changes in the mouth, tongue, and the human vocal tract, including the larynx. Compared with the larynx of a variety of primates, the human larynx has lost vocal fold membranes, a simplification that enabled speech. Because of uncertainties in when language evolved, the timing of this change remains uncertain, and thus it is difficult to discriminate whether genetic and morphological changes or behavioral changes occurred first. The most plausible scenario is that gesture and speech coevolved with biological changes through gene-culture coevolution.[15]

Cooperation

Imagine that it is spring and time to clear and repair the *acequia* (irrigation ditch) serving the village fields. If most villagers are related then cooperating with this task makes sense, just as cooperating with kin often makes good evolutionary sense. Such kin selection occurs in many groups of animals. But if I am not related to most of the villagers, why should I cooperate? I could plead a bad back or make some other excuse to avoid the work, safe in the knowledge that others would show up and clean the ditch, and my fields will be irrigated come summer. If being thought a freeloader was not a concern, then avoiding such communal obligations is sensible. But most people do not cheat, and this prosocial behavior raises questions about the evolution of human social behavior. Most social animals cooperate only within small groups and can be highly antagonistic toward outside individuals. Biologically, no increase in fitness is achieved by supporting people to whom we are not closely related.

We are a highly cooperative species, and many feel an obligation to support others. This cooperative nature is an essential evolutionary novelty as well as a potentiating factor for the growth of social complexity. Prosocial behavior, behavior that is intended to benefit others, often leads to cooperation among large numbers of unrelated people. Prosocial behavior is not transactional ("Vote for me and I will get a road/school/airport built for you") but reflects empathy. Feelings of obligation and cooperation beyond kin are a unique cultural attribute of humans, and they have been developed through cultural evolution to foster increased social complexity.[16]

Some cultural anthropologists frame such behavior as a question about the generation of public goods. But recall that public goods are those that are non-excludable (I cannot keep you from using the same good) and non-rivalrous (many of us can use the same good without significantly depleting the resource). But water from an *acequia* is a club good, not a public good, because cheaters could be excluded from using the resource. Many other purported public goods are in fact club goods. Nonetheless, human cultures do generate some public goods, and at the end of this chapter we return to cooperation, cheating, public goods, and niche construction. I tend to agree with Rob Boyd that this may be more of a challenge for anthropologists than some have recognized.[17]

Most readers of this book are WEIRD, belonging to Western, educated, industrialized, rich, and democratic cultures. Psychology was founded on studies of WEIRD people (well, WEIRD college students), even though we represent a small fraction of the populations of the world. WEIRD people share cultural attributes that foster prosocial interactions. We are more likely to trust individuals beyond our immediate kin, we expect politicians and public servants to be honest and not to reward family members and cronies with jobs and contracts (well, until recently), citizens not to cheat on their taxes, jurors to rely on the evidence presented in the courtroom rather than family or ethnic ties, and so forth. We see ourselves as individuals rather than nodes in familial and social networks and see individual rights as universal. We are analytical and favor identifying general principles or rules that do not depend upon family connections. The WEIRD are more likely to be racked by personal guilt rather than shame for having dishonored their family or clan. Anthropologist Joseph Henrich traces these collective psychological attributes to the social norms of Christianity, particularly those associated with family and marriage that arose after 700 CE. Despite having sketched this apparent dichotomy between WEIRD and non-WEIRD people, this is a continuum. None of us can be easily packaged, and we all have multiple dimensions of psychological complexity.[18]

The attributes of WEIRD people help clarify one of the conundrums of prosociality and the diffusion of a suite of novelties. Whether in villages in Africa or on Wall Street, markets function effectively because traders either trust those with whom they are trading or believe that institutions exist that ensure the fairness of the market. Such institutions can be as informal as the group running your local farmer's market on the weekend or highly structured entities, such as the US Security and Exchange Commission, with enforcement powers. People within rural villages often exhibit generosity and cooperative behavior among the family and clan networks in which they are embedded, a case of interpersonal prosociality. When a society grows large enough that shaming recalcitrant members is no longer effective, either prosociality breaks down or means of impersonal prosociality must arise based on principles of fairness, honesty, and conditional cooperation. Thus prosociality is, like language and cumulative cultural evolution, one of the general-purpose evolutionary novelties that has fueled human innovations.

Collective Intelligence

Many of us were raised on stories of the heroic inventors (usually white and male) who created our modern world: Johannes Gutenberg, Benjamin Franklin, Ada Lovelace, and the Wright brothers. Younger generations have Sergey Brin, Lonnie Johnson, and Grace Hopper. Each of these individuals deserves their accolades, but focusing on individual achievement is similar to Chomsky's claim that language was due to the sudden appearance of Merge, or according primacy to individual cognitive capacity over the effects of social learning (the cognitive niche hypothesis). But the long list of simultaneous inventions described in chapter 3 is evidence that context was often as important as individual creativity.[19]

Human creativity reflects our collective intelligence, the mutual cognitive dependences underlying many aspects of human culture and technology. Our brains are adapted to social interactions, seeking solutions to problems from others and transmitting new ideas and new solutions. Using the internet to find a solution to your clogged drain was an expansion of this approach, and the introduction of artificial intelligence promises even more creative developments. Collective intelligence is a dynamic where novelty and innovation emerge at the population rather than the individual level (however much individuals might be credited for collective activities; see also Thomas Edison, for example).

The rate and effectiveness of novelty generation in human societies reflect at least four factors. First, larger and more interconnected societies will have a larger number of potential models for social learning and greater capacity for fortuitous recombination. The relationship between the number of patents and population size shows increasing returns to scale, or supralinearity, meaning that the number of patents awarded (a proxy for inventiveness) increases more rapidly than the size of the population. Not only are you more likely to find a mechanic in a big city who knows how to fix your 1974 Land Rover 90, you are more likely to be able to find the necessary parts. Larger cities accommodate more ethnic restaurants, more diverse cultural opportunities, and a greater range of medical specialists. This is the key to the success of Silicon Valley or Route 28 around Boston, and why governmental attempts to replicate such inventiveness are doomed to failure. What is sometimes missed by the scaling literature (see chapter 4) is that inventiveness is not simply a function of population size but one of the interconnectedness of social networks. Networking allows the generation of shared behavioral standards or expectations among members of a group, or norms. Social or behavioral norms can eventually lead to stronger patterns preserved as institutions.[20]

But population size and social connectivity alone are insufficient to explain patterns of innovation. Environmental context also plays a role, particularly predictability and the extent of cultural-niche-constructing activity. Anthropologist Laurel Fogarty has built computational models showing that rates of novelty (albeit defined differently than used here) increase in more unstable environments. This is intuitively plausible, as fluctuating environments pose more challenges. Fogarty's models, however, also show that dominant styles of learning play an important role. In some cases, social learning may preserve traits that are not of current use or that can inhibit the spread of novel behaviors.[21]

A third factor promoting collective intelligence is the fidelity of information transfer during social learning. In telephone or Chinese whispers, a phrase is whispered from one individual to the next. The final person in the chain announces that they heard "Send three and fourpence, we're going to a dance," before the instigator reveals that the phrase began as "Send reinforcements, we're going to advance." This simple children's game demonstrates the difficulties of high-fidelity transmission of even simple information and how readily errors can propagate. The inevitable loss of information during social learning is magnified as our brains tell stories to make sense of incomplete information.

In chapter 4 I described the expansion of the number of possible recipes as the number of ingredients grows as a path to novelty. Food is not the only example of recombination in culture. As cultural diversity increases the space of possible recombinations expands even more rapidly, so that cultural diversity breeds diversity. And just as genomes have devised structures to limit recombination, societies have done the same. For example, vitally important cultural information can be sacred to a specific clan, limiting transmission to others in the same community. Schimmelpfennig and colleagues have incorporated the importance of diversity within their concept of "cultural evolvability," or the ability of societies to adapt through cultural evolution. Evolvability balances stability and efficiency against creativity and diversity.[22]

Cumulative culture or cumulative intelligence is shared among some group of people and reflects contributions from many individuals over time, with results that no one individual, however skilled, could have achieved. Language, mathematics, and writing are among such cultural public goods. But languages, for example, can enhance communication among one group while limiting the access of others, as shown by the Navajo code-talkers used by the US military to communicate between islands during World War II. Navajo was then unwritten, and the Japanese had no one who could understand these messages. In this case the Navajo language represented a club good rather than a public good, since access to the language was limited to those growing up in Navajo communities in Arizona, New Mexico, and Utah. Writing is similarly a club good, as are products of cumulative cultural evolution that require considerable social learning or can readily be restricted: learning how to haft an arrowhead onto a shaft, the best times to plant, or recipes for detoxifying harmful plants. Technological novelties may often be public goods, which led to the development of patent laws to allow inventors to capture some of the profits of the invention.

To a paleontologist the origin of cumulative cultural evolution and language seems straightforward: humans are embedded within cultures that accumulate over time. Neither language nor such extensive cumulative cultural evolution is characteristic of our closest relatives. Since each is a property of the species, they must have arisen before the last common ancestor of all living humans. For most species, and particularly for the invertebrate clades I study, we rarely possess the granularity of population-level information that we have for humans. This high level of resolution provides an unrivaled view of the dynamics of human novelty and innovation. Moreover, cumulative culture and language might have appeared among the many fossil hominins any time after our lineage split from chimpanzees. But cultural evolution introduces the possibility of decoupling between the acquisition of the biological potential for a novelty and the cultural realization of that novelty, as discussed for aspects of mathematics in the introduction to this chapter.

Here I present a scenario derived from the work of a number of scholars, principally psychologist Michael Tomasello and his colleagues and philosophers

Ronald Planer and Kim Sterelny. Both groups have considered the timing of cumulative cultural evolution, language, and collective intelligence (or "collective intentionality," for Tomasello). Some aspects of this scenario are not well constrained, and specialists debate the quality of the evidence and the reliability of the inferences described here. At the end of this section, I briefly discuss potential pitfalls.[23]

Social learning among chimpanzees differs from that in humans. Chimpanzees seem to learn from others only when they cannot readily solve a problem themselves. Such limited social learning and individual innovation likely characterized early hominins through the pre-Oldowan (over 2.5 million years ago), Oldowan (before 1.7 million years ago), and Early Acheulian tool traditions. Such tools are found throughout Africa, into Europe, and across Asia with relatively little change in basic form. Large gaps are present in the record of pre-300,000-year-old tools. Tomasello finds the earliest evidence for cumulative cultural evolution with Late Acheulian stone tools (750,000 to 125,000 years ago), as they seem to require skill beyond individual learning or passive observation. Individual innovation continues in this phase, but with flexible social learning and thus teaching.

There is evidence for use of fire, cooperative hunting of large game animals, and obligate cooperative foraging by the time of *Homo heidelbergensis* (about 800,000 to 130,000 years ago). For Sterelny, the protolanguage of *H. heidelbergensis* was limited to gestures and mime, despite impressive skill in toolmaking and evidence for cooperative hunting along with use of fire. Planer and Sterelny suggest that by 500,000 years ago these hominins were "language-ready," but their social structure limited the extent of their cumulative culture. The complexity of language expanded as hominins acquired progressively better technology that required teaching. Tomasello shares this view, suggesting that by 400,000 years ago collaborative activities and intentional teaching were occurring, demonstrating the presence of an effective cultural ratchet. Although the effectiveness of this cultural ratchet would have depended on population density and structure, as with the spread of genetic changes, and thus varied regionally. Archaeologist Steven Kuhn, in studying the evolution of Paleolithic technologies, reaches a similar conclusion, pointing out that the appearance and spread of technologies is less about the origin of a particular capacity than about a collective decision to exercise that capacity in the repeated generation of a technology.[24]

Modern humans were present by about 200,000 years ago, and by 125,000 years ago more complex tool kits reflect collaborative innovations and cumulative culture. In O'Madagain and Tomasello's scenario, this marks the transition from joint intentionality among cooperating individuals to a collective intentionality among members of a cultural group, with the latter a prerequisite (and thus a potentiation) for the sustained generation of cultural and technological novelties. Finally, detailed, causally accurate reasoning appears as the final stage in the scenario. As an example, O'Madagain and Tomasello cite explaining not just that sinew is used to attach arrowheads to a shaft, but why a particular pattern of wrapping will ensure the arrow shaft will not split on impact. The ability to identify salient causal factors was essential to the slow spread of an evidence-based understanding of technology. They suggest that such reasoning likely arose independently at different times in different regions between 40,000 and 10,000 years ago. Progress was likely episodic, marked by occasional widespread reversals in rationality.

But whether learning and cumulative cultural evolution appeared as early as suggested by this scenario is uncertain. After the appearance of our species about 300,000 years ago, tools evolved rapidly, with extensive regional variation. Archaeologist John Shea notes that archaeologists can readily identify the region and approximate time of origin of stone tools. He suggests that early hominins were not obligatory tool users but, rather, picked up stones occasionally. Learning to craft stone tools was a skill parents passed on to their children. Thus, evidence of obligatory tool use and the formation of regional traditions of different types of tools indicates social learning and cumulative improvement. The absence of such evidence before 300,000 years ago suggests that Tomasello's timeline may overestimate the extent of social learning and collaboration among *Homo heidelbergensis* and other early *Homo*. Instead, if language was required for conveying tool-building skills, Shea suggests that effective speech must have evolved by 200,000 to 300,000 years ago, far younger than other estimates, but an interval that would encompass Neandertals and Denisovans as well as early *Homo sapiens*. Other paleoanthropologists have examined the use of fire (another technological novelty), hunting, or other skills.[25]

One challenge to reconstructing the diffusion of human technology and cumulative cultural evolution more broadly is an unhealthy obsession with identifying the "oldest" evidence of tool use, fire, music, art, or other cultural attributes. In the past this led to claims of an "Upper Paleolithic Revolution," signifying a sudden increase in human cognitive abilities and technological skill. But if the expansion of human cumulative culture and language was cultural, we should also expect that it was likely episodic and discontinuous. New capacities appeared and persisted for some time before disappearing, only to re-evolve again somewhere else. In general, new technologies appeared long after the origin of the responsible species of *Homo*. The persistence of any novel technology is a function of population size and the interconnections among populations, which allow the technology to diffuse and eventually form an innovation (or not). As Sterelny put it, "Informational capital is, then, vulnerable to demographic attrition." Behavioral lags among species of *Homo* mirror the macroevolutionary lags described in earlier chapters.[26]

In Daniel Everett's fieldwork with the Pirahã in the Amazon Basin, he recognized that their language lacked recursion, the attribute that Chomsky viewed as the critical event in the origin of language. This discovery has been taken as strong evidence against Chomsky's views and in support of language as a culturally generated phenomenon. But all we really know is that this language lacks recursion, for this cannot reflect preservation of an ancestral condition. Evolutionary biologists know that loss of characters, even seemingly critical characters, is widespread through the history of life. Recall that the intestinal parasite *Giardia* manages quite well without mitochondria.[27]

Here I have emphasized scenarios that explicitly recognized changes in the transmission of cumulative cultural knowledge over the evolution of our species. The capacity for cumulative cultural evolution may have been nascent in early hominins, but the space has expanded through the past few million years, and particularly in the past few hundred thousand years, and continues to expand today. Standardized intelligence tests show a steady increase in mean scores, an increase that is evident across different groups and different ages. This is known as the Flynn effect, for James Flynn who first identified the pattern. Flynn proposed that the most likely explanation was a steady expansion in conceptual and abstract thinking. This scenario captures the expansion of social learning and cumulative cultural evolution, but fails to fully incorporate feedbacks from cultural change to the genome (gene-culture coevolution), or how our ancestors modified their own environment through niche-constructing activities. As with other innovation events, niche construction has played a significant role, with the application of fire and the spread of agriculture being only the most obvious examples. Next, we turn to cultural and technological novelties that have arisen within this expanded opportunity space.

Culture Builds Technology

Today we often consider technology a by-product of the Industrial Revolution or of science and engineering, but technology began when early members of the genus *Homo* began shaping Oldowan stone tools over 2.5 million years ago. Technology is the application of practical knowledge, knowledge that has been refined, preserved, and passed on by social learning and cumulative cultural evolution. In human cultures innovation is invention plus social learning. By this definition, technologies are by-products of cultural evolution. Animal artifacts long preceded humans, of course. Termite mounds, beaver dams, and bird nests are obvious examples. The challenge to separating culture from technology is illustrated by differing views of whether language and writing are technologies; some would restrict technologies to human-made artifacts. But human technology is a product of cumulative cultural evolution to solve a practical problem, whether in hunting and fishing, cooking, clothing, transportation (from pots and kayaks to planes), or a myriad of other human activities. Earlier sections of this book have somewhat disingenuously distinguished three domains: biology, culture, and technology. But technological novelty and innovation are inseparable from social and cultural evolution. Nonetheless, this artificial dichotomy is useful for two reasons. First, as discussed in chapters 2 and 3, while cultural anthropologists and archaeologists have long encompassed technological change within their remit, economic historians consider human artifacts distinct from culture. Much of the literature on technological innovation is largely independent from anthropology. Second, aspects of technology from the domestication of plants and animals to more recent products of the Industrial Revolution raise distinct questions about niche construction and economic growth.[28]

In this book I have invoked the concept of biological homology in defining novelty. Recall that the emphasis on individuation in biological novelty arose from concerns over identifying new homologous characters. But recognizing novelty through individuation seems more complicated in culture and technology because of rampant borrowing. Recombinations likely represent the major avenue to human and cultural novelties, just as for novel proteins. For recent technological inventions, the advent of patent law has rendered the problems of cultural individuation far more tractable. The US Patent and Trademark Office establishes a patent as legal protection for "any new and useful process, machine, manufacture, or composition of matter, or any new and useful improvement thereof." While humans are promiscuous borrowers, so too are microbes, and thus I am unconvinced that recognizing novelties via individuation, recombination, and deep transformation, as discussed in chapter 5, is an insurmountable problem with culture and technology.[29]

There has been much ink (and a little blood) spilt over whether technology "evolves." If *evolve* is a synonym for *change*, then technology evolves, but if evolution requires replicators and a means of inheritance, then the classical viola did not "evolve" into the ergonomically improved Rivinus "Pellegrina" viola (look them up—they are wonderfully bizarre). Rather, motivated by chronic injuries among many violists, luthiers took up the task of adapting classic violas to human anatomy. Like all technology, violas are artifacts of human cultural processes, just as trace fossils are artifacts of the behavior of animals. One could compare the "fitness" of different technologies by examining the decline in the number of telegraph stations and the increase in the number of telephones, followed by the slow decline in landline telephones as cell phones exploded in popularity. But technological evolution should be understood as a shorthand for the evolution of cultural artifacts.

Here I consider different cases of cultural and technological novelty (or invention) and innovation: domestication and agriculture, finance, and radar and transistors, which illustrate the application of this conceptual framework. Domestication occurred repeatedly in many places via deep transformations of plants and animals through human interactions. Although the rise of agriculture has commonly been linked to the rise of cities and states, more recent studies have revealed a more complex picture. The Mongol Empire under Genghis Khan, for example, was arguably a state, even if it did not conform to definitions developed for Mesopotamia, Egypt, India, China, and Central America. Money, banking, and finance have been a continuing source of novelty for centuries and are tightly linked to the growth of institutions. Double-entry bookkeeping, joint stock companies, hedge funds, and mortgage-backed securities represent new approaches to tracking profit and loss, managing risk, and making more money. Finally, the development of radar during World War II illustrates technological lags and the important role of combinatorics.

Domestication and Agriculture

Some innovations have greater impact than others, often because they involve creation of vast possibility spaces for follow-on inventions. Steam power, railroads, electricity, and information and computer technologies are generally ranked as the most important general-purpose technologies, although one could add language and mathematics. Equally significant was what has been termed the "Neolithic Revolution," beginning about 11,700 years ago, when hunter-gatherers in the Levant and Fertile Crescent settled into a less nomadic lifestyle as they increased their management of plants and later animals. These changes developed after the end of the Pleistocene glacial interval and during a general amelioration of climate. Between 10,000 and 3,000 years ago, sedentary

Neolithic societies also show population growth, aggregation into towns and eventually cities, irrigation, pottery, specialization, and division of labor.

When Darwin wrote *The Variation of Animals and Plants under Domestication* in 1868, he described the morphological changes that accompanied domestication, which have become known as the "domestication syndrome," with the implication that a similar pathway was responsible across many different species of plants and animals. No domestication of any plant or animal was a sudden, transformative event, at least on the timescales of anthropologists, nor was there a single route to domestication, since it was not a goal-directed process, but rather a gradual coevolutionary accommodation between humans and future domesticates.[30]

Domestication is often considered mutualistic (benefitting both species), arising from coevolutionary interactions as humans generated environments, allowing them to manage and benefit from the propagation and reproduction of domesticated species. This definition is specific to humans but can readily be generalized to encompass domestication by other species, such as the domestication of fungi by ants. As domestication is an outcome of a coevolutionary process, it is distinguished from mere management, which involves manipulation of the environment or conditions under which plants or animals grow. Selective breeding or other changes are not a part of management but arise in pathways leading to domestication. If I drop some sunflower seeds near a village in the hope that they will grow the following summer and be easier to harvest, I am managing the sunflowers, but not yet domesticating them. Domestication might involve preferentially planting seeds from plants with the strongest stalks, ensuring seedlings have sufficient water the next spring, or ensuring the plants are not trampled by dogs, children, or something else. Finally, agriculture is a human social and cultural innovation encompassing a suite of domesticated plants and animals that supply necessary food and resources. In other words, agricultural societies draw much of their resources from domesticates. Although management (sometimes described as cultivation), domestication, and agriculture represent a continuum, domestication does not necessarily lead to agriculture, any more than management of a species ensures that it will eventually become domesticated.[31]

Today, several dozen animal species and hundreds of different plants have been domesticated. My colleague Melinda Zeder distinguishes among domesticates that "adopted" humans by moving into human habitats before a more mutualistic interaction developed, species that were initially a target of humans through hunting, fishing, or harvesting, and finally domestication because of

310   CHAPTER 9

intentional human activity. The first two paths involve species that likely already possessed features that made them suitable for domestication (which I might describe as potentiation), whereas the final, directed pathway often involves species for which some barrier to domestication needed to be overcome.[32]

Cultures that adopted dairy farming boosted the nutrition of children (and sometimes adults), so the adoption of dairy in the Middle East, Africa, South Asia, and Europe is hardly surprising. Adequately digesting lactose, the key sugar in milk, requires a change in the enzyme lactase. It was once thought that the nutritional advantages of milk established intense selection and led to the rapid spread of people with this change. If so, this involved only a simple mutational change, and thus no true novelty. But a recent study of about 7000 pottery sherds from 554 sites across Europe has forced a drastic, but interesting, reassessment of the traditional view. Dairy fats are preserved in sherds (as long as archaeologists did not wash them), allowing reconstruction of the timing of dairy farming. There were two important results. First, people consumed milk for almost 1000 years before the persistence of lactase in adults was widespread. This is a classic example of a behavioral shift generating the environment for subsequent adaptive genetic shift leading to an evolutionary novelty. But a second point is of equal interest. Dairy farming did not increase steadily across Europe, but rose and fell, with contractions during times of famine. The authors postulate that lactose intolerance may have been detrimental during times of starvation and malnutrition, with greater survival of those with the mutation.[33]

One of the striking facts about domestication is that although it occurred on every continent except Antarctica (although Australia seems to have possessed few species amenable to domestication), there is no sign of domestication before about 12,000 years ago, with the end of the great Pleistocene climatic variability. This correlation has led many to link domestication to climatic amelioration. While we can be reasonably confident that agriculture would have been challenging and probably short-lived during the Pleistocene, that does not necessarily mean that the end of the ice ages was causally connected to the success of management, domestication, and agriculture. There seems to be no reason that management could not have occurred during previous interglacials. Moreover, climatic amelioration cannot be the only factor enabling domestication and agriculture. Although domestication first appeared after 11,500 years ago, new domesticates have continued to appear. And recall that agriculture often lagged domestication by 1000 to 2000 years. Population growth or greater population density may also have played a role, consistent with evidence for lags between initial cultivation of species and progress to communities dependent upon agriculture for much of their diet. A third scenario advanced by archaeologist Bruce Smith implicates niche construction, suggesting that domestication involved "deliberate enhancement in resource-rich environments." These are not mutually exclusive hypotheses, and indeed each may have contributed. With the large number of transitions from wild to cultivated species, it seems possible that the driving forces may have differed regionally and from species to species. I cannot help but note, however, how similar these three driving factors are to environmental change, ecological networks, and niche construction, as discussed in chapters 7 and 8.[34]

Today the niche-constructing aspects of agriculture are evident, whether in extensive artificial selection of cultivars or carefully tending crops or domesticated animals. How such systems arose, however, is more conjectural, relying on archaeological evidence and insights from studies of modern small-scale societies. Based on archaeological evidence from eastern North America and South America, Smith proposed that domestication likely began in small settlements with resource-rich river bottoms, lakes, and estuaries, where a high density of managed species could be easily maintained. Those species that responded readily to management were encouraged, resulting in rounds of positive feedback and sustained utilization. Through this process, human intervention selected useful traits and eventually led to domestication, but just as significantly, humans would have discovered appropriate conditions for sustaining the proto-domesticates. In doing so, these small-scale societies modified their environment to perpetuate both managed "wild" species as well as more domesticated forms through progressive rounds of niche construction.[35]

These small-scale societies exhibited low-level food production, a mix of traditional plants and animals along with a contribution from domesticates. Increasing the contribution of domesticates to food production likely necessitated adjustments to social and cultural practices, including systems for managing and maintaining the domesticates, expanding cooperation, and regulating access to resources. Who controlled where people planted, how crops were distributed and stored, or how seeds were stored for the next season? Such societies faced new coordination problems requiring resolution before they could adopt (if they did) full-scale agriculture. Cultural inheritance facilitated the transmission of knowledge and practices between generations.

Agricultural societies created new opportunity spaces for downstream novelties and innovations. Indeed, the conception of a Neolithic Revolution assumed linked innovations, with agriculture furthering changes in social complexity by altering the nature of work and fostering divisions of labor, progressive rounds of new technology for farming, storing crops, and transporting them to market and new economic opportunities. Although the first cities appeared before agriculture, concentrations of people led to changes in housing styles, creating niches for dogs, rats, cats to catch the rats, and opportunistic vermin from cockroaches to termites. Human diseases requiring person-to-person transmission thrived. And these new opportunities, whether in farming technology, transportation, or housing, triggered further rounds of novelty and innovation. Because combinatorics is fundamental to human generation of novelty, agriculture transformations percolated to expand opportunities in other spheres.[36]

The consequences are sufficient justification for including domestication and agriculture, but they also reveal insights into the dynamics of novelty and innovation that may not be evident from events deeper in the past. First, the origins of agriculture were only 12,000 years ago. This would appear in the fossil record as an almost instantaneous transformation, albeit one that was significantly accelerated by cultural evolution and learning. Second, the evolution of maize from teosinte, a grass found in the Oaxaca region of Mexico, was a deep morphological transformation but one largely accommodated by the genetic variation available within ancestral teosinte. In other words, no genetic or developmental novelties appear to have been required. This leads to the third point, which is that agriculture represents innovations combining adaptive evolution of domesticated plants and animals with cultural evolution of the human populations that adopted agriculture. Domestication and agriculture demonstrate that network transformations associated with significant innovations do not necessarily require novelty. Finally, successful transitions to agriculture are inherently combinatoric, requiring an appropriate combination of plants and animals to provide sufficient nutrition to sustain social groups shifting from hunting and gathering to agriculture. For example, beans, squash, and turkeys provided substantial nutrition in the US Southwest, along with corn. Although single domesticates have often been the foundation for agricultural societies (rice across Asia and India, corn in Mexico and the American Southwest), they are usually accompanied by other domesticates.[37]

Institutions

The successful domestication of plants and animals and the diffusion of agricultural economies required institutions to ensure property rights to both land and agricultural products, the ability to construct and maintain common-pool resources, principally irrigation systems, and eventually the storage of surpluses. The traditional account of the transition from hunting and gathering to settled agriculture often links irrigation systems of large size to increasing political control and hierarchy, the generation of agricultural surpluses to some people specializing in craft industries and ceremonial or religious activities. Under this paradigm, these processes led to stratified societies and the first cities, including Çatalhöyük in Anatolia, followed by others in Mesopotamia, China, the Indus Valley, the Andes, and Mexico. Despite the widespread doubts about causal linkages between domestication, cities, and social stratification, it is nonetheless true that this transition required the development of new institutions that likely grew out of prosocial interactions among hunter-gatherer societies long before the rise of agriculture.

Today we think of institutions as governments, universities, large companies like General Electric or Apple, or perhaps the Catholic Church. These are formal institutions, bound by laws and regulations that ensure the persistence of an organized network of interactions over many human life-spans. We might consider the establishment of formal legal systems as foundational novelties that permitted the elaboration of specific institutions. More broadly, however, institutions are clusters of social norms that help govern relationships between people. The effectiveness of an institution often depends upon the ability of norms to encourage or ensure compliance. Norms arise from cultural practices that arise and are spread by social learning so that they become culturally inherited. Norms influence the foods we eat, who one can or cannot marry, whether men and women can see each other during religious services, when one begins to plant or to harvest a crop, and even the shape and painting of a pot. When similar social norms are woven together, an institution may arise, hence the colloquial phrase "the institution of marriage," which captures social norms about behavior, offspring, property, and related values. In many Westernized societies marriage represents a social pair-bond reflecting not just affection but romantic love, but the consensus among anthropologists and historians is that this view of marriage emerged in sixteenth-century Europe, as the Roman Catholic Church began requiring a priest to sanctify the union. But the origins of the institution lie in Roman common law regulating the inheritance of property and sexual activity.[38]

Social norms underpinning institutions require standards to judge when someone has violated the norm and means to rectify violations. Many institutions are self-enforcing because individuals who follow the norms benefit, as do those who participate in sanctioning individuals who violate norms.

Among the Hopi and other pueblo peoples in the southwestern United States, a group of "clowns" at ceremonial summer dances make fun of village members who violated social norms. Such public shaming reinforces community traditions. More recently the clowns have been deployed to make fun of Western visitors, particularly those surreptitiously taking notes or photographs. But as populations expand, public shaming becomes less effective and people become less closely connected. This may be one factor leading to the formalization of institutions through rules, laws, and similar practices. Once established, institutions enable large-scale interactions and exchanges that do not depend upon kinship relationships. Formalization helps ensure the stability of institutions as well as the means to ensure compliance with institutional norms. Whether informal or formal, institutions represent an important source of cultural novelties.[39]

Institutions are unique to human societies. Although social interactions such as reciprocal exchanges and social learning are found in many other species, no other species regulates social interactions through arbitrary, socially constructed rules, enforces the rules, and perpetuates them through cultural evolution. Together with language—for bargaining over the rules and their enforcement—prosocial behavior was a necessary precondition for establishing institutions. But a dichotomy exists, for example, between those who view institutions as underlain by rules and laws, and those who view them as stable patterns of behavior. For our purposes it is sufficient to recognize that institutions are the means by which we accomplish tasks beyond the capacity of a single individual, a family, or a group of related individuals.[40]

Thus, our collective ability to form institutions is a significant cultural innovation that permits a broad range of downstream cultural novelties and underlies increases in social complexity. As described in chapter 3, anthropologists once invoked stereotyped levels of complexity from bands and tribes to chiefdoms and states. Today most anthropologists recognize the great diversity in social complexity. Primary states need not be founded on agriculture and irrigation, or even cities. The number of hypotheses proposed to explain expanding complexity ranges over the rise of agriculture, provisioning of public goods such as irrigation, cities, trade, social stratification, religion, economic inequality, and conflict between groups. Studies by ecologist turned cultural evolutionist Peter Turchin and his associates have used a large-scale database of social attributes to test alternative scenarios. They concluded that agriculture and adoption of new military technologies (such as cavalry) jointly provide the best explanation. I was intrigued, however, to see that their analysis identified a significant lag, averaging about 2000 years, between the adoption of agriculture and the formation of large-scale states. Further explorations of how cultural novelties underpin expanding social complexity will be an intriguing project for the future.[41]

Some anthropologists and political scientists have favored scenarios in which state formation depended on acquiring a monopoly on the use of force and coercive powers. In anthropologist Robert Carneiro's classic work on the origin of the state, he contrasted coercion with volunteerism and concluded that only the former provided a realistic explanation. The resulting autocracies were associated with increased social and economic complexity. Collective solutions to public goods problems had been largely discounted as ineffective. Free-riding, corruption, and the seeming inevitability of aggrandizement seemed insurmountable obstacles. Since Carneiro's study, work in anthropology, political science, and economics, notably that of Nobel Prize–winning economist Elinor Ostrom, have explored the requirements for successful collective action institutions. At Monte Alban in Oaxaca, Mexico, for example, anthropologists Linda Nichols and Gary Feinman have used broad surveys of the surrounding valleys to support a view of what they describe as "quasi-voluntary compliance," in which widespread social cooperation produced institutions that furthered trade and social interactions. In this case, even hierarchical societies may have relied upon collective action rather than coercion, possibly even adopting democratic traditions independently of ancient Greece.[42]

One challenge to understanding social complexity is that individuals participate in multiple networks, and in doing so they connect institutions into complex interaction networks. As political scientist John Padgett analyzed the social dynamics of Renaissance Florence, he reconstructed economic networks of companies and goods, kinship networks of marriages between families, and political networks of interactions among competing Florentine factions, tracing the positions of individuals across these different networks. Men might also belong to military networks, and families could advance religious ties as sons joined the priesthood or took holy orders, or through confraternities. Elaborating this multinetwork perspective from the medieval period into early modern Europe in Italy and France enabled Padgett to examine feedback between early capitalism and the formation of modern states, capturing the rise of merchant banks with international networks, and their displacement by the earliest joint stock companies and central banks associated with the Dutch Revolt in the late sixteenth and early seventeenth century.

Such changes were necessary but not sufficient preconditions (a potentiation) for spillovers and network reconstruction. Indeed, Padgett redefined Schumpeter's creative destruction as "the breakup of old networks and their replacement by new ones." Counting bills of exchange or marriage contracts, for example, would have been insufficient. Only by examining economic, kinship, and political networks and the multiple roles of individuals could Padgett reveal the interactions among these institutions.[43]

Hierarchical, autocratic social organizations require a suite of institutions, as do social systems organized around the more collective form of quasi-voluntary compliance advocated by Nichols and Feinman for Monte Alban, but the nature of the institutions differs. Our unique capacity to build institutions was initially expressed through informal structures governing behavior. These were elaborated into more formal structures that continued that influence into the complexity of our social, economic, political, and cultural lives. The strength of institutions varies globally between cultures and regions, leading economists Daron Acemoglu and James Robinson to conclude in *Why Nations Fail* that political and economic institutions underlie differences in economic success among nations. They marshalled data from ancient, historic, and modern settings to argue that institutions enabling prosperity generate positive feedback loops, or a "virtuous circle." Abandoning the typological progressivism of increased social complexity acknowledges the likelihood of a far greater range of social institutions and interesting possibilities for future work in cultural novelty and innovation.[44]

Radar and Patents

World War II forged two of the most important inventions of the past century, radar and the digital computer, and was indirectly responsible for a third, the transistor. That radio waves could be deflected by metal objects was recognized by German physicists in the 1880s, and by the 1930s early steps toward radar for finding ships and aircraft had been taken. During the Battle of Britain in the summer and early fall of 1940, the British suffered terrible losses as Germany prepared for a cross-channel invasion, but existing radar lacked sufficient power to direct anti-aircraft weapons or fighter aircraft toward incoming aircraft. As the Battle of Britain was underway, physicists John Randall and Henry Boot greatly increased the power of radar by inventing the cavity magnetron. This new radar combined the magnetrons used in earlier radars with a klystron tube to amplify microwaves to produce higher-energy radars operating on microwave frequencies. After the British revealed the prototype to the American military, the cavity magnetron was rapidly put into production and made incredible contributions to the war effort.

The history of microprocessors, the key components in our computers, cell phones, and increasingly cars, refrigerators, and other devices illuminates the sometimes-complicated relationships between novelty, innovation, and adaptation in technology. Modern electronics are based on the transistor, which replaced earlier vacuum tubes but also expanded the range of possible technologies. Conceived by Julius Lilienfield in 1926, transistors were implemented by a team from Bell Labs in 1947. As is so often the case with inventions, in 1948 a group in France independently made similar discoveries. Eventually silicon-based transistors were packaged into integrated circuits and microprocessors.

Before Gordon Moore cofounded Intel Corporation, he was the director of research at Fairchild Semiconductor, where in 1965 he wrote an essay estimating that the number of transistors per chip would double every year. A decade later he revised this to a doubling of the number of transistors every two years, a more realistic estimate that held true for five decades. Moore's law, as this relationship became known, translated into increased processing power. There was nothing preordained about such progress, however, despite popular misconceptions. Rather, it was a deliberate strategy of the semiconductor industry, which enabled coordination of the many allied technologies needed to achieve more-capable chips. Coupled with an increase in clock speed—the rate at which instructions were executed—computers increased steadily in power and decreased in size. This enabled the regular appearance of new technologies, from minicomputers in the late 1960s to personal computers in the 1980s to laptops, smartphones, tablets, and so forth. The realities of physics brought this trend to a halt. Clock speeds were capped in 2004 because of excessive heat, and quantum effects came into play as circuits became smaller. By the early 2020s progress in microprocessors had reached its limit, and engineers and computer scientists are seeking new strategies.[45]

The initial transistor from Bell Labs was a novelty, and a general-purpose technology that would generate innovations across information technology and beyond. The steady progress charted by Moore's law is the technological equivalent of adaptive evolution, aided by the development of allied technologies (some of which may have qualified as inventions in their own right, of course). Here the trend was facilitated by industrial coordination. Major advances in microprocessor technologies were not themselves novelties or inventions in the sense that I am using the terms (although they likely qualified for multiple patents), but they enabled further technologies, from printers, laptop sleeves, and word-processing and spreadsheet programs to apps and other software. These ancillary technologies represent innovations, as new ecosystems developed for successive phases of technology. Gaming and video streaming led computer designers to offload graphics-intensive activities onto specialized graphics chips, which soon acquired new uses.

Technological Spaces

Technology is often viewed as an expanding possibility space of combinations. This perspective has been widespread in the business and technology literature, in part through the views of Stuart Kauffman (chapter 4). Viewing technological novelty as an expanding combinatoric space is important but leaves out two significant considerations. First, as in biology, many technologies are highly modular. If you imagine putting together the mechanism of an analog watch, trying to keep all the gears, springs, and whatnot together to fit into the case is a recipe for disaster. Far better to design and build the watch as a series of modules assembled as separate components that fit together to form the completed mechanism. The evolution of technology involves combinations of combinations of combinations in an unending recursion. Economist Brian Arthur identifies building blocks, or primitives, as the foundation of technology. Within a particular domain of physical phenomena, such as chemistry, optics, or electronics, these primitives are combined into functions, tested, and if valuable the combinations are encapsulated (like the modules of a watch). Rounds of combinations and testing generate a library of functions that can be used for a new round of combinations. Each step is important: forming a new combination, selection or testing, encapsulation, and the formation of a new library of building blocks. Arthur explicitly used biology as a metaphor for the evolution of technology, and as in biology, Arthur's view of technological change generates modular, hierarchical structures.[46]

The second conundrum involves the origin of novelty in a combinatoric space. For Arthur, novel technologies arise from new ideas or concepts that can be used to achieve some goal. Arthur believes that combinatorics suffices for all technological inventions. The introduction to a fascinating study of novelty among logic circuits begins: "New technologies are never created from nothing. They are constructed—put together—from components that previously exist; and in turn these new technologies offer themselves as possible

components—building blocks—for the construction of further new technologies." I agree that this is mostly true, but not, I think, entirely true. New
combinations help explore an existing space, but do they construct new ones?
Just as Schumpeter noted that stringing together horse-drawn wagons did not
create a railroad, multiplying vacuum tubes did not create a transistor, although early transistors served the same function in an electrical circuit, any
more than the early magnetrons produced the cavity magnetron so critical to
the development of effective radar during World War II. As with biological
evolution, true novelty ultimately depends on expanding the supply of building blocks to a library, as was true of the first transistors. The transistor opened
a new opportunity space for electronic devices that would have been unimaginable in the age of vacuum tubes. Similarly, when John Randall and Harry
Boot invented the cavity magnetron, they did not imagine its eventual co-
option to serve as the foundation for microwave ovens. Indeed, today radars
involve new technology, so that most existing cavity magnetrons are found in
kitchens, with little sign of their early history in the nose of fighter planes.[47]

The domains of technology have expanded through human evolution as we
have developed the capacity to harness new physical processes. For our ancestors, technology was limited: stone tools for hunting and butchering; hafting
blades or axes to a handle (requiring twine or other adhesives); and hooks,
nets, weirs, and related technologies for fishing. Some cultures constructed
bone flutes and other musical instruments. The next major increase in technological domains came with domestication and the spread of agriculture, as
well as the wheel and the use of horses for transportation. With agriculture
came new domains of technology for food production, cooking, and the storage of food. Agriculture also led to increasing social aggregation and the spread
of cities, with increasing specialization of social, economic, and political roles.
Eventually pottery, iron, and other technologies spread with increasing exploration of new physical domains, particularly since the Industrial Revolution. The
expansion of each domain allowed inventors to exploit new opportunities. As
Arthur observed, these domains differ greatly in extent, with innovations in
some fields suddenly transforming a different domain. In the case of agriculture, crop improvements had long depended on artificial selection by farmers.
State extension services and commercial agricultural suppliers systematized
crop improvement, and biotechnology has expanded opportunities again.
That we can debate the merits of new opportunities in chemical pesticides,
gene transfer, and global homogenization also reflects changes in political opportunity spaces. Inventions require subsequent refinement and improvement

before they are sufficiently useful to become widespread. Arthur describes this as "structural deepening," while others have used different terms, but the point is that once a basic principle has been established, cheaper, faster, better means to achieve the same end will almost certainly be discovered.[48]

Patent records and product spaces can help test the relative importance of search and combinations of existing products versus the introduction of new technologies that provide the basis for additional rounds of search and combinatorics (or exploitation versus exploration). The records of the US Patent and Trademark Office date to 1790 and provide a consistent series of technology codes. Since patent applications are required to cite related patents, networks of patent citations have been constructed, revealing patterns of interaction. Most patents cite only a few other patents, while a small number cite many, generating a nested hierarchy of related technologies. Examining these records allows a more quantitative view of the invention and evolution of technologies. Most new patents arise through search near existing patents via local optimization. Local search depends upon a sphere of expertise, within which an individual inventor or a company can generate patents. Patents also represent proven technologies, allowing them to serve as modules for recombination into larger technologies. But such analyses need to distinguish between patents awarded for the discovery of new principles and those associated with structural deepening. By analyzing the combinations of codes listed in patent applications, a recent study confirmed Arthur's claims, with combinatoric search through existing technologies accounting for most patents. A small number of patents represent novel combinations of technologies, or exploration. The combinatorics of product spaces composed of English words, recipes, cocktails, and software has also been analyzed, with similar results.[49]

Patent records have produced interesting insights into inventive activity, but many patents never have any economic impact and thus do not contribute to innovation. In the case of radar, most histories of World War II in Europe mention its impact (although hardly a systematic study, radar seems to feature less frequently in accounts of the war in the Pacific, although it was equally important). As with biological innovations, examining the network effects is essential. Within a technology a novel subassembly may require modifications to surrounding subassemblies, just as a new technology may require adjustments within social or economic networks. The current push for cleaner energy is triggering such network effects: the buildout of charging stations for electric cars, for example. Similarly, our electric grid was built to move electricity from large power stations to cities, and neither the economics nor the network was designed for a distributed system where increasing numbers of buildings have solar panels. The expansion of options markets after the Black-Scholes-Merton model was published similarly triggered extensive downstream adjustments. Both the innovation and diffusion phases feature network adjustments, but the expansion of a network approach is required to better understand the process.

In an 1869 essay, Alfred Russell Wallace wondered why human capabilities were so much greater than needed when we diverged from other primates. The discoveries of *Australopithecus*, *Homo habilis*, and *Homo erectus* only partly resolves Wallace's dilemma. Our abilities in building societies; creating art, music, and technology; elaboration of a mathematical view of the universe; and even learning to pitch a four-seam fastball seem far beyond those plausibly required by our ancestors at the origin of *Homo sapiens* some 300,000 years ago. Whether one describes our capabilities as cumulative cultural evolution, as collective intelligence, as a collection of cognitive gadgets, or by some other rubric, the early steps of human evolution endowed our species with potential, with capacity, far beyond what was utilized by the earliest members of our species. Early humans were not engaged in quantum physics or constitutional law. Nor had they yet domesticated plants and animals, acquired writing or mathematics, or built universities. The acquisition of cumulative cultural evolution and all that it entails was ultimately one of the most transformative evolutionary novelties in the history of life, a general-purpose technology that rivals the last common eukaryotic ancestor.[50]

I have argued that human uniqueness is collective, through social learning and niche construction to preserve and transmit information between generations. Although the ability to generate social, cultural, and technological novelties and innovations is rooted in biological novelties, these new opportunity spaces are fundamentally cultural, not biological. But one lesson from this book is that most lineages are overdesigned (in an engineering sense) for their current life habits, and humans are not unique in this regard. Eukaryotes and their gene regulatory processes were absurdly overdesigned for life two billion years ago, as were the earliest animals, or the fish ancestors who possessed the necessary regulatory machinery for life on four limbs.

The final chapter reconsiders the nature of novelty and innovation and returns to the questions posed in chapter 1 about a common framework for novelty and innovation across the three domains of biology, culture, and technology. Chapter 10 also considers some broader matters: the prospects for a general theory of novelty and innovation, the extent of determinism and the role of contingency and evolvability, and whether evolution is open-ended, before concluding by addressing how the evolutionary process has itself evolved through the history of life. Although Darwin's view of evolution was deeply embedded in his understanding of the history of life, the Modern Synthesis view is essentially ahistorical: the framework of quantitative genetics largely ignores changes in the process over time. The emergence of cumulative cultural evolution is perhaps just the most obvious example of ways in which the evolutionary process has changed. Over the past several decades, however, there has been growing interest in how the possibility spaces for evolutionary change and evolutionary dynamics have changed over time.

10

Toward a Theory of Evolutionary Novelty and Innovation

FRENCH HISTORIAN Jacques Le Goff decried periodization, the seemingly innate drive among historians to segment the past into ages, epochs, and periods, often separated by various revolutions. Some revolutions have been transformative: the Glorious, French, American, and Russian changed internal political and social dynamics and influenced later revolutions. In his final book, Le Goff argued that many of these transformations were more apparent than real, serving as a convenient segmentation for historians themselves. Today historians see more continuity between the Middle Ages and the Renaissance than when I studied European History, for example.[1]

Historians are not the only scholars ennobling discontinuities more apparent than real. Archaeologists periodized the Stone, Bronze, and Iron Ages, the fall of the Classic Maya (the Maya are still living in the Yucatan), and debated pseudo-evolutionary trajectories of social complexity from tribes to chiefdoms to states. Geologists constructed the geological time scale by recognizing changes in the dominant fossils in European rocks. Indeed, when English geologist John Phillips introduced the Palaeozoic, Mesozoic, and Cainozoic, grouping periods into more encompassing eras, he believed that each era represented a separate origin of life. Later geologists recognized these profound changes in animal life as the end-Permian and end-Cretaceous mass extinction events. By the 1980s paleontologists distinguished between mass and background extinctions.

Some divisions have a whiff of desperation about them. Does anything noteworthy distinguish the Silurian from the Early Devonian? In 1972 Stephen Jay Gould and Niles Eldredge brought discontinuities to the origin of species with their theory of punctuated equilibria, which metastasized to all manner of events having nothing to do with speciation. Similarly, human consciousness is held to be something different in kind from the behavior of other animals, and we separate geniuses from the rest of us only to have physicists distinguish "ordinary geniuses" from "magicians." Perhaps the most pernicious discontinuity in the history of ideas was philosopher of science Thomas Kuhn's recognition of paradigm shifts as intellectual revolutions that changed normal science. The term *paradigm* became so overused by those seeking to hype their manuscripts that, mercifully, its use has been banned by many scientific journals.

This inquiry into another of life's discontinuities, the idea that novelties and innovations represent different evolutionary dynamics from other evolutionary processes, may be subject to the criticism that Le Goff leveled at his colleagues. I see the message as more nuanced, however. I have argued that the only reliable definition of evolutionary novelty is individuation of new characters, but that these may arise in different ways. The conceptual framework I have advocated poses challenges both for those who have identified unique processes associated with the generation of evolutionary novelty and those hewing to a more adaptationist paradigm.

Chapters 2 and 3 described the history of work on these apparent discontinuities. Although the pattern was widely recognized, scholars differed over the salient features: Were novelties distinguished by more rapid rates of change? By the apparent morphological (and later genetic or developmental) discontinuities? By the involvement of different evolutionary mechanisms? Günter Wagner and his collaborators have identified a variety of character homology identity mechanisms (ChIMs) as generating the individuation of evolutionary novelties, as discussed in chapters 5 and 6. But ChIMs could be viewed as either a process or a consequence. If we take recursively wired gene regulatory networks as an underlying feature of regional patterning in developing embryos (as kernels by the late Eric Davidson and me) and of cell types (as ChINs by G. Wagner), these may have been gradually assembled and consequential only after they are formed. Alternatively, it might be that the process of generating such subcircuits is somehow distinct from other evolutionary processes. As we saw, human language was enormously consequential once it evolved, but all evidence is that it was assembled over time (however rapid it may appear to some paleontologist far in the future), and I suspect this is likely to have been true of most evolutionary novelties. Once assembled, however, novelties may represent evolutionary ratchets that influence the range of available options.

Even as biologists and paleontologists accepted the idea of descent with modification after 1859, they differed over the efficacy of natural selection, with many (particularly in Germany and the United States) favoring orthogenesis or other evolutionary drivers. Goldschmidt was just the most famous of those who proposed alternative genetic or developmental mechanisms for the generation of the apparent "jumps" seen in the fossil record or evident in the gaps between existing higher taxa, imputing a decoupling of evolutionary mechanisms as an explanation for apparent discontinuities. In seeking to eliminate both Osborn's orthogenesis and Goldschmidt's systemic mutations, Simpson's *Tempo and Mode in Evolution* shifted the argument. Underlying Simpson's adaptive radiation model was the assumption that the cumulative effect of small mutations over time, coupled with ecological opportunities, could generate the patterns we observe in the fossil record.

This final chapter examines the nature and ongoing challenges of novelty and innovation, issues surrounding individuation, and search versus construction. With this we return to a question posed in chapter 1: Is a general theory of novelty and innovation possible? If biological, cultural, and technological evolution are open-ended as suggested by both theoretical and empirical evidence, then construction of new evolutionary spaces makes an important contribution to novelty and innovation. In making this claim I do not reject novelty by search, although I suspect that the most fundamental innovations are those associated with construction of new evolutionary spaces. Parallel and convergent events in biology, culture, and technology have been interpreted as evidence that the scope or potential of novelty is limited rather than open-ended. Interrogating this problem raises issues of evolvability, contingency versus determinism, growth, complexity, and the always-fraught topic of progress. Evolutionary possibilities have expanded through the history of life, as have cultural and technological possibilities more recently. But has the evolutionary process itself evolved as well?[2]

Discontinuities, Decoupling, and Developmental Capacity

For Darwin, jumps were signs of a fragmentary fossil record. Huxley, Pictet, and others rejected Darwin's view, concluding that discontinuities were fundamental and required explanation. By the 1950s when Mayr and Simpson participated in the Modern Synthesis, they acknowledged that at least some of the discontinuities might be real. But for transformationists like Goldschmidt, discussed in chapter 2, and advocates of macroevolution, discussed in chapter 3, these discontinuities revealed changes in evolutionary rates or distinct evolutionary processes.

Previous chapters have examined some examples of real discontinuities. But in the cases where we have sufficient information to judge, these episodes involved potentiation; the foundations of a new novelty, whether genomic, developmental, morphological, or cultural; refinement of that novelty; environmental changes; and the construction of new niches to facilitate successful innovation.

The early expansion of developmental capacity provides a new perspective on the assembly of evolutionary novelty. Whether this is generally true is a question for empirical study. This view is emphatically *NOT* an argument for "hopeful monsters"—sudden transitions with no connection to what came before—but nor is my view a validation of microevolutionary views that fail to recognize the importance of potentiation, of establishing newly individuated characters, or that rely upon adaptive radiation scenarios. Moreover, early developmental capacity poses challenges to the use of evolutionary rates as a metric of novelty. This is the main reason there has been so little discussion of rates after the introductory chapters. Although implied rates have featured in discussions of novelty and innovation since the nineteenth century, potentiation and the decoupling of novelty and innovation suggest that much evolutionary change will be missing from the fossil record. Reconstruction of a temporal framework for novelty and innovation requires fossil evidence in a robust phylogeny and trustworthy molecular clock dates of divergence times, integrated with comparative molecular and developmental studies.

If my claims for developmental capacity and potentiation are correct, then lineages with such capacity may be "poised" for morphological novelties, allowing for a more rapid morphological transition than would be the case if the genomic and developmental capacity were acquired during the novelty event itself. Calculating rates of morphological change during such events without considering potentiation will often lead to overestimation of the extent of the evolutionary change. I was certainly guilty of such overestimation in earlier work on the Ediacaran-Cambrian diversification. Similarly, failing to discriminate between novelty and innovation complicates rate estimation if the underlying novelties substantially precede the ecological success of the clade. As others have recognized, decoupling between taxonomic diversification and ecological and evolutionary impact raises questions about the widespread reliance upon diversification as a metric for innovation. Going forward, estimates of rates of macroevolutionary change need to specify underlying assumptions and what the rates are meant to imply about the evolutionary process.

Nature of Novelty and Innovation

How an evolutionary biologist views apparent discontinuities is just one of the intangible expectations associated with any definition of novelty: Is all evolutionary change gradual? Was I trained in an intellectual tradition that embraced distinct macroevolutionary processes? If I refuse to be tied down about the nature of novelty, how do I differentiate novelty from adaptation? Do novelties defined based on process (new gene regulatory networks, for example) disappear if I shift to a definition based on character individuation, or some other criteria? Does discontinuity in morphology or other aspects of the phenotype require discontinuities in developmental processes or in the genome to qualify as an evolutionary novelty?

This book has advanced a view of novelty based on individuated characters, deep character transformations, or new character combinations. Although an emphasis on character homology has been important since 1990, other approaches have enriched our understanding and remain informative. As a reminder, these different perspectives are:

- Darwin and Mayr's view of novelty as a new structure with a new function;
- novelty structured by ecological opportunity, as championed by Dobzhansky;
- the primacy of adaptive radiations in generating novelty, with Simpson as the primary proponent, but a view widely adopted by later evolutionary biologists;
- higher taxa (phyla, classes, and orders), and after the phylogenetic revolution, major clades, as a metric of novelty and innovation;
- novelty as the formation of new, individuated homologous characters, growing out of work by Gerd Müller and particularly advanced by Günter Wagner;
- a suite of related combinatoric approaches. including Kauffman's adjacent possible, with Andreas Wagner as the most prominent recent advocate;
- proposals that behavioral changes initiate the evolutionary changes that result in evolutionary novelty, particularly from Mary Jane West-Eberhard.

In chapter 5 I developed the idea that the key differences between these approaches lie in whether novelty differs from evolutionary adaptation (and thus was even a question worth worrying about) and in differing perspectives on the relative significance of novelty or innovation, which in turn reflected whether novelty was considered as a radical departure in phenotype, generative of new taxa, or consequential in terms of ecological or economic impact. Each of these seven approaches has been scientifically fecund, but collectively they suffer from several challenges. Incorporating macroevolutionary lags and potentiation requires distinguishing novelty from innovation, and thus a new scenario. Here I briefly enumerate some of their limitations, before examining remaining challenges with novelty and innovation.

These seven approaches overlap and are to some degree complementary. Many founders of the Modern Synthesis, including Mayr, Dobzhansky, and Simpson, and later evolutionary biologists viewed mutations as sufficiently frequent that they focused attention on ecological opportunity as a driver for evolutionary innovations. (Schumpeter independently arrived at the same view of technological invention and innovation.) Novelty reemerged as an important issue with the explosion of comparative studies of developmental processes after 1990, but these studies sought explanations among the mechanisms generating novelty, rather than their success (or failure) as innovations.

Although the authors of first three approaches acknowledged novelty as an interesting evolutionary phenomenon, the scenarios are adaptive and microevolutionary. Coupling structure and function is at least partly character based. Simpson's work is most linked to generating new species or promoting more consequential evolutionary impacts, and thus the subsequent success of an evolutionary novelty, but emphasis on the consequent history of novelties also infects both the Darwin/Mayr approach and Dobzhansky's emphasis on opportunity. Later workers decoupled structure and function, discriminating between novelty (new structure) and innovation (new function). Behavior-first models also focus on the structure-function dichotomy but invert the polarity, suggesting that new functions arise from opportunities and only later become instantiated as new structures.

The emphasis on new higher taxa as a metric of novelty was long central to paleontological work and influenced other evolutionary biologists. Some events linked to new higher taxa are critical to understanding novelty and innovation, including the origin of eukaryotes and the Cambrian Explosion of animals, as well as the evolution of specific clades such as turtles, birds, and whales, among others. But focusing on higher taxa conflates novelty with innovation, the appearance of a clade with the origin of specific morphological novelties, and emphasizes consequences. While discussions of new higher taxa may have sufficed in pre-cladistic days, before the ability to deconvolve many of these issues, they seem difficult to sustain today.

Combinatoric approaches have figured prominently, particularly in Kauffman's adjacent possible and A. Wagner's conception of novelty as an issue of the accessibility of regions of pre-statable spaces. While interesting insights have emerged from these studies, a central theme of this book is the expansion of the evolutionary process over time, in the sense that new mechanisms of regulatory control have created new possibilities for evolutionary novelty. Other new possibilities include cumulative cultural evolution and language in humans, while new ecological and cultural structures have expanded opportunities. While aspects of this view are captured by Kauffman's expanding possible, I have argued that some interesting possibility spaces are not pre-statable. That opportunity space for novelty and innovation has expanded over time is a view that contrasts sharply with the idea that search through combinatoric possibilities is a sufficient explanation for innovation. Combinatoric views have not distinguished novelty from innovation, and while theoretically interesting, they are not character-based. Finally, combinatoric approaches in general represent what Calcott described as lineage-based explanations, focusing on the history of a particular line (for example, a mutating DNA or RNA sequence). Such a typological framework is hard to reconcile with the population-based focus of much of modern evolutionary biology.[3]

I have argued that novelties often require potentiating evolutionary changes to establish the proximate foundation upon which the changes in regulatory patterning occur. Among the potentiating events were those associated with the origin of animals, aspects of vertebrate limb development found in early fish, and development of social complexity in ants. Absent potentiating changes the novelty itself would never arise. But critically, potentiating changes are no guarantee that the novelty will arise or that it will occur in a timely fashion.

In the conceptual framework developed here, potentiating events are not limited to the genetic and regulatory changes that facilitate novelty but may also contribute to the success of innovations. This is demonstrated by the environmental and ecological control of macroevolutionary lags, which determines whether novelties become ecologically and evolutionarily significant. In many cases innovations have required the construction of new habitats, new opportunities, by the activities of other organisms.

Challenges in Defining Novelty

Whales, razor clams, the anomalocarids of the Burgess Shale, and the oaks outside my office each represent groups of species, discrete from related groups in the sorts of spaces described in chapter 4. Other discrete things in the world include genes, cell types, tissues, and some ecological communities, while languages, institutions, and rituals are examples of discrete cultural entities. Organisms form nested clumps, each whale species is one clump, followed by clusters of fin, humpback, killer, and other whales, which are divided between those whales with baleen and those with teeth, and then all whales form a great cluster near clusters of dolphins. Yet examined in detail each of these seemingly discrete entities has blurry boundaries. The late Harvard geneticist Richard Lewontin once described explaining such persistent clumpiness as one of the great challenges for biologists. But I fear that the challenge is even greater than Lewontin believed. The nature of these entities is a continuing problem for biologists and for philosophers of biology.[4]

If we accept individuation as the most defensible definition of novelty, does this resolve our problems? Individuation of new characters has been a key factor in the definition of novelty used here, but some readers will have concerns about two substantive problems: individuation and homology.

The approach to historical kinds adopted here differs from two related views, which may be more familiar to some readers. The first is an essentialist, or typological, view positing that kinds possess an essential physical property. The atomic number (the number of protons) is the essential physical feature defining the properties of an element. Essentialist arguments have been fundamental to discussions of natural laws by philosophers for several hundred years, and can often be traced back to Plato's views of eternal forms. Essentialists have viewed species as imperfect reflections of such forms. Mayr's objection to body plans as an evolutionary problem, as discussed in chapter 2, was grounded in this sense of types or body plans.

Questions about the nature of species led to an ontological definition of historical kinds in which each species had a defined beginning (speciation) and a defined end (extinction) and thus represented an individual historical trajectory. Philosopher of biology David Hull advocated this view, and the implications of species as individuals has been explored by many philosophers and biologists, with philosopher Peter Godfrey-Smith's "Darwinian individuals" among the most prominent recent examples. Viewing species as Darwinian individuals but not natural kinds avoids the essentialist critiques of Mayr and others. Discrete evolutionary individuals subject to selection have also featured in discussions of the major evolutionary transitions and multilevel selection.[5]

The view of individuation invoked in studies of evolutionary novelty is distinct from both essentialist and ontological views of historical kinds but is an outgrowth of the latter, a point raised in chapter 1. Here historical kinds are defined by the establishment of stable interaction networks, whether regulatory networks at the genetic level, cell types at the tissue level, or laws and procedures for states and institutions. Individuation is a consequence of stable, recurring processes or mechanisms that, within certain environmental or cultural bounds, ensure a consistent outcome. In the framework of complex adaptive systems, these represent basins of attraction. The generation of cell types is a developmental process that ensures reproducible formation of the cell type and stable patterning of embryos. Regulatory states in plants and animals are mechanisms that ensure stable production of organisms. Similar mechanisms ensure that once novelties have formed they will persist, while accommodating the necessary adaptive changes elsewhere within networks of gene, protein, or developmental interactions. The definition of innovation used here views them as a consequence of the construction of new, stable ecological networks.[6]

If novelties represent natural kinds defined by stable interaction networks, this raises question about how such networks form and how novelties at one level create scaffolds for novelties at other levels. One ontological approach to species was philosopher Robert Boyd's development of the idea of homeostatic property clusters. Homeostasis involves processes that maintain stable internal dynamics of living things—for example, stable internal temperatures among mammals. Boyd extended the idea of homeostasis to species, suggesting that species are natural kinds based on a cluster of properties and mechanisms shared by all individuals, which distinguish them from individuals in other species. Boyd suggested that more-inclusive monophyletic clades might represent natural kinds by extension of the principle of homeostatic property clusters. Since Boyd's early work the idea of homeostatic property clusters has been extended to settings from cell types to culture. I view the stable regulatory networks underlying developmental and morphological novelties as a mechanistic underpinning for Boyd's homeostatic property clusters.[7]

Biologist Armin Moczek's comment "Novelty begins where homology ends" encapsulates the view that recognizing a novel character is an explicit acknowledgement of the absence of homology. Homology is important to understanding novelty because of the challenges of distinguishing between different character states and new characters, as discussed in chapter 3, and because issues of homology challenge our understanding of the appropriate focal level of evolutionary novelties, a problem that is at least partly resolved by focusing on individuation. Through much of this book I have treated the issue of identifying newly individuated, homologous characters as relatively straight-forward. As many readers will be aware, this is anything but the truth. Chapter 6 presented one example of the challenges to identifying homologous characters, where insect and vertebrate limbs are obviously not homologous, yet they share developmental pathways. The discussion of the origin of insect wings also raised the still-unresolved issues of correctly identifying homology.[8]

Novelty has been tied to the generation of recursively wired gene regulatory networks, which would effectively decouple that character from previous characters. If so, the underlying regulatory structure helps define the newly individuated structure, neatly circumventing the sorites paradox. Subsequent research showed that new features, such as a new cell type, are not necessarily associated with recursively wired regulatory networks. For me, the critical issue is the individuation of new characters, not the mechanism of individuation. In later work, Wagner and his colleagues generalized his view to character identity mechanisms (ChIMs). Underlying these character identity mechanisms are stable interaction networks, which may or may not incorporate recursively wired gene regulatory subcircuits. These networks might represent gene interaction, or they might represent the outcome of a network of interactions, in which individual pieces of the network change but the outcome remains stable. As I emphasized in chapter 6, our developing understanding of the evolution of gene regulatory networks shows that what persist over evolutionary time are networks of interaction, not the specific regulatory genes that make up the network.[9]

The cases in chapter 6 involve individuation, but whether via formation of a new character or deep transformations of existing character states (as with the transition from fins to limbs) is often contested. Moreover, recognizing new characters makes no assumption about the rate of transformation. Tempo is dependent on context, and apparently sudden morphological novelties associated with the origin of new characters may mask more gradual transformations at the genomic or developmental level.

Challenges in Defining Innovation

If problems with defining novelty center on debates over individuation, the challenges with innovation are entirely different. Many novelties have played no role in innovations. Indeed, this decoupling between novelty and innovation underpins macroevolutionary lags and latent innovations. The contingent nature of innovations was illustrated in chapters 7 and 8. While innovations have traditionally been seen as a source of new taxa or new clades (generative), I have focused here on network effects: whether the adoption or spread of novelties restructures relationships sufficiently that removal of the novelty (or suite of related novelties) triggers changes in network structure. I acknowledge that this approach is not fully worked out, and future work may provide new frameworks for evaluating innovation.

The contingent nature of innovations extends to their ecological, evolutionary, and environmental impact. The spread of oxygen-producing photosynthesis after 2.4 billion years ago irreversibly changed the geochemistry of the planet and eventually produced the oxygen-rich world in which we live. Flowering plants have had a pervasive influence on terrestrial ecosystems over the past 65 million years, as have human cumulative cultural evolution and language. But other innovations have been less far-reaching. Is the differing impact of evolutionary innovations a completely contingent phenomenon, depending on the opportunities faced by evolutionary novelties, or can we find any useful logic? The phenomenon of macroevolutionary lags, which has featured so prominently in this account, favors the argument that innovation is largely a result of historical contingency. At this point, our understanding of innovations remains too undeveloped to offer a definitive answer. But I believe that considering innovations in the context of the evolutionary spaces reviewed in chapter 4, and particularly the distinction between search and construction, might suggest that certain classes of innovation may be more impactful than others.

Recall that approaches to novelty and innovation differ in whether they involved search through definable opportunities (Kauffman's "pre-statable space") as the dominant mechanism or the construction of new opportunity spaces. Intermediate between these were alternatives in which evolution occasionally identified new methods of exploring a space, or "lucky leaps." As noted in chapter 4, past discussions of "leaps" have often foundered on failures to recognize that apparent leaps at one level (phenotype) may be achieved by straightforward search at an underlying level (genotype). This divergence was beautifully illustrated with the RNA folding model, which has a clear connection between sequence and the folded, two-dimensional structure. In this case, search through series of single changes in the pre-statable RNA space permits sudden shifts in the folded structure.[10]

The concepts of search and construction operate for both novelty and innovation. Exploring existing opportunities through search involves local optimization of an existing functional form, and this is likely to be the most common form of innovation. Combinatorics extends the reach of search, particularly when already-optimized modular assemblies are involved. New operators for searching or expanding evolutionary spaces as disparate as alternative splicing of proteins and the combinatorics of language increase the range of new combinations and thus the rapidity of search through an evolutionary space.

The construction of new spaces for innovations encompasses new ecological opportunities (hence Simpson's focus on adaptive radiations), such as those allowing access to new environments. Five hundred million years ago during the early Paleozoic, the open ocean was sparsely populated by animals, but over the next few hundred million years a variety of different clades, including fish, nautiloid cephalopods, and others, populated the space. New spaces for cultural and technological innovation range from the most fundamental cumulative cultural evolution and language, and the invention of the process of innovation during the eighteenth and nineteenth centuries, to the construction of various technological spaces. In these cases, spaces associated with the generation of public goods, whether atmospheric oxygen in the case of oxygenic photosynthesis or the multitude of spillovers from culture and language, are often likely to have the greatest long-term impact.[11]

Prospects for a General Theory of Novelty and Innovation

"A fox knows many things, but a hedgehog one important thing." What the ancient Greek poet Archilochus meant in this tiny fragment is anyone's guess, but from these meager foundations the philosopher Sir Isaiah Berlin constructed an essay about Tolstoy's *War and Peace* and the nature of history. Berlin grouped thinkers from Plato and Lucretius to Friedrich Nietzsche as hedgehogs, viewing the world through the lens of a single, encompassing idea. Aristotle, Erasmus, William Shakespeare, and James Joyce were among the foxes, deeply concerned with chance and contingency. Natural selection is one of the most transformative in the history of ideas, but Berlin counted Charles Darwin as a paradigmatic fox, or perhaps a fox who wanted to be a hedgehog. *War and Peace* is a theory of history masquerading as a novel (if you made it through the book, you may remember that near the end Tolstoy abandons his characters entirely for an extended riff on his main interest). In analyzing Tolstoy's theory of history, Berlin used the conceit of the fox and the hedgehog to argue that Tolstoy was a fox captivated by a hedgehog's view of history.[12]

To Berlin's discomfort, distinguishing foxes from hedgehogs quickly became a parlor game for the intelligentsia. It seems easy (too easy, in fact) to
divide those who view historical events as mediated by broad laws, the hedgehogs, from the foxes who recognize contingent outcomes as the fundamental
dynamic of history. Among scientists, theoretical physicists surely nest with
the hedgehogs, while fieldwork-loving biologists, geologists, and archaeologists are happiest among the foxes. Berlin's comparison informed political
scientist Philip Tetlock's brilliant exploration of the accuracy of forecasts,
philosopher Ronald Dworkin outed himself in his argument for the supreme
importance of value in *Justice for Hedgehogs*, while Stephen Jay Gould argued
for the critical role of contingency in the history of life in *Wonderful Life*.[13]

Berlin's parlor game masks a deeper and persistent tension between the
nomothetic versus the idiographic, between searching for general laws and
cataloging contingent complexities. Many scholars have decried idiographic
approaches: in 1892 William James suggested that psychology was "only the
hope of a science," while in another context Berlin criticized historians' failure
to identify general laws, or even "moderately reliable maxims." Geneticist
Theodosius Dobzhansky opened his 1937 contribution to the Modern Synthesis
with a comparison of natural history to the more nomothetic field of genetics
and its quantitative account of evolutionary dynamics. Whereas Dobzhansky
argued that quantitative genetics was the proper foundation of a nomothetic
evolutionary theory, in 1980 Gould argued for a strong nomothetic component
in evolutionary biology, with the ability to quantitatively explore long-term
trends in the history of life. Although he acknowledged the critical contributions of historical, idiographic components of paleontology, Gould argued
that the study of the fossil record "resides in a continuum stretching from idiographic to nomothetic disciplines." Indeed, much of his 2002 opus is an inquiry into nomothetic versus ideographic approaches to the history of life, with
an extended epilogue contrasting his views with Stuart Kauffman's relentlessly
nomothetic approach. Gould concluded that human history would remain
"recalcitrantly idiographic" because it deals with a single species, a claim challenged by archaeologists, economists, sociologists, and some historians seeking
general patterns.[14]

This tension between ideographic versus nomothetic understanding infuses this book as well. Chapter 1 laid out a suite of possibilities for a general
theory of novelty and innovation spanning biological, cultural, and technological domains, ranging from a quantitative, mathematical theory to the
natural history of novelty, invention, and innovation. Two irreconcilable

perspectives are found among hedgehogs: strict Darwinians deny that novelty can be distinguished from adaptation (many of them will have flung this book across the room long before reaching this chapter), and a very different group embrace novelty and innovation as singular features of evolution but argue that a general, quantitative framework is at hand. But claims for a generalized theory of novelty require a very limited definition of novelty, as discussed further below. Foxes, on the other hand, reject claims for a general theory, a view that has been widely adopted, particularly by those writing on invention and the history of technology. In the details one could easily argue that no novelty is really like any other. But the long argument of this book is for conceptual unity in the origins of evolutionary novelty and innovation across domains. One might consider this the views of a fox envious of hedgehogs, unconvinced they have a firm grasp of reality but viewing success as requiring both idiographic and nomothetic approaches. But the other possibilities are hardly false alternatives. Thus, it is worthwhile clarifying my concerns about the other scenarios, particularly as others may favor a different conclusion.

Berlin's hedgehogs will favor a single, deterministic theory covering all three domains. One such theory already exists and is quite well developed, at least within the biological and cultural domains. Quantitative population genetics and allied methods constitute such a theory, albeit one that downplays any need to distinguish novelty from the consequences of natural selection and drift. Gene-culture coevolution, championed by Boyd and Richerson, extended this approach to cultural evolution, building off workers going back to Darwin who have generalized ideas from biological evolution to the social, cultural, and technological realms.

Generalized Darwinism extends variation and selection from biology to psychology, history, economics, and computer science, and in different forms has been championed by psychologist Donald Campbell, philosopher Daniel Dennett, business theorist Eric Beinhocker, and economic historian Joel Mokyr, among others. While this book is very much in this generalizing tradition, I am more interested in models incorporating sources of variation that fuel the path to novelty. Moreover, I have drawn inspiration from Schumpeter and Romer, among others in economics, who provide useful concepts to incorporate into biological thinking.[15]

Much as the strength of population genetics relies on an agnostic view of the gene, some encompassing theories are similarly agnostic about the nature of innovation, applying to whatever one wishes to define as novelty or innovation. While there is a certain logic to this position, it makes it difficult or impossible to examine failed innovations, the contributions of lags, or the sensitivity of innovation to ecological, environmental, or economic context. Moreover, where feedback and decoupling occur between different levels, as between genes, development, and morphology in their interactions with the environment, or in the construction of ecosystems, contingent outcomes preclude successful general theories.

Pre-statable spaces are a more fundamental challenge to efforts to craft a general theory, as is the difference between search of pre-statable spaces versus the construction of a new space. Recall that in pre-statable spaces the full range of a possibility space (genes, morphology, or technology) can be identified independently of any occupation of the space. The universe of all RNA sequences of 100 nucleotides and that of protein sequences of 300 amino acids are pre-statable spaces, as are morphospaces for logarithmically coiled shells. A formal or quantitative model can generally be constructed for pre-statable spaces. But as discussed earlier, these are a fraction of the interesting systems in which novelty and innovation arise.

Novelties and innovations can affect evolutionary spaces in three ways: First, they can facilitate access to previously inaccessible regions of a space. Some regions of protein space are accessible only because heat shock proteins and other chaperone proteins ensure that other proteins fold into their proper three-dimensional geometry. Absent chaperones vast regions of protein space would not yield stable and useful proteins. Second, they can generate new means of exploring evolutionary spaces. Combinatorics is a key example of this, with alternative splicing of proteins allowing evolution to rapidly sample different regions of protein space by recombining already-tested shorter protein domains. Arthur's emphasis on the importance of recombining "primitives" in technological evolution is essentially the same process. Finally, and perhaps most significantly, novelty and innovation have generated entirely new spaces in which biology, culture, and technology can play. The transistor created spaces for a vast range of digital devices. Although the first two types of changes occur in both pre-statable and non-pre-statable spaces, my intuition is that the most interesting sorts of novelties and innovations are those that could not have been foreseen, and thus could not be covered by a formal theory. Intuition is not science, of course, but this seems a promising avenue to prove me wrong.

Of course, novelty and innovation do occur within pre-statable spaces where search can be an effective method for uncovering new opportunities. The critical issue is the relative frequency of novelty via search versus construction, and particularly whether the exploitation of new general-purpose technologies and their biological equivalents is effectively achieved via search.

If contingency is pervasive, then any attempt to identify a general framework for novelty and innovation may be doomed, and the best we can hope for is natural history. The claim that no common threads exist across different domains goes too far, in my view. I have identified commonalities that structure a conceptual framework for evaluating novelty and innovation. Potentiation need not occur before all cases of novelty and innovation, and I have no doubt that some cases of novelty will collapse to key innovations that generate adaptive radiation as proposed by Simpson. Many novelties have not led to innovations, and this does not mean that such novelties were not an evolutionary success. Rather, not all novelties necessarily result in a restructuring of ecological networks. Some novelties are latent, a phenomenon that allows macroevolutionary lags.

This conceptual framework is just the start of the project, however. Conceptual frameworks can be important steps in understanding. Some scientific perspectives become instantiated as mathematical models. But building such models requires specifying interactions among variables. Quantitative models can be incredibly complex: Modern climate models contain over a half a million lines of computer code, and even on the fastest computers they can consume considerable computer time. But even such complex simulations remain a cartoon of reality, albeit often an incredibly useful cartoon. Moreover, whether studied analytically or through simulations, models can readily be tested against empirical data and refined or discarded as necessary.

Other conceptual frameworks, such as plate tectonics, explain how the world works without a mathematical framework. The view I have developed here is that pursuing a single model or suite of interrelated models that span biological, cultural, and technological domains is unlikely to be fruitful. It does not follow from this conclusion that modeling and simulations cannot provide significant insights, however. Climate models cannot predict whether it will be raining a month from now in Mid-Coast, Maine. I can look up when the tides will be right for kayaking along Penobscot Bay, but I cannot know this far in advance whether doing so would be enjoyable. The utility of models, including simulations, depends upon the resolution of the model results and the questions asked, so I anticipate that properly constructed models could explore the dynamics of various types of novelty and innovation, yielding insights beyond a compilation of case studies.

Two final issues are of primary interest to philosophers of science: theory reduction and philosophy of practice. Theory integration, or theory reduction, was once a goal of some scientists and philosophers, particularly the reduction of chemistry, biology, and even geology (!) to physics as *the* goal of science. Grounding biological processes in chemical or physical principles represents one avenue of theory reduction, while cognitive science seeks to integrate neuroscience, linguistics, psychology, and philosophy into a more rigorous understanding of cognitive function. Evolutionary biology increasingly serves this integrating role across organismic biology. This project is certainly motivated as an inquiry into the possibility of conceptual unification across the domains of biological, cultural, and technological innovation, but the conclusions outlined here firmly reject the possibility of theory reduction. Philosophers Alan Love and Ingo Brigandt embraced a range of definitions of novelty, consistent with their interest in the practice of science: how scientists work, rather than prescribing how scientists should work. Some view such polyvalency as a strength, reflecting the diversity of ways that the terms *novelty* and *innovation* are currently used. Unlike Love and Brigandt, however, this project focuses on the lessons about novelty and innovation that can be drawn from a broad comparative study, and thus common definitions of these terms across domains is essential.[16]

Is Evolution Open-Ended?

If the environment stopped changing, would evolution come to a halt? In 1973 evolutionary biologist Leigh Van Valen took data from fossil vertebrates to argue that the extinction probability of a genus or family was independent of its age: at any point in time, older and younger taxa were equally likely to become extinct. Under Van Valen's hypothesis species are constantly adjusting and adapting to other species and to changes in their environment, just as the Red Queen in Lewis Carroll's *Through the Looking Glass* said: "It takes all the running you can do, to keep in the same place." Among the questions sparked by the Red Queen hypothesis is whether evolutionary novelties and innovations are required to generate evolutionary change, or whether, in the absence of environmental change, further novelty and innovation would die out, leaving a relatively static configuration of species and ecological interactions. The Red Queen hypothesis helps frame two broader questions that have puzzled evolutionary biologists since Darwin: Is evolution open-ended or unbounded, and is there a sense in which evolution leads to progress? Similar questions about economic growth have long concerned economists, but many have reached very different conclusions from biologists. Over the past several decades, anthropologists and others concerned with cultural evolution have largely declined to engage with such questions because of the long history of using them to advanced narratives of Western, Eurocentric superiority. This section engages with the related issues of growth, progress, evolvability, and determinism versus contingency, before returning to open-endedness of evolution.[17]

Growth and Progress

Individuals grow, as do populations, cities, and national economies. Biologists measure growth in the number of cells, by the number and abundance of species in ecological communities, and by the amount of carbon fixed by plants available to other organisms (net primary productivity). Economists see invention and innovation as an essential component of economic growth. This growth comes in two flavors, "Smithian" and "Schumpeterian." Smithian growth, named for Adam Smith, results from trade, increased specialization, and better protection of property rights, and is characterized by institutional changes and diminishing returns over time as resources become limited. Absent a change in the system, the rate of Smithian growth will eventually decline as trade, property rights, and other institutions become efficient, and resources are effectively allocated.

Schumpeterian growth derives from increases in useful knowledge from discoveries in science and technology, new products, and improved production techniques. Some of those exploring Schumpeterian growth have described it in evolutionary terms. At least in principle, Schumpeterian growth is not subject to diminishing returns over time unless there is an ultimate limit to scientific and technological knowledge. But this form of growth does require the destruction of previous innovations to free resources for new innovations. Modern economies are commonly driven by both types of growth, although archaeologists have argued that Smithian growth may have predominated in the distant past, while economic historians Joel Mokyr and David Landes, among others, have argued that the explosion of economic growth that began with the Industrial Revolution was dominated by new technology.[18] Philosopher Alfred North Whitehead captured the essence of the Industrial Revolution in remarking:

> The greatest invention of the Nineteenth Century was the invention of the method of invention. A new method entered into life. To understand our

epoch, we can neglect all the details of change, such as railways, telegraphs, synthetic dyes. We must concentrate on the method in itself; that is the real novelty, which has broken up the foundations of the old civilization.[19]

Each quarter, the Bureau of Economic Analysis of the US Treasury releases estimates of the gross domestic product (GDP), the most widely reported measurement of economic activity, as do the financial ministries of other well-run countries. Measures of GDP incorporate the total value of all finished goods and services, added to an economy by private consumption and investment, and government spending. Since economist John Maynard Keynes established the modern definition of GDP during World War II, this measure has become the standard metric of economic production and prosperity. Comparisons of GDP estimates among countries serve as markers of differing levels of prosperity, and GDP has been widely used to structure economic advice for developing countries. Yet the utility of a single number as a metric of growth and prosperity has been challenged almost since its inception. Among other problems, GDP does a poor job of integrating technological innovation. Consider light. Since 1800 we have shifted how we light our homes and offices from candles to whale-oil lamps and kerosene to carbon filament, then tungsten filament bulbs and now compact fluorescent lights and light-emitting diodes. Economist William Nordhaus showed that modern lights are orders of magnitude more efficient than their predecessors and can be acquired for a small fraction of the labor cost than was the case two centuries ago. From the sesame oil lamps of ancient Babylon to nineteenth-century candles, lighting efficiency increased by 0.04 percent per year. Since then, progressively more sophisticated technologies have resulted in a 900 percent improvement in efficiency, or 3.6 percent per year. But despite these innovations that provided lighting at lower cost and far greater efficiency, using the standard means of calculating GDP the cost of lighting increased by three to five times. When I grew up in Los Angeles during the 1960s, we had maybe a dozen television channels. Today hundreds of cable channels are available via high-definition digital signals. Similarly, digital cameras in smartphones have replaced film cameras. Yet little of this change is captured by standard economic metrics such as GDP. But pity the economists, for comparing candles to compact fluorescent lights is far easier than comparing a horse to a subway for commuting, fountain pens and paper to email, or a battle ax to a battle tank.[20]

Biologists and anthropologists generally abjure claims of progress. Into the 1980s, claims of progress, as in culture, tended to view Western human culture as the apogee of evolutionary development. Repeated mass extinctions and other biotic crises over the past half-billion years have led paleontologists to question any claims of long-term progress in the history of life. Certainly, many clades have disappeared (trilobites, ammonoids, and non-avian dinosaurs, for example), but one can argue that these disappearances were due to environmental crises that pruned fauna and flora without apparent regard to Darwinian selection. But many of us harbor an intuition that there has been progress in the history of life and in the course of human affairs. Improvements in human technology have increased standards of living over the past few centuries. The transformation in China since the 1980s is unparalleled in human history. The traditional metric for these economic changes is per capita GDP (all of these have come at considerable and now irreparable environmental cost, however).

In chapter 5 I described several views of progress in the history of life. The major evolutionary transitions, from the origins of life through human cumulative cultural evolution and language, focuses on repackaging of information to generate novel evolutionary individuals and new opportunities. Some have viewed these evolutionary transitions as steps toward greater complexity, which could be interpreted as a sort of progress. I also presented Vermeij's argument that successful lineages are not simply lucky, but have greater power (roughly, the ability to acquire and utilize more energy per unit time), which is similar to anthropologist Leslie White's views of human societies. Although I have discussed both energy and information many times, my approach is open to criticism for not making either of these the primary focus. The acquisition of energy has certainly been an important attainment of some novelties and innovations, but as I discussed earlier, I think this is too narrow a focus. Information could be deployed in an alternative narrative about novelty and innovation, by developing metrics of information content. Statistical mechanics has a wealth of tools to support such an approach. My principal reluctance to pursue either of these avenues, however, is that in most cases they tend toward an inherently progressivist view of the history of life.

Among the many challenges in rigorously addressing the issue of progress is the difficulty of distinguishing between an expansion of an evolutionary opportunity space from a shift in the mean. Paleontologists Andrew Knoll and Richard Bambach took an ecological approach with their megatrajectories, and Dan McShea has spent much of his career investigating patterns of increased biological complexity. Both scenarios are more rigorous than the Great Chain of Being described in chapter 2, yet they share some of the same progressivist outlook. McShea articulated this distinction as the difference between passive and driven trends. The expansion of evolutionary opportunity over the past three billion years has been partly through the generation of entirely new evolutionary spaces. The eukaryotic cell, various types of complex multicellularity, cultural evolution, and human cumulative culture are among the most obvious examples of such new evolutionary spaces.

One might call this progress, but I hesitate for two reasons. First, the idea of progress has an inherent valence of a linear Great Chain of Being, which seems all too convenient. If we used the Eocene-Oligocene boundary as the cutoff, removing humans from the scenario, would we also claim evidence for progress? Yes, if our metric was the number of cell or tissue types, information (DNA content), energy flow, or ecological complexity, but these metrics are too easily conflated with expanding evolutionary spaces. A second concern arises from the evidence for contingency in both novelty and innovation. The idea of progress often assumes a driven evolutionary trend. Yet if novelties are often contingent upon prior potentiation, and if macroevolutionary lags intervene before equally contingent ecological and evolutionary changes deliver successful innovations, then the role of contingency seems too pervasive to support claims of a driven trend. Claims for evolutionary progress may be best grounded in performance metrics within clades, as in Nordhaus's analysis of improvement in light. Novelty still plays a role, but there is perhaps a greater role for adaptive evolutionary change. Paradoxically, novelty may play little role in the most rigorous scenario for establishing a sense of progress.

Evolvability

Tardigrades are tiny animals with eight pairs of stubby legs and a smashed-in face. Most are less than a millimeter long, and the 100-plus modern species are found from Antarctica to tropical forests. Little of the half-billion years of genome evolution since the appearance of the clade has translated into obvious morphological change, and Cambrian tardigrades are almost indistinguishable from modern ones. Compare tardigrades to their cousins the arthropods. Arthropods include millions of species from crabs to insects thriving in every imaginable environment on Earth. Similar divergences in evolutionary scope are found when comparing horseshoe crabs to other clades of chelicerate arthropods (for horseshoe crabs are not crabs), between sharks and bony fish, or between animals as a whole and their close cousins (the Holozoa). Such differences in evolutionary history between sister clades can be explained by the dynamics of mutation, variation, and development of the lineages, and in differing responses to evolutionary opportunity. An alternative avenue of explanation, known as evolvability, focuses on differences in the generation of available variation. Using the ideas of evolutionary spaces from chapter 4, evolvability reflects how much of a phenotypic space is available over evolutionary time.

Evolvability is the capacity of different systems to generate useful variation upon which selection can act. Population geneticists, evolutionary developmental biologists, and paleontologists differ in how they define and analyze evolvability. Paleontologist David Jablonski defines evolvability as "the differential (phenotypic) ability to take advantage of, or respond to, opportunity," while others focus on an adaptive result. Although living organisms provide incomparable insight into the genetic basis of evolvability, the fossil record is at least as important. Consider tuataras. This New Zealand "lizard" is the closest living relative of snakes and lizards, which number about 8000 living species. One might assume that tuataras, the sole species within the order Rhynchocephalia, have low evolvability. But living tuataras are a remnant of a once-proud clade of Mesozoic reptiles with two dozen or so genera and a wide range of feeding strategies. So, comparing the tuatara to lizards and snakes provides a very misleading view of the evolvability of the squamates versus rhynchocephalians. More broadly, the evolvability of any clade can be assessed only with a well-studied fossil record and a detailed phylogeny, because the current context may provide only limited insight.[21]

More striking than contrasts in evolvability between different clades are shifts in evolvability within a single clade. Today the echinoids within the phylum Echinodermata encompass not just sea urchins but also sand dollars and heart urchins. These might seem variations on a theme of spines surrounding a sphere of sutured plates, but what is remarkable is the plethora of modern echinoids relative to the paltry variety of Paleozoic sea urchins. One effect of the Permo-Triassic mass extinction 251 million years ago was an increased developmental capacity for the clade. Although not yet analyzed quantitatively, a similar pattern is found among marine gastropods (snails). Paleozoic gastropods had a great variety of shell forms, but relatively few were predators, and they lacked the spines, knobs, and elaborate protuberances that decorate post-Paleozoic gastropods. As with echinoids, gastropod evolvability increased after the Permian.[22]

Enhanced modularity is linked to evolvability. The many morphological modules of arthropods (segments, appendages) facilitate independent novelties and adaptations. The wings, treehopper helmets, and beetle horns discussed in chapter 8 illustrate how the morphological and developmental modularity of arthropods leads to enhanced evolvability. The independence of arthropod segments permits flexible responses to differing evolutionary opportunities. The evolvability of arthropods is further enhanced by differentiation between larval and adult phases of the life cycle, in which different phases can adopt distinct adaptations, pursue different feeding and ecological strategies, and acquire novelties. In chapter 6 we explored the combinatoric structure of enhancers in gene regulatory networks and their role in co-option of subcircuits to generate novelty. Beyond gene regulatory networks, modularity provides an avenue toward combinatoric novelty.

The generation of modularity is only one facet of the connections between novelty and evolvability. Potentiation represents a form of evolvability that increases variability that may lead to novelty *if* the specific mutations required for the novelty later arise. It only increases the likelihood of subsequent novelty; it does not guarantee that it will arise. Sister clades with the same potentiation may differ in whether they succeed or fail to acquire the mutations needed to construct novelties. Simpson's connection between novelty and invasion of new adaptive zones was a recognition that novelties can enable new avenues of variation (although Simpson did not use the term *evolvability*). In addition, successful novelties often enhance evolvability of the clade to which they belong. Enhancers again serve as a useful example, for their widespread deployment in bilaterian clades has allowed reuse and co-option of genes, keeping genome sizes low while permitting remarkable diversity in development, morphology, and behavior. I suspect that this work has only begun to explore the relationships among novelty, innovation, and evolvability.[23]

Contingency and Determinism

In his book *Wonderful Life*, Stephen Jay Gould deployed the exquisitely preserved animal fossils of the Cambrian Burgess Shale (chapter 8) to illuminate his argument that at many points the history of life could have taken a very different path. Today the Earth might have been dominated by intelligent crabs. Recall that the fossils of the Burgess Shale preserved a wealth of anatomical detail normally hidden to paleontologists, including the brain, nerves, and beautifully preserved eyes. The fossils reveal architectures that soon vanished from the fossil record. *Opabinia* is one of these iconic "weird wonders": about five inches long, with a long proboscis ending in a claw extending forward of the mouth, and possibly five eyes, it clearly belongs to the Panarthropoda but otherwise looks like nothing else in the history of life. Priapulids are another example: carnivorous, burrowing worms that greatly outnumbered annelid worms in the Cambrian but are relatively insignificant in most ecosystems today. Gould argued that other clades might have had greater evolutionary success during the Cambrian: lobopodians instead of arthropods, and priapulids instead of annelids, and the history of the past 500 million years could have looked much different. Why did annelids triumph while priapulids became insignificant? Why did the giant (for the Cambrian) predator *Anomalocaris* disappear along with *Opabinia*, marine lobopodians like *Aysheaia*, and many others? As Gould put it in *Wonderful Life*, if we "replay[ed] the tape of life again," would history have been different?[24]

Gould's argument for contingency was challenged immediately. Some paleontologists questioned the pattern of contingent decimation he identified (leading to the development of the quantitative methods of characterizing disparity mentioned in chapter 4). Biologists rejected the importance of macroevolutionary patterns and contingency as significant factors in evolution. And philosophers critiqued the rigor of Gould's claims. Philosopher John Beatty pointed out that Gould used the term *contingency* in different ways. Beatty and other philosophers identified at least five different types of contingency and have explored broader issues of chance and contingency in the history of life. Phylogenetic methods to reconstruct evolutionary relationships have found homes for most of the "weird wonders" Gould discussed. *Opabinia* and *Anomalocaris*, for example, are cousins within the Radiodonta and allied to true arthropods within the broader panarthropod clade.[25]

Any historical process involves some path dependency, so necessarily involves contingency, whether biological evolution, the cultural change of the ups and downs of technology, or the behavior of galaxies. In path dependency, initial conditions and points along a trajectory help determine the outcome. Many different paths may be accessible from any point in history. In the context of evolutionary spaces, this means that multiple alternative trajectories thorough a space may be possible depending on selection, environmental events, or chance. Some of these paths may lead to the same outcome (equivalent to a basin of attraction for complex systems), while others will take a lineage along different evolutionary trajectories. For Gould, in some alternative Phanerozoic history arthropods do not become successful on land and sea, and other bilaterian clades dominate modern ecosystems.[26]

Those opposed to Gould's claims of contingency have invoked the frequency of convergence as evidence that the trajectories through evolutionary spaces are limited, and even when different paths are available, they will lead to a similar outcome. With this argument that many aspects of evolution are predictable, arthropods and vertebrates always win, mollusks always displace brachiopods, and the survival of a few species of echinoids through the end-Permian mass extinction to become dominant predators was not fortuitous but fate. Moreover, life elsewhere in the universe must share many of the same characteristics we see around us, including brains at the front of the body, bilaterality, and even intelligence. It is certainly true that physical and biological factors can limit and even direct the evolutionary opportunities of very different clades, as is the case with sharks and other open-ocean predators, and venoms, but is convergence widespread across distantly related clades? If so, this would be an important and fascinating discovery about the nature of evolutionary spaces, for it would reveal that spaces of different clades are somehow deeply entangled.[27]

Sensitive dependence on initial conditions and degree of subsequent unpredictability represent two major aspects of Gould's contingency. Separating them reveals four domains (figure 10.1). Gould's domain is in the lower right, while the domain of physics is in the upper left, where predictable outcomes enable laws and generalizations. Isiah Berlin's hedgehogs imagine they live in a world in the upper left, while the foxes belong in the lower right. In my view, Conway Morris and Vermeij believe that the history of life largely lies in the two domains in the upper half of the figure, with little sensitivity to initial conditions generating highly deterministic or mixed histories but little contingency.[28]

This argument over convergence and determinism is critical to understanding novelty and innovation, for if Conway Morris and McGhee are broadly correct, then the patterns described in the past few chapters, while interesting, reveal little about evolution. Extensive convergence can only mean that the scope of evolution is limited, and that the creativity of novelty and the opportunities for innovation are constrained at all levels, from DNA sequences to morphology and behavior. Arguments that evolutionary novelty is essentially a process of search among a library of potential solutions to evolutionary problems are similarly deterministic. In most arguments for the primacy of search, initial starting conditions are important when they limit the accessibility of some regions of a possibility space, but there is generally little role for contingency.

	Low unpredictability	High unpredictability
Low sensitivity to initial conditions	Much of physics Highly deterministic	Mixed
High sensitivity to initial conditions	Mixed	Gould Highly contingent

FIGURE 10.1. Sensitivity to initial conditions and unpredictability allow discrimination of different domains of contingent and deterministic behavior. After Erwin (2015c).

Similar limits to technological and intellectual creativity have been inferred from patterns of simultaneous discoveries. On June 18, 1858, Darwin received a letter from Alfred Wallace conveying his manuscript about natural selection. This is just one of hundreds of inventions and discoveries that occurred independently yet nearly simultaneously. Patterns of simultaneous discovery fascinate historians and sociologists for what they might reveal about the dynamics of human creativity. In chapter 2 I described William Ogburn and Dorothy Thomas's catalog of nearly-simultaneous inventions. Ogburn and Thomas view their compilation as a vast underestimate of the frequency of independent innovations.

Two interpretations of such accounts of convergence have been advanced, whether of biology or technology. One possibility is that the space of evolutionary options is limited, and many inventions must be largely inevitable. Thus, the independent origin of states in China, Egypt, India, Mesopotamia, Mexico, and Peru might support the argument that the range of human political institutions is small. Ogburn and Thomas favored an alternative explanation, arguing that cultural and technological discoveries reflect the cultural settings of their time: "And the elements of the material culture at any one time have a good deal to do with determining the nature of the particular inventions that are made." In other words, the environmental setting (biological, cultural, economic, technological, or intellectual) generates the conditions under which novelties or innovations become possible. Far from being a sign of limited evolutionary opportunity, simultaneous discoveries document the vital role of potentiation.[29]

In my view, Ogburn and Thomas's evaluation of simultaneous discoveries provides a salient resolution to the ongoing conversation over the roles of contingency and determinism. Their view of history seems to have been one that was largely contingent until certain prerequisites had been met (material culture), at which point the probability of a suite of discoveries increases, becoming more deterministic. I hope I am not imputing too much to their work, but it seems highly congruent with my argument: potentiation helps generate the possibilities for novelty, but in contrast to adaptive radiation scenarios there is no cause-and-effect relationship between a "key innovation" and occurrence of a novelty, or of the success of the novelty. Macroevolutionary lags may occur between novelty and innovation when the biological, cultural, or technological conditions are not yet sufficient.

Endless Possibilities?

Will novelty and innovation eventually come to an end, gradually dribbling away into some stable constellation of diversity that will suffer gradual attrition through extinction? Or are evolutionary possibilities so endless that both novelty and innovation will continue indefinitely? The contingent nature of evolution and the infinite spaces for genomes and proteins discussed in chapter 4 suggest that it would take billions of years to sample even a reasonable fraction of most evolutionary spaces. But recall that constraints and degeneracy require that many different genotypes will produce the same or similar protein structures, and the requirements of two- and three-dimensional protein folding further limit the number of actual proteins. The space of possible proteins is vast, but the array of venoms is limited because the venoms must be effective in disrupting muscle or nervous system activity. While questions of contingency and determinism are related to those about the open-ended nature of evolution, there is a subtle difference as well. Many-to-one relationships between hierarchical levels raise the possibility that evolution may be open-ended at the level of genomes and proteins, for example, but severely limited for morphology or behavior. This eliminates any simple answer to the question of whether evolution is open-ended.

Economists have differed over whether growth is open-ended or whether increasing resource limitation will force growth to stagnate, with detrimental consequences for those in the poorest countries and the less fortunate members of rich countries. Were technological innovation to come to a halt, so too would economic growth, and with it improving living standards for hundreds of millions of people. For support of these ideas, we need look no further than the remarkable changes in China since Deng Xiaoping began economic modernization in 1978. Between 1978 and 2005 China's economy grew at an estimated rate of 10 percent annually. Statistics from the World Bank reveal that between 1990 and 2016, the number of Chinese living in poverty plunged by about 750 million people, two-thirds of the population. Such a transformation in human well-being is unprecedented in human history. Similar economic transitions in the West in the wake of the Industrial Revolution reduced infant mortality, childhood diseases, and malnutrition. Steven Pinker of Harvard published two well-received if controversial books that charted declining violence, both as wars between states and as individual violence, and increasing prosperity over the past several centuries. From such data many have drawn the conclusion that progress in human well-being is tied to economic growth, and economic growth is closely tied to innovation. Indeed, although the term *innovation* does not appear in the index of Pinker's 2018 book *Enlightenment Now*, the volume is a paean to the role of knowledge in driving human progress.[30]

Evolution may appear limited within restricted regions of an evolutionary space when trajectories are constrained by adaptation and innovations by search rather than construction. But if many evolutionary spaces are manifolds, where the local region is Euclidean but becomes increasingly less Euclidean with distance, the appearance of determinism is dependent upon the local domain. If some novelties and innovations construct new evolutionary opportunities, then it seems likely that evolution is open-ended rather than constrained. But that is supposition, and this is another issue that will reward further consideration by biologists, philosophers, and students of cultural evolution.

Evolution Evolving

One of my favorite cartoons shows Leonardo de Vinci struck by inspiration and building the first personal computer. It's an old cartoon from the 1980s, so the monitor is a big, boxy thing. The cartoon works because Leonardo could have designed a personal computer. Why not? He was an endlessly creative inventor, a brilliant artist, a contributor to what today are at least a dozen different disciplines, while inventing technologies well ahead of their time, including a helicopter. But in the last frame of the cartoon Leonardo is looking around for an electrical socket to plug in the computer. Only in the nineteenth and twentieth centuries has our technological and economic capacity, our product space, caught up with Leonardo's inventive mind. To me this cartoon perfectly captures how novelties can change possible futures.

Since Charles Lyell's *Principles of Geology* first appeared in 1830, generations of geology students have parroted his mantra, "The present is the key to the past," a claim incorporating two ideas: First, that we can understand and explain the geological history of the Earth based on the processes occurring today without recourse to the catastrophic explanations favored by Lyell's contemporary Georges Cuvier (see chapter 2). The second claim, also due to Lyell, is that these processes operated in the past at the same rates observed today. Uniformity of rate and process. Lyell summarized his argument in a letter to a fellow geologist:

> [The book] will not pretend to give an abstract of all that is known in Geology, but will endeavour to establish the principles of reasoning in the science, & all my Geology will come in as illustration of my views of those principles . . . [that] no causes whatever have, from the earliest time to which we can look back, to the present, ever acted but those now acting, & that they never acted with different degrees of energy from that which they now exert.[31]

When Louis Alvarez and colleagues proposed that the end-Cretaceous mass extinction was cause by the collision of a bolide with the Earth, many geologists dismissed it because such a catastrophic explanation conflicted with the uniformitarianism we had been taught since our first geology courses. (Alvarez was a physicist, although his colleagues included several geologists.) Today geologists acknowledge the limitations of Lyell's uniformitarianism, but the same assumptions underlie other fields. Biologists face a uniformitarian conundrum over the applicability of modern experimental studies throughout the history of life. Indeed, uniformitarian assumptions are so deeply ingrained in biology that few have questioned them. When the eminent biologist John Tyler Bonner challenged this assumption, he admitted: "I published the beginning of these thoughts in two places . . . expecting to be burnt at the stake, but silence was the only response." John Maynard Smith and Eörs Szathmáry recognized that their major evolutionary transitions represented nonuniformitarian transitions in the evolutionary process, and Rob Boyd, Kim Sterelny, and Michael Tomasello have each discussed how cultural evolution has changed over human evolution.[32]

That the evolutionary process has evolved over time is indisputable, providing a nonuniformitarian component to evolutionary dynamics. But there are at least four distinct, but nonexclusive, modes in which evolution could evolve: the expansion of opportunity spaces; new levels of selection and the origin of new evolutionary individuals, as envisioned by Maynard Smith and Szathmáry; new types of inheritance, including niche construction and cultural inheritance; and systematic changes in the efficacy of recombination, selection, and drift.

First, the portions of an evolutionary space that can be occupied may change over time. As the Earth evolved from a state with essentially zero oxygen in the atmosphere and oceans before 2.4 billion years ago, through a low-oxygen and likely low-productivity interval from 2 billion years ago to perhaps 800 million years ago, followed by increasing oxygenation of the oceans and atmosphere, the evolutionary opportunities have expanded manyfold. The opportunity space for evolution successively expanded again as plants, insects, and some vertebrates, mollusks, and annelids moved onto land, with the rise in flowering plants from the Cretaceous, and with cumulative cultural and technological evolution. Opportunity spaces have contracted as well, most notably with the collapse of ecosystems during mass extinction events. The part of arthropod morphospace occupied by trilobites was largely abandoned in the early Paleozoic and then slowly dwindled away until the last trilobite went extinct in the late Permian. But do such patterns reflect systematic differences in the probability of novelty and innovation, or different types of novelty, through the history of life?

A similar pattern can be traced in economic growth, particularly since the onset of the Industrial Revolution. In both cases, progressive changes in conditions opened new possibilities for evolutionary novelties or cultural invention, and opportunities for exploitation by innovations. If the capacity to generate novelties and inventions or the opportunity for them to be successful as novelties is controlled by features of the environment, then changes in Earth's habitability may serve as a primary control. Although increased oxygenation is a long-term, directional trend, episodic events such as mass extinctions and their aftermath could also regulate novelty and innovation.

Evolution also can evolve through the intrinsic generation of new potential or capacity. While the previous mode focused on extrinsic, environmentally driven changes in the possibility space, chapter 6 investigated whether genetic and developmental processes could affect the ability to generate novelties, including sexual recombination with the origin of eukaryotes, effective symbiosis, distal enhancers, and modular structures in gene regulatory networks that increase the possibility of effective co-option.

Second, evolutionary biologists have recognized that selection can act at multiple levels. Darwin's natural selection influences individuals within populations, shifting the mean of their fitness, but Darwin also recognized that selection could operate on sexual attributes. Selection requires variation and heritable differences in fitness. At least in principle, selection can operate on a variety of levels beyond individual selection, from cells and chromosomes to species and clades. One of the generating ideas for work on the major evolutionary transitions was recognition that selection could operate within individuals, as between the eukaryotic host cell and the symbiotic mitochondrion and chloroplast, leading to conflict. This conflict was at least partially ameliorated by establishing new evolutionary individuals with selection largely focused on that level.

Altruistic behavior is an additional puzzle and led to the idea that groups could be selected. In the 1960s theorist William Hamilton convincingly showed that kin selection was a more plausible explanation for altruistic behavior than selection on unrelated individuals in a group. Although kin selection is now widely accepted, issues of cooperation and cheating persist as challenges to claims of selection at other levels. Proposals for selection on multiple levels, including cultural evolution, recognize that the nature of the evolutionary process changed when new evolutionary individuals were formed. I know of no systematic analysis of changes in the number of potential levels of selection through the history of life, but simple inspection reveals that they have increased. Thus, a corollary of multilevel selection is implicit recognition that evolution has itself evolved, although this may not have been apparent to many advocates of such ideas (although it was explicitly noted by Maynard Smith and Szathmáry). This is not the venue for adjudicating the multilevel selection debate. I simply note that to the extent that selection *does* operate on multiple levels, this may represent one avenue of change in evolutionary dynamics.

New types of inheritance represent a third mode by which evolution has evolved. Niche construction generates structures that persist for long periods of time and thus influence the selective regime. Ecological inheritances produced by niche construction or ecological engineering can be traced back at least two billion years to increased atmospheric oxygen and some of the earliest microbial reefs, although their extent may be more pervasive in the past half-billion years. Cumulative cultural inheritance adds interesting complexities to evolutionary theory. In both cases nongenetic changes convey information between generations independently of the intergenerational transmission via DNA. Although seemingly more limited in duration, epigenetic inheritance of modification of DNA via sequence-specific methylation further expands the types of inheritance. Niche construction and cumulative cultural inheritance are of particular interest because of their potential to generate public goods that can spill over to influence other species. As novelties have aided in the establishment of new mechanisms of inheritance, the dynamics of evolution are modified. The cumulative generation of new modes of inheritance has increased opportunities for both novelty and innovation, depending on the type of inheritance.

Finally, systematic changes are possible in the efficacy of primary evolutionary mechanisms such as recombination, selection, and drift. The efficacy of selection and drift are closely related to population size. Many species of plants and animals have relatively small population sizes or have passed through restrictive bottlenecks during their history (during biodiversity crises, for example). Low effective population sizes increase the possibility that genetic, developmental, and morphological attributes may become fixed within populations by drift alone. Geneticist Michael Lynch has suggested that significant developmental innovations such as the adoption of Hox genes for anterior-posterior body plan formation in animals may have arisen through drift rather than selection.[33]

Horizontal gene transfer plays a major role in archaeal and eubacterial lineages, allowing the spread of beneficial genes and retarding the buildup of harmful mutations (known as Muller's ratchet). Eukaryotes adopted sexual recombination to the same end, and a recent analysis concluded that the larger genome size and increases in the number of repeated sequences among early eukaryotes may have limited the ability of horizontal gene transfer to purge deleterious mutations. This shift may also have permitted the growth in regulatory complexity in eukaryotes and the expansion of morphological and ecological diversity.[34]

The extent to which these four different modes have expanded the scope of evolutionary dynamics and their impact on novelty and innovation are ripe for future work. Have these added to the complexity of life? Have they made novelties and innovations more likely, particularly what I have described as "general-purpose technologies"?

Epilogue

ROBERT GORDON'S 2016 book *The Rise and Fall of American Growth* is sobering. With a relentless stream of data and analysis, Gordon contrasts the transformative innovations between 1870 and 1940 that improved the quality of life of most Americans, including electricity, urban sanitation, chemicals and pharmaceuticals, the internal combustion engine, and modern communications, with the changes of the past few decades. Electric lights, flush toilets, automobiles, and the telephone replaced candles, outhouses, horses, and mail (except, as Gordon points out, in many parts of the rural South). These technologies were based on inventions of the late nineteenth century, followed by refinement and exploitation during the twentieth century, and their diffusion across most of the country. Gordon could identify relatively little impact on productivity from the information technology "revolution" of the late twentieth century. Personal computers, smartphones, and the internet have been significant, but less transformative than earlier changes (how extensively artificial intelligence will challenge Gordon's claims is, of course, unknown). For Gordon, the pulse of growth from 1870 to 1940 may have been a one-time event, because these changes eliminated the grinding, back-breaking work of earlier generations and introduced far easier conditions at work and at home.

Gordon's argument is a tonic against rampant "techno-optimism." The naïve sense that the next big new technology is just around the corner is part of our national DNA. Abundant clean energy from fusion power has been just 10 years away since I was about 10 years old. And still is. Thus, the view that the innovation party is largely spent challenges deeply held views about progress. If, as Gordon suggests, living standards are likely to be stagnant for the coming decades, this raises some troubling questions for social and political dynamics. But I am a geologist, not an economist, so my time horizon is broader. Even if Gordon is correct about conditions in the United States, economic growth may be driven by changes elsewhere. From the 1960s onward, economic growth in Europe was driven not by indigenous innovation, but by the importation of new ideas, processes, and products from the United States. After the pandemic, concern has turned back to how to balance cooperation and competition between two economies both driven by invention and innovation.[1]

Yet Gordon's vision might be too optimistic. Recent work by economist Chip Jones begins with analyses suggesting that human population growth may soon turn negative. The steadily dropping fertility rates over the past half-century have reached a point where in many countries the rate is only slightly above the replacement rate (which would yield a steady population). A lower human population would have undoubted benefits in reduced environmental damage, but Jones explores a different impact: a drop in creativity and innovation. If the generation of new ideas is, at least in part, a function of population size, a large drop in human population could lead to a similar decline in the flow of new ideas, and thus in innovation and economic growth, leading to a stagnant economy. There are countervailing forces, not least of which is that an increased focus on education, particularly in Africa, could increase the supply of ideas even as global population declines.[2]

This book is not a fully articulated theory of novelty and innovation, because I do not believe that such a theory is possible, unlike the theory of adaptive radiations, infinitely malleable to cover all possible situations, or theories of novelty that require pre-statable spaces but downplay the interesting ways evolution has found to construct new evolutionary spaces. My hope is that this exploration of potentiation, novelty, and innovation, the relative impact of search versus construction, and related ideas will expand the conceptual space for such problems.

One thread through this book has been connections between biological innovation and issues of economic growth and technological innovation. But my goal was emphatically not to identify "insights" from evolution that might be useful for the business community, or for society more generally. The question of commonalities among biological, cultural, and technological dimensions of novelty and innovation is intriguing. Yet it should be obvious that the conclusions of this book are unlikely to generate a TED talk or discussion among "thought leaders." But I will indulge myself in some final observations.

First, Schumpeter was right: novelty (or invention) is a different problem from innovation. Tackling novelty and innovation as distinct problems is essential to understanding their mechanisms. This is just as true for understanding the history of life as it is for understanding technology. Making a new type of toothpaste is no more a novelty than is a change in coloration of the wing of a fly. As is evident from the general conceptual model of potentiation, novelty, adaptation, and innovation presented here, adaptive change (whether in biology, culture, or technology) is a vital component, but empirical results have convinced me that the small-scale adaptive changes championed by the transformationists discussed in chapter 2 are insufficient to explain the pace and nature of novelty and innovation. Patent records can provide insights into invention but precious little information about innovation. Solutions to improve the flow of inventions, for example, will not necessarily translate into the economic growth generated by innovation.

Second, I have argued that no single definition of novelty is sufficient, yet individuation of new features provides a unifying framework that allows us to bound the problems of interest in a useful way. At the same time, I have resisted the temptation to claim that all novelties are the same. Returning to the theme of evolutionary spaces, novelties that expand the space of possibilities potentially have greater impact than those allowing exploration of an existing space. Chapters 7 and 8 identify public goods and niche construction as key components of the success of innovations. In a sense, the most generative novelties are those that induce positive feedback to advance their own success and the success of follow-on novelties and innovations. I identified those novelties that generate public goods as among the most successful examples available. This creates a challenge for business but an opportunity for the public sphere, as the most useful public goods, by definition, are those that are most difficult for private enterprise to monetize. Whether the interstate highway network in the United States or the internet, public goods provide necessary potentiation for economic growth; they rarely specify the nature of the innovations that will follow. Indeed, many public goods are agnostic about further developments, building a possibility space for a wide range of potential innovations.

This brings me to my third observation, that the distinction between novelty and innovation and the range of driving factors provide multiple points of leverage. Searching for novelty within an existing product space may not be trivially easy, but it is generally more straightforward than generating a new space or envisioning a new means of searching a space. Recall that the invention of distal enhancers at the origin of animals appears to have been a critical novelty in the structure of developmental gene regulation. Distal enhancers greatly expanded the developmental possibilities for animals, and thus the range of morphologies they could explore and the complexity of ecological interactions. But as far as we know, this was a single novelty for animals. Schumpeter's argument against the importance of invention (stringing together horse-drawn coaches not producing a railroad) is worth remembering, but I have argued here that both novelty and innovation may be important, even if the relationship between the two is neither linear nor deterministic.

Finally, generation of novelty and innovation is a hard problem. If, as I have argued (following Huxley, Simpson, Schumpeter, and many others), the discontinuities between novelty and innovation versus adaptive or microevolutionary changes are often real, then the idea of a simple, all-encompassing theory or model of novelty and innovation is a chimera. No simple answer can encompass the origin of eukaryotes, the Cambrian Explosion of animals, or the cultural shifts associated with our own history. Understanding the mechanisms that generate new cultural spaces in which novelty and innovation can flourish, such as the corporation or the nation-state, may not necessarily generate insights into the production of new products. The only means of generating a general model is through a definition of novelty that is so broad as to encompass virtually any type of change. While some may find this approach useful, I disagree. Yet at the same time I have argued that sufficient commonalities exist to warrant a general conceptual framework. This echoes Carl von Clausewitz's view of the utility of theory for training one's intuition and thinking about problems in a way consistent with a view of history in which contingency plays an important role.

During his reelection campaign in July 2012, President Barack Obama delivered a speech about entrepreneurship in which he described the extensive government support for economic growth through education, infrastructure, and public investment in basic scientific research. This became known as the "you didn't build that" speech, with *Fox News* and the Romney campaign capitalizing on some inartful phrasing. Obama was channeling earlier comments by Elizabeth Warren:

> There is nobody in this country who got rich on their own. Nobody. You built a factory out there—good for you. But I want to be clear. You moved your goods to market on roads the rest of us paid for. You hired workers the rest of us paid to educate. You were safe in your factory because of police forces and fire forces that the rest of us paid for. You didn't have to worry that marauding bands would come and seize everything at your factory.[3]

Government investments and public goods are the foundation for economic growth. Thomas Jefferson sent army captain Meriwether Lewis and William Clark across the continent to explore the Louisiana Purchase, continuing a tradition of supporting scientific research and exploration that had begun in Europe by the sixteenth century. In the United States, federal and state governments began building roads and canals in the eighteenth century, provided generous allotments of land for railroads in the nineteenth century, and expanded airports and the internet in the twentieth century. Corporate research labs began in the late nineteenth century, led by German chemical firms. The most notable American examples include Thomas Edison's research laboratory in Menlo Park, New Jersey; the Bell Laboratories founded by Alexander Graham Bell; and Xerox PARC; each of which has generated major technological advances. Today, Alphabet's X (the "moonshot" factory) is attempting to replicate Bell Labs in their prime. But today too few public companies invest heavily in basic research—corporate research and development are largely development. A study by researchers at the Duke Business School found substantial reductions in basic research activities and increasing reliance on research conducted outside the corporation. Yet federal support for research is dwindling as well. Corrected for inflation, federal investments in non–Defense Department research and development have been essentially flat since about 2003 (aside from a one-time infusion after the 2008 recession) and in 2018 were a bit less than double what they were in 1976. In the 1960s total federal R&D was about 12 percent of the budget and today is less than 4 percent (although in fairness, in 2017 the non-Defense R&D budget was larger than the Defense portion for the first time since 1980).[4]

Gordon realized that the factors responsible for the relative decline in innovation after 1940, and particularly after 1970, have involved complex social and economic forces. How much of this decline should be apportioned to a drop in invention and novelty and how much to a drop in innovation is unclear, but whatever the causes the United States is no longer investing as heavily in the public goods and research needed for continued economic growth. He also published his book before the impact of the artificial intelligence revolution began in earnest (although its foundations were laid back in the 1940s). It is too soon to speculate whether AI represents a transformative innovation on the scale of earlier innovations and one sufficient to reverse the declines in American innovation.

ACKNOWLEDGMENTS

This book has had a very long gestation, so I am deeply indebted to many colleagues for their insights, discussions, patience, and occasional disagreements. Several people require particular thanks: Discussions with Jeremy Jackson, David Krakauer, Kevin Lala, Manfred Laubichler, John Odling-Smee, Carl Simpson, Rachel Wood, and the late Eric Davidson were enormously helpful at different times. Jerry Sabloff provided helpful feedback on an early draft of chapter 9. Günter Wagner's 2014 book on evolutionary novelty was published at just the right time to clarify my thinking, and I greatly appreciate subsequent discussions. My colleagues Bill Dimichele and Conrad Labandeira at the National Museum of Natural History, and former postdoc Marc Laflamme and former student Sarah Tweedt provided critical insights in refining the structure of the book. I am particularly grateful to several philosophers and historians of biology who invited me to meetings and helped sharpen my thinking: John Beatty, Alisa Bokulich, Ingo Brigandt, Max Dresow, Jim Greismer, Chris Haufe, Alan Love, and David Sepkoski. When Jim Valentine, my former PhD supervisor, and I decided to write our book *The Cambrian Explosion*, I put off an earlier version of this book, but in the end this project benefitted enormously from the additional time and our exchanges on that project. Jim's perspectives on evolution permeate this volume, although there are doubtless many sections with which he would disagree. I am likewise very grateful to the three reviewers for Princeton University Press for the deep engagement with two drafts and the perceptive suggestions for improvements (not all of which I accepted).

Science is a deeply collaborative enterprise, and I benefitted from taking part in several critical collaborative efforts. These include participation in the MIT/Harvard node of NASA's National Astrobiology Institute over three successive grants, which funded much of this research, led by Andy Knoll of Harvard and then Roger Summons of MIT. Alan Love and Bill Wimsatt were thoughtful enough to include me in an exciting collaboration on generic versus genetic approaches to evolution.

As will be evident to some, the book was profoundly shaped by colleagues at the Santa Fe Institute through various workshops and working group meetings, at lunch, at tea, and in casual discussions. Since the late 1990s I have spent considerable time at SFI, and those fortunate enough to be familiar with the institute will recognize that its interdisciplinary flavor permeates this book. Too many people associated with SFI influenced this book to name, but particularly important were the late Linda Cordell, Doyne Farmer, Jessica Flack, Walter Fontana, Michael Hochberg, David Krakauer, Jose Lobo, Scott Ortman, Jerry Sabloff, Eric Smith, Debbie Strumsky, Jessika Trancik, Andreas Wagner, Geoff West, and the inimitable Cormac McCarthy. Discussions with participants at the 2014, 2015, and 2017 SFI Innovation workshops organized by Jose Lobo were valuable in clarifying my thinking. Chapter 4 reflects discussions with Brian Arthur, Walter Fontana, and Chris Wood. On the cultural side, John Padgett and Woody Powell included me among the biologists in a fascinating project on history and evolution, funded by the Social Science Research Council and SFI.

This research project would not have been possible without the librarians at various branches of the Smithsonian Institution Libraries, particularly at Natural History and the Dibner Library at the National Museum of American History. Of equal importance has been Politics & Prose in Washington, DC, our beloved local independent bookstore. As this book extended into areas of culture, technology, and economics, I found many paths to follow among the books at P&P. And thanks too to the bakers of the late Red Fox (adjacent to P&P) and Livin' the Pie Life in Arlington, Virginia: their brownies are very different but equally delicious.

At Princeton University Press, Alison Kalett was a persistent and encouraging sounding board during the overly long gestation of the project. Jenny Wolkowicki oversaw production of the final book, careful copyediting by Maia Vaswani saved me from many errors, and Dimitri Karetnikov greatly improved my crude original drawings.

This project, like most of my recent work, would not have come to fruition without the love and endless support of my wife, Wendy Wiswall. She has lived with this project, in different forms, almost since we met and has always been happy to listen to my thoughts and fears about this book. Perhaps most importantly, her devotion to the book sometimes exceeded my own.

NOTES

Chapter 1

1. Strömberg 2005, 2011. As most of the topics discussed in this chapter receive more detailed coverage in later chapters, the notes for this chapter are limited to specific citations.

2. The story of Atanasoff and ENIAC has been described in many places, but particularly well by Gleick (2011).

3. Simpson 1944, expanded in Simpson 1953b.

4. Simpson 1960, p. 162. Adaptive radiations and the examples here will be discussed in detail in chapters 3 and 6.

5. See Newton 1999, p. 794. The role of parsimonious explanations is discussed by Siskin (2016). For extrapolationist views, see Futuyma (2015).

6. For evolutionary biology: Moczek 2008; G. Müller and Wagner 1991; Shubin et al. 2009; G. Wagner 2014. For business: Christensen 1997. For economics: W. B. Arthur 2009; Romer 1990. For culture: North 2013.

7. Schumpeter 1935. There is a perceptive biography of Schumpeter by McCraw (2007). Schumpeter used the term "development" for what today most economists term "growth" (I am indebted to a reviewer for noting this point). Here I have used Schumpeter's term but later will use "growth" in its modern sense. According to Cambridge University records, Malthus used the name Robert, although historical accounts generally call him Thomas, his first name.

8. There is also the possibility that he entire enterprise of seeking a general theory of novelty and innovation is so obviously delusional that anyone pursuing it should be strongly medicated.

9. Maynard Smith and Szathmáry 1995; Szathmáry and Maynard Smith 1995. Some of the controversy has been summarized by Calcott and Sterelny (2011). Szathmáry recently updated his views (2015), and there is an insightful overview by O'Malley and Powell (2016).

10. Universal Darwinism is very much a reductionist attempt to extend a gene-centered view and should not be confused with generalized Darwinism. On universal Darwinism: Dawkins 1983; Dennett 1995.

11. Minsky 2006.

12. S. Wright 1932, 1982a, b. The paleontological view is articulated by Jablonski and Bottjer (1990b), who discuss macroevolutionary lags.

13. Blount et al. 2012; G. Wagner 2014.

14. Gould 1989; Monod 1972. On the deterministic perspective: Conway Morris 2009; McGhee 2007, 2011, 2016. Conway Morris and his colleagues edited an issue of the Royal Society journal *Interface Focus* entitled "Are There Limits to Evolution?," published December 6, 2015 (vol. 5, issue 6).

Chapter 2

1. Wooton 2015.

2. On the growth of evolutionary ideas: Bowler 1996; Mayr 1976, 1982; Ruse 1996; Russell 1916.

3. Neo-Lamarckianism remains a persistent undercurrent in evolutionary thought, mining new discoveries in search of support for its views (Gissis and Jablonka 2011). Strictly speaking, orthogenesis was killed off by the Modern Synthesis, but a curious (and largely inchoate) mélange of religion and internalist views known as "emergent evolution" persists. This perspective dates to George Lewes in the 1870s and focuses on the discontinuous origins of variations and the appearance of greater organismic complexity. Psychologist C. Lloyd Morgan's *Emergent Evolution* (1923) represented the acme of such thinking before the Modern Synthesis, but the ideas have recently reemerged, as discussed by Erwin (2007b).

4. Ernst Mayr cast those concerned with the concept of the type as Platonic essentialists whose work lost any relevance with Darwin's introduction of population-based thinking. There are many discussions of the concept of a type but see discussions by Amundson (2005) and Farber (1976). Winsor (2006) is a wonderful corrective to the classic story equating typological thinking with Plato's essentialism. On Goethe: Levit et al. 2014; Riegner 2013.

5. Lovejoy 1936; Ruse 1996, ch. 1.

6. Lamarck 1809, p. 122, cited in G. Müller and Wagner 1991, p. 229.

7. On Lamarck: Burkhardt (1977) 1995; Ruse 1996, esp. ch. 2. Burkhardt points out that although many historians have suggested that Lamarck had developed a more dynamic view of the scala naturae, this is nowhere clear in his publications. Ospovat (1981) notes that in Darwin's 1838 essay on natural selection he shared Lamarck's view that organisms were perfectly adapted to the environment, and only subsequently developed a more dynamic view of natural selection.

8. Two useful discussions of the Geoffroy-Cuvier debate and the differences between them are by Appel (1987) and Le Guyader (2004). Le Guyader's translations of many of the relevant papers include the caption to figure 2.1. The classic history of the debate over form and function, and between Cuvier and Geoffroy, is by Russell (1916; see also Russell 1936). In this abbreviated telling of the Geoffroy-Cuvier debate I have had to omit many fascinating details of the development of their ideas, the political setting, and the implications for evolutionary history.

9. Le Guyader (2004) argues that Geoffroy was clearly thinking about evolution, but a considerable gap existed between his ideas and Darwin's.

10. For consistency, in this book I characterize this distinction as transformationist vs. saltationist, but other authors have used a variety of terms for the same conceptual divide. Riedl (1978) used the terms *transformationist* vs. *emergentist*, which is also used by Rieppel (2017), while Amundson (2005) adopted *adaptationist* vs. *structuralist*. While *emergentist* does have some historical validity, it has more recently been championed by some who would harken back to this earlier work (e.g., Reid 2007). The transformationist-structuralist distinction also roughly corresponds with Mayr's championing of the population view versus what he described, always pejoratively, as the typological view. This might also seem to correspond to the distinction between microevolution and macroevolution. However, as discussed in the concluding chapter, while microevolution does correspond to the transformationist viewpoint, much of modern macroevolution focuses on distributional explanations of large-scale evolutionary patterns and thus corresponds more closely with what I term *innovation* rather than the generation of novelty. For other discussions of this controversy: Bateman and DiMichele 1994, 2002; Turner 1983.

11. For Naturphilosophie and its approach to biological change: Gould 1977.

12. Baer 1828. The idea of recapitulation in developing embryos was widespread during the late eighteenth century and first half of the nineteenth (Richards 1992).

13. Richards 1992.

14. Darwin 1859, p. 471.

15. Fishburn 2004.

16. Buffon 1778; Hutton 1788; Lewis 2000.

17. Huxley to Darwin, November 23, 1859, in T. Huxley and Huxley 1900, p. 189, cited in Di Gregorio 1984, p. 66; Huxley, "The Origin of Species" (1860), in T. Huxley 1894, p. 22.

18. S. Lyons 1995.

19. Pictet 1860, cited in D. L. Hull 1973, pp. 144–145.

20. Cited in D. L. Hull 1973, p. 147.

21. Cited in D. L. Hull 1973, p. 152.

22. Letter from Darwin to Asa Grey, April 3, 1860, quoted in Hull, 1973, p. 142.

23. Darwin 1872; Mivart 1871.

24. Mivart 1871, p. 35.

25. Suzuki 2017.

26. Phillips 1841, 1860; Chamberlin 1909. Credit for the theory of diastrophism also goes to the Austrian geologist Eduard Suess, who discussed simultaneous transgressions of marine waters onto land, and regressions, in his monumental *Das Antiliz der Erde* (*The Face of the Earth*) (Suess 1904).

27. Bowler 1992.

28. Romanes 1895, quote vol. 1, p. 164. On exaptation: Gould and Vrba 1982.

29. Romanes 1895.

30. De Vries 1906.

31. Beatty 2016, p. 675, emphasis in the original.

32. Laubichler and Niklas 2009. There were Darwinian evolutionary biologists in Germany as well, of whom the most prominent was Bernhard Rensch, whose views are discussed later.

33. The German zoologist Wilhem Haacke introduced the term *orthogenesis* in 1893, probably based on ideas circulating among other German biologists. On orthogenesis and Haacke: Levit and Olsson 2006. Haacke had been a student of Haekel's, but after Haacke introduced orthogenesis Haekel described him in a letter as "a talented but irresponsible scientific con-man" (translation of letter from Haekel in Levit and Olsson, pp. 98–99).

34. Jaekel 1902, translated and abridged in Rensch 1959, pp. 98–99.

35. The views of many German paleontologists are discussed by Rensch (1980), and analyzed by Laubichler and Niklas (2009), Reif (1986), and Reif et al. (2000).

36. Osborn 1897, 1933.

37. J. Huxley 1942, ch. 9, quotes at pp. 486 and 487.

38. Simpson 1944, 1953b; quotes Simpson 1944, p. 3.

39. Simpson 1953b, p. 162.

40. Simpson 1953b, p. 340.

41. Simpson 1953b, p. 349.

42. Simpson 1960.

43. Mayr 1954, 1960; Nitecki 1990, Provine 2005.

44. Mayr 1942, p. 298.

45. Mayr 1963, pp. 600–607.

46. Mayr 1954, p. 176. By "continuous system" Mayr means a broadly distributed species with continuous gene flow.

47. Mayr 1954, p. 176.

48. Stebbins 1974, p. 301.

49. Rensch 1959 (German editions in 1947 and 1954; English edition 1959), quote p. 271. Rensch's views in the context of Schindewolf and other German evolutionary biologists are discussed by Reif (1983).

50. Futuyma (2015) makes a similar case.

51. Levit and Meister 2006; Reif 1983; Schindewolf (1950) 1994.

52. Schindewolf (1950) 1994, pp. 105, 106; Schindewolf 1936, p. 85, translated and quoted in Rieppel 2017, p. 96; see generally Schindewolf (1950) 1994.

53. On Goldschmidt: Dietrich 2003; Goldschmidt 1940; Gould 1982.

54. Schindewolf (1950) 1994, p. 202.

55. Mayr, 1960, pp. 354 (quote), 356. Curiously, however, Mayr had cited Schindewolf more positively in his 1954 paper, where he discussed "typostrophic variation" involving something "entirely new." After considerable effort the Society for the Study of Evolution was founded in 1946, and its journal *Evolution* began publication the following year. The vertebrate anatomist Rainer Zangerl introduced the terms *morphotype* and *structural plan* to English-speaking evolutionary biologists in a paper in volume 2. The consolidation of the Modern Synthesis made that the last such paper in *Evolution*, but the broader issues presented by apparent discontinuities associated with evolutionary novelties persisted (Cain 1994; Love 2003; Zangerl 1948). On the debate over the "creativity" of natural selection: Beatty 2016, 2019.

56. Walcott 1910; Yochelson 2006.

57. On diastrophism: Chamberlin 1909; Chorley 1963; Simpson 1944, p. 3.

58. Cloud 1948. Cloud, like Simpson, would never have been described as avuncular. Imperious perhaps. In typical Cloud fashion he objected to explosive evolution: "The term 'explosion' has been used for this phenomenon but as the so-called explosions may be in process during millions of years and probably did not make a loud noise, it is not an altogether felicitous designation . . . and the word eruptive—defined by Webster's Dictionary as a 'breaking forth from restraint'—is a more expressive term for the phenomenon in question" (p. 342). I spent my graduate school years housed in the Preston Cloud Laboratory at the University of California, Santa Barbara, built by NASA during the Apollo project for his studies of moon rocks, and where Cloud had an office in retirement. Many UCSB undergraduate students thought we made clouds inside the building (I am not making this up).

59. The papers were published in 1952 in the *Journal of Paleontology*, along with comments and discussion from several other geologists (Henbest et al. 1952).

60. Mayr 1960, p. 351.

61. Mayr 1960, p. 351.

62. Mayr 1963, p. 602.

63. Gould and Vrba 1982; Mayr 1960.

64. Darwin 1872. The importance of changes in function was proposed by Darwin and elaborated by the German evolutionary morphologist Anton Dohrn (1875).

65. Mayr 1960, p. 367.

66. The papers are in volume 14, issue 4, of *Systematic Zoology* (now *Systematic Biology*), published in December 1965 and based on a symposium held in 1963 (Schaffer and Hecht 1965).

67. Eldredge and Gould 1972; Gould and Eldredge 1993; Jablonski 2008; Stanley 1975.

68. De Vries 1906. Here De Vries is quoting a comment by "Mr. Arthur Harris in a friendly criticism of my views" (p. 826). But the term has a longer lineage. Branch (2015) identified Harris and traces the origin of the term to books by Jacob Schurman, a professor of Christian ethics and president of Cornell University. The phrase "arrival of the fittest" continues to be used, notably in a paper by Fontana and Buss (1994) and in the title of A. Wagner's 2014 book on innovation: *Arrival of the Fittest*.

69. Gould 1977.

70. Riedl 1977, 1978; Schoch 2010; G. Wagner and Laubichler 2004. I first read Riedl's 1977 book early in my time as a graduate student. Some of my German-speaking colleagues have viewed my interest in Riedl with amusement, insisting that one must read him in the original German to have any hope of understanding his ideas.

71. Gould 1977, pp. 337–338.

72. Filipchenko 1927. An excellent history of these events is Jenner's (2022).

73. Valentine 2004; Kemp 2016.

74. Kemp 2007a, b, 2016, quote p. 29.

75. An insightful discussion of the long history of interest in "the new" is North's *Novelty* (2013).

76. Malthus (1798) 1926.

77. Pomeranz 2000. On technology and the Industrial Revolution: Landes 1976; Mokyr 1992; Rosenberg 1982.

78. Pearl and Reed 1920. The addition of Alaska as a state significantly increased the total area of the country but added little agricultural output and seems unlikely to have materially affected the estimates of Pearl and Reed.

79. A. Smith 1776, bk. 1, 1.5.

80. On Marx: Bimber 1990; Frison 1988; Usselman 1995. The nature of novelties and issues of discontinuity versus continuity do not appear to have been a concern for either Smith or Marx. Each recognized that technological innovations (including the organization of factories) could have a significant impact on output.

81. Boyd and Richerson 1985; Cavalli-Sforza and Feldman 1981; E. Wilson 1975.

82. On Darwinian versus Spencerian approaches: Currie and Mace 2011; Dunnell 1980; Kuhn 2021; Pearce 2020. Although their views influenced several generations of researchers, theirs were not the only views on culture during the late nineteenth and early twentieth centuries. Degler chronicled the shifting views over the primary source of human nature from biology to culture and back to biology (1991).

83. On Spencer's views on organism-environment interactions: Pearce 2020.

84. On the history of approaches to sociocultural evolution: Graeber and Wengrow 2021; Harris 1968; Trigger 1998.

85. Pitt Rivers 1906. This volume was a posthumous collection of previously published essays and includes a useful introduction by Henry Balfour describing the implications of some of Pitt-Rivers's work. See also discussion by Trigger (1998, pp. 97–98), which concludes with Trigger's assessment that Pitt-Rivers made the most significant contribution to understanding cultural change of the Victorian evolutionary anthropologists.

86. L. Morgan 1877. Although Morgan is little remembered today, he was prominent in scientific circles during the late nineteenth century, helping to found the University of Rochester and attempting to start a companion women's college. He and his wife left their estate to the University of Rochester to endow a women's college.

87. Mason 1902.

88. Sahlins and Service 1960; Spencer 1997; White 1959. White summarized his views as culture = energy × technology.

89. Godin 2017; Kirch 2010. I am indebted to Godin's excellent analysis of the history of ideas on technological innovation for the insights about the role of the anthropological controversies.

90. Harris 1968.

91. Barnett 1953, p. 7.

92. Graeber and Wengrow 2021.

93. Ogburn and Thomas 1922; there is a more thorough discussion of "multiples" by Merton (1961).

94. Gales of destruction is from Schumpeter (1942). McCraw's *Prophet of Innovation* (2007) is an excellent biography of Schumpeter, with extensive discussions of his views on innovation; Swedberg's *Schumpeter* (1991) is also useful. By World War II Schumpeter's work was eclipsed by the success of John Maynard Keynes and other economists, but his insights into the importance of technological innovation for economic growth have been validated by later work.

95. Schumpeter 1935, pp. 4, 6. Although Schumpeter traditionally gets the credit for the distinction between novelty and innovation, a tradition I have followed in this book, Godin (2017) cites E. M. Rogers (1978) as crediting at least two predecessors to Schumpeter: Ogburn (1922) and Linton (1936), to which Godin adds others. Adjudicating this issue is unnecessary here, as it is Schumpeter's distinction that made an impact on subsequent generations.

96. Schumpeter (1912) 1954. Lest I leave the impression that Schumpeter was the only economist interested in these issues during the interwar years, it is worth recognizing that several others were engaged in similar research.

97. The distinction between Schumpeterian and Smithian growth is from Parker (1974), as clarified and expanded by Mokyr (1992, 2016).

98. Maclaurin 1950, 1953; Godin 2017.

99. From the 1930s into the early 2000s, thinking about innovation was dominated by five sequential models. This account relies on Rothwell (1992), who first recognized these models, as well as Albu (2017), Scoville (1951), and Godin (2017).

100. Godin 2017.

101. Godin 2017; Schmookler 1966. Economic conditions were an important factor for Schmookler; see particularly his chapter 4 "The Use of Important Inventions as a Cause of Further Inventions." Schmookler did recognize that knowledge could be considered an economic good. On p. 176 he discusses the conditions under which knowledge could be considered a consumer (private) good and those where it could be considered an economic good (if it had value for "economic, medical or military utility"). See discussions of types of economic goods at the end of chapter 3.

102. Kline 1985. On Schmookler: Godin 2017, pp. 112–113.

103. E. Rogers 1995.

104. Knight 1967.

105. Smil 2023.

106. Godin 2017, references at p. 214; Knight 1967. Kline (1985) largely eliminated invocation of any linear model, whether driven by basic science, invention, or demand/need.

107. Those familiar with these disciplines will have realized that each of these is an example of metrification gone mad: counting something because it is easy rather than because it is informative. None of these is a particularly easy thing to count, whether species or GDP. For an enlightening view of the fallacies of GDP: Gordon 2016.

108. For the neo-Schumpeterian approach: Aghion et al. 2021. On the Great Divergence: Pomeranz 2000.

Chapter 3

1. Strömberg 2005, 2011.

2. On the transition to grasslands and C_4 plants: Edwards et al. 2010. On their impact on climate: Pagani et al. 2009.

3. In 1988 Matthew Nitecki, a curator at the Field Museum of Natural History in Chicago, organized a symposium on evolutionary novelty (Nitecki 1990). The papers from that meeting provide a perspective on widely accepted views just prior to these changes. In his introduction to the volume, Nitecki largely accepted Mayr's 1960 definition, with novelties allowing clades access to new adaptive spaces. Paleontologists David Jablonski and David Bottjer took a broad view of evolutionary innovation, viewing it as involving a functional threshold; for them a novelty is simply any derived trait.

4. The phylogenetic affinities of *Tullimonstrum* continue to be debated, but the latest work suggests a vertebrate affinity (McCoy et al. 2020). Evidence for the notochord was presented by McCoy et al. (2016), analysis of the eyes was by Clements et al. (2016), and the critique of these results was from Sallan et al. (2017); see also C. Rogers et al. (2019).

5. This distinction has been advanced by Love (2006) and G. Wagner and Lynch (2010). For discussion: M. Donoghue and Sanderson 2015; Love 2003; Wainwright and Price 2016.

6. Wainwright and Price 2016, p. 480.

7. Wainwright and Price 2016.

8. Pigliucci 2008b, p. 890.

9. Andersson et al. 2015; Kassen 2019; Nasvall et al. 2012.

10. Davidson and Erwin 2010; Moczek 2008; G. Müller and Wagner 1991.

11. Panchen 1992; Russell 1916. On fish glycoproteins: L. Chen et al. 1997.

12. For more detailed discussions of the concept of homology: M. Donoghue 1992; A. Larson 2014; Wake 2003.

13. G. Müller 1990; G. Müller and Wagner 1991, p. 243.

14. Moczek 2008, p. 436.

15. G. Wagner 2014.

16. Benson et al. 2018. G. Wagner describes the deep transformations leading to type II novelties as the origin of a novel variational modality. This was initially developed by G. Wagner and Altenberg (1996) and is closely related to the concept of developmental modularity (Raff 1996; Raff and Sly 2000).

17. Bonner 1982; Gould 1977. Some of the early excitement is captured by Raff and Kaufman (1983), which was partly motivated by Goldschmidt's intuitions. For a retrospective on the 1982 meeting with papers by some of the original participants as well as by younger biologists, historians, and philosophers of science: Love 2015.

18. G. Wagner 2007, 2014.

19. Davidson 2006; Davidson and Erwin 2006, 2010. G. Wagner's expanded view of character identity networks: DiFrisco et al. 2020.

20. G. Wagner 2014, ch. 9; Prum 2005.

21. These definitions evolved through a series of papers: Travis et al. 2010; Peterson and Müller 2013, 2016. Here I have adopted those from the 2016 paper.

22. Simakov and Kawashima 2017.

23. For general discussions: Alberch 1989; Dobzhansky 1937; Mayr 1960. On novelty as overcoming phylogenetic and developmental constraints: Hallgrimsson et al. 2012, p. 502; and a critique by Peterson and Müller (2013).

24. Galis (2001) and Hunter (1998) chronicle the many definitions of key innovations.

25. Coddington 1988; Galis 2001; Heard and Hauser 1995; Lauder 1981; Liem 1990; Rabosky 2014, 2017.

26. Cracraft 1985, 1990; Erwin 1992, 2015a; Rabosky 2017. The study of island adaptive radiation is by Miles et al. (2023). In 1992 I identified four different types of evolutionary diversifications: novelty events, involving high morphological disparity but not necessarily a great increase in the number of taxa; broad diversification events, in which the number of species and genera increases in many different clades owing to expanded ecological opportunities; economic radiations due to a more limited ecological opportunity that can be exploited by a single clade, or a group of related clades; and finally a more restrictive definition of adaptive radiations. While this classification resolved some problems, I did not use novelty and innovation in a consistent fashion.

27. M. Donoghue and Sanderson 2015, p. 265.

28. S. Wright 1982a, b.

29. Erwin et al. 1987, with data from Sepkoski 1982. For Valentine's earlier studies: Valentine 1973, 1980, 1989.

30. Jablonski 2005; Jablonski and Bottjer 1990a, b, 1991; Jablonski et al. 1983, 1997; Miller and Sepkoski 1988; Sepkoski and Miller 1985. Not all of these studies relied entirely on higher taxa as proxies for evolutionary novelty. For example, Jablonski and Bottjer (1990a, b) relied on specific novelties within clades of bryozoans, and Jablonski et al. (1997) combined analysis of

specific morphological novelties in two clades of bryozoans with study of the environmental context of their first appearance.

31. An advocate of the gradualist position is Futuyma (2015).

32. G. Müller and Wagner 1991; Riedl 1978.

33. On body plans: Amundson 2005; Hall 1996, 1999; Valentine 2004. Quote Hall 1999, p. 99.

34. M. Donoghue and Sanderson 2015. These authors were principally interested in events that increase species diversification rates (or reduce extinction rates), rather than evolutionary novelties, but here I apply their ideas in the context of novelty as well. The correlated progression scenario suggests that a series of new characters arises, each generating the opportunity for the next. In one sense this might appear to be a series of key innovations, but the correlated progression model proposes a greater degree of functional integration among the new characters, a view shared by others (Kemp 2007a; Rosenzweig and McCord 1991).

35. Fusco and Minelli 2010.

36. Reader and Laland 2003a.

37. West-Eberhard 2003, 2005, 2008.

38. West-Eberhard 2008, pp. 899, 903.

39. Moczek et al. 2011. The role of phenotypic or developmental plasticity in evolution has been discussed for decades as the Baldwin effect, genetic assimilation, and genetic accommodation. General discussions: Fusco and Minelli 2010; Moczek et al. 2011; Schlichting and Pigliucci 1998; West-Eberhard 2003, 2005, 2008. For papers specifically focused on the role of plasticity in generating evolutionary novelty: Levis et al. 2018; Moczek et al. 2011.

40. Standen et al. 2014. The early bird scenario: M. Wang et al. 2018.

41. On the plasticity-first model: Levis et al. 2018; Levis and Pfennig 2018. For criticisms: G. Wagner et al. 2019.

42. G. Wagner et al. 2019. Nedelcu and Michod (2020) extended this model to propose that the distinction between germ and somatic cells in the evolution of multicellularity may also have been a response to stress, drawing on evidence from *Volvox*, a green alga, as well as social amoebae. On ricefish: Hilgers et al. 2022. For the relationship between eye evolution and stress: Oakley and Speiser 2015; West-Eberhard 2003.

43. The power of kitchen combinatorics is from Fink et al. (2017), combining ingredient lists for 56,498 recipes from allrecipes.com, epicurious.com, and menuplan.com, which required a total of 381 ingredients.

44. This example is from W. Arthur (2009, p. 175). For n building blocks, and removing noncombinations, the number of combinations increases as $2^n - n - 1$. ($-n$ for singletons and -1 for the null case.)

45. Oakley 2017. Combinatorics poses significant, although not insurmountable, problems for phylogenetic approaches. Phylogenetic methods trace the vertical descent of characters from one generation to the next, or one species to the next. Borrowing features from a different species (technically, horizontal transmission) can confuse some methods.

46. The RNA model for evolutionary novelty was developed by a group of theoreticians including Peter Stadler (now in Leipzig), Walter Fontana (Harvard), and Günter Wagner. More recently, Andreas Wagner at the University of Zurich and his research group have taken this approach further. Stadler, Fontana, and Günter Wagner were all students in the theoretical biology group in Vienna; Stadler and Fontana with Peter Schuster, Wagner with Rupert Riedl (see chapter 2).

47. W. B. Arthur 2009. Salazar-Ciudad (2006) discusses a suite of basic developmental patterning mechanisms, including those involving patterning changes due to changes in cell state or fate, including stripes, spots, and so-called "French-flag" patterning; patterning changes due to changes in cell location; and some of the variety of developmental patterns that can be produced by combinations of these basic mechanisms. Salazar-Ciudad's work has focused on better

understanding genotype to phenotype mapping, as was also the case in the work of Stadler, Fontana, and Wagner.

48. Maynard Smith and Szathmáry 1995; Szathmáry and Maynard Smith 1995. Related work includes: Bonner 1988; Buss 1987; Michod 1999; Michod and Roze 2000.

49. That major evolutionary transitions constitute a form of evolutionary novelty was recognized by G. Müller and Wagner (1991).

50. Calcott and Sterelny 2011; O'Malley and Powell 2016; Queller 1997; Szathmáry 2015; S. West et al. 2015.

51. Knoll and Bambach 2000. Issues of large-scale trajectories in the history of life have been a long-standing interest of some paleontologists and evolutionary biologists and are intimately connected to issues about whether the history of life exhibits something that might be termed progress. These issues are explored in: Gould 2002; McShea 1998; Rosenzweig and McCord 1991; Ruse 1996; Stebbins 1969; Valentine 1973.

52. Vermeij 1998, pp. 1444–1445.

53. Vermeij 1998, p. 1445.

54. Canfield 2005; Catling et al. 2005; Johnston et al. 2009; Lenton and Watson 2011; T. Lyons et al. 2014, 2015. A readable introduction to the debate is Canfield's (2014).

55. The estimates for the available energy supplied by each transition are from Lenton and Watson (2011).

56. Lenton and Watson 2011.

57. Useful introductions to the panoply of approaches include those of Laland and Brown (2011) and Mesoudi (2011). Some may decry the absence of memes from this discussion. Dawkins's attempt to define a cultural analog to genes (1989) attracted considerable attention from some (e.g., Dennett 1995) and remains a darling of thought leaders in Silicon Valley, but it has largely been ignored by those working in this area because there is no productive research program. See discussions by Lewens (2015) and Mesoudi (2011). Universal Darwinism is a related attempt to apply standard population genetic thinking to culture and a variety of other areas. Although the idea predates Dawkins's application of the term (Campbell 1960; Dawkins 1983), and is derived from the views of early pragmatist philosophers such as Charles Pierce and John Dewey, it retains many adherents, particularly Geoffrey Hodgson in economics (Aldrich et al. 2008; Hodgson and Knudsen 2010).

58. On Darwin's and Simpson's views of what is now known as niche construction: Darwin 1881; Simpson 1953b.

59. Kolodny et al. 2015. On magnetrons: W. B. Arthur 2009.

60. Older estimates from Sterelny (2016); younger estimates from Tattersall (2018).

61. Berwick and Chomsky 2016; Bolhuis et al. 2014.

62. Lieberman 2015; Pinker and Bloom 1990; Sterelny 2016.

63. Campbell 1960; Simonton 1999, 2003. On the relationship between models of cultural evolution and creativity: Fogarty et al. 2015. Simonton's (1999) remains one of the most incisive analyses of human creativity.

64. Phelps 2013.

65. Romer 1990; also 1986. Although the distinction between different types of economic goods has a longer history, Romer credits Cornes and Sandler (1986) for developing the distinctions between different goods. An accessible introduction to Romer's contributions is by Warsh (2006).

66. For more on endogenous growth theory: C. I. Jones 2002, 2019. On the tragedy of the commons: Hardin 1968.

67. There will be more to say about biological public goods later. James McInerney at the University of Nottingham developed similar ideas and has cogently argued that "tree-thinking" can sometimes blind biologists to other ways of thinking about evolution. See: Erwin 2008b, 2015b; J. McInerney and Erwin 2016, 2017; J. McInerney et al. 2011b.

68. Sterelny 2014, 2016; Tomasello et al. 2012.

69. Christensen 1997. After Christensen's death the *Economist* described him as "the most influential management thinker of his time" ("Clayton Christensen's Insights Will Outlive Him," January 30, 2020, https://www.economist.com/business/2020/01/30/clayton-christensens -insights-will-outlive-him).

70. Lepore 2014.

71. Brigandt and Love 2012.

72. Hochberg et al. 2017. This is an introduction to a suite of papers from a 2014 workshop at the Santa Fe Institute, where I discussed Schumpeter's distinction between invention/novelty and innovation.

73. Application of the sorites paradox to the terms *novelty* and *innovation* is discussed by Hochberg et al. (2017). On the paradox: Hyde and Raffman 2018. *Sorites* comes from the ancient Greek *soros*, meaning "heap or mound," hence the heaps of wheat in most discussions of the paradox.

74. Pigliucci 2008b.

75. Schumpeter 1935, p. 4.

76. Love (2006, 2008) discusses form and function, levels of hierarchy, and generalizability.

77. Losos 2010.

78. Brigandt and Love 2012; Laubichler et al. 2018.

Chapter 4

1. Borges 2007. Some readers will realize that Spanish contains 27 letters, not 22 (ñ is added to the 26 of English; some also include the double letters *ch*, *ll*, and *rr* for a total of 30 letters). In his earlier story "The Total Library," Borges discussed the history of language and letters. In this he derived the 22 by rejecting the double letters, ñ, *k*, and *w* because they are used only in loan words from other languages, *q* because it is "superfluous," and *x* as an abbreviation. Borges never lacked a point of view.

2. Among other differences, McCarthy's version of this book would be shorter, with no semicolons.

3. Eco 1983; Nolan 2014.

4. Block 2008. Borges reveals that there are letters on the cover of each book but does not indicate how many. Including these symbols would further enlarge the size of the Library.

5. Similar ideas of combinatorics long predate Borges. Gottfried Leibniz explored combinatorics in his doctoral dissertation in 1666, later published as *Dissertario de arte combinatorial* ("Dissertation on the Art of Combinations"; Leibniz (1666) 1923). He discusses the number of possible books of a fixed length, as did Borges, in *De l'horizon de la doctrine humaine* (Leibniz (1715) 1991); Leibniz-translations.com, consulted August 31, 2018. Leibniz's interest in combinatorics is described by Antognazza (2009). Dennett 1995; Dennett's inspiration was Dawkins (1986).

6. L. Carroll 1879.

7. There is a large and often complex literature on adaptive and fitness landscapes. Among the works that have helped in writing this section are: De Visser and Krug 2014; Dietrich and Skipper 2012; Fragata et al. 2019; Greenbury et al. 2022; Serreli 2015. Also particularly helpful were chapters in Svensson and Calsbeek's (2012) *Adaptive Landscape in Evolutionary Biology*. A useful recent summary of the variety of landscapes is by Bataillon et al. (2022).

8. Maynard Smith 1970.

9. F. Arnold 2011; Romero and Arnold 2009.

10. S. Wright 1932. Strictly speaking, Wright's description began with multiple different alleles at different gene loci to illustrate the combinatorics. But the simplified fitness landscape he introduced features different adaptive values for gene combinations. The first representation

of a fitness surface evidently dates to Armand Janet in 1895, but this had little impact, while Wright's paper introduced the concept to evolutionary biology (Dietrich and Skipper 2012). General introductions to adaptive landscapes: Serrelli 2015; Svensson and Calsbeek 2012. Wright was sufficiently taken with the landscape metaphor that he applied it to at least two very different situations (Provine 1986). The examples in the text are based on the paper from his 1932 presentation, in which the landscape represents the possible gene combinations to be found in different organisms, while the height (generating the surface) reflects the fitness of an individual with that combination of genes. But Wright also applied the concept to a setting in which a point in the landscape corresponds not to an individual but to a population with the genes at the frequencies of that point.

11. Dobzhansky 1937. Dobzhansky's expansion of Wright's adaptive landscape also features in later editions of the book in 1941 and 1951.

12. Dietrich and Skipper 2012; Simpson 1944.

13. Kokko et al. 2017.

14. On N-K models: Kauffman 1993; Kauffman and Levin 1987; Kauffman and Weinberger 1988. In these models, N is the length of a binary string and K is the number of interactions. As K varies the nature of the landscape changes from a Mount Fuji landscape to one that is completely random. Theoretical ecologist Simon Levin at Princeton University coauthored the first paper with Kauffman, but Kauffman went on to extend and elaborate these models. Although epistasis is important, and N-K models have an important role in thinking about novelty and innovation, they also have significant limitations: they assume evolution is governed entirely by selection; neutral mutations are either entirely absent or insignificant, and population sizes are infinite (Fragata et al. 2019). Since these assumptions are unlikely to obtain in most real-world events, these models are primarily of heuristic interest.

15. Examples of recent models of the adjacent possible as applied to issues of novelty: Loreto et al. 2016; Tria et al. 2019.

16. Gavrilets 2010; Pigliucci and Kaplan 2006. The wide distribution of holey landscapes is demonstrated by Dochtermann et al. (2023).

17. Pigliucci and Kaplan 2006.

18. On neutral landscapes in general: Fragata et al. 2019.

19. Fontana 2006. Recent work has shown that the genotype to phenotype mapping for short RNAs is more complex owing to plasticity in the phenotype, which makes the mapping non-deterministic: Garcia-Galindo et al. 2023.

20. Dennett 2017, p. 118, emphasis mine. Dennett also explores the Darwinian space constructed by philosopher Peter Godfrey-Smith (2009), which is a cube that ordinates Darwinian, quasi-Darwinian, and proto-Darwinian phenomena. Dennett applies this to both the origin of life and to human culture, and to how the Darwinian nature of these events changed through time.

21. Louis 2016. Ideally, we would like to know the mapping between sequence and the number of 3-dimensional structures, but solving this problem remains computationally challenging.

22. A. Wagner 2011, 2014. For two excellent reviews of these spaces and the complexities of genotype to phenotype mapping: Aguirre et al. 2018; Manrubia et al. 2021.

23. Fontana 2006; Fontana and Schuster 1998.

24. Pigliucci and Kaplan 2006.

25. The activities of a species modifying its own environment, and often the environments of other species, was the topic of Charles Darwin's last book, on earthworms (Darwin 1881). Today this is commonly known as niche construction or environmental engineering and will be discussed further in chapter 8. Pigliucci and Kaplan (2006) credit Lewontin (1978) for the rubbery landscape metaphor. Lewontin described flaws in traditional adaptive thinking in papers in the late 1970s and 1980s (esp. Lewontin 1978, 1983).

26. I am indebted to my Santa Fe Insitute colleague David Krakauer for this example.

27. Chatterjee et al. 2014.

28. On extensions of the protein space: Brunette et al. 2015. On fitness spaces: Fragata et al. 2019. These constraints on how proteins can evolve are another example of epistasis.

29. Starr et al. 2018.

30. Bogorad and Deem 1999.

31. A. Wagner 2011, 2014a.

32. On Waddington's developmental or epigenetic landscapes: Baedke 2013; Waddington 1940, 1957. On more recent developmental spaces: Bhattacharya et al. 2011; Felix 2012; Fusco et al. 2014; Huang 2012. Related are network spaces of metabolic activities (Raymond and Segre 2006).

33. On developmental abnormalities: Alberch 1989. On polydactyly at Chaco Canyon: Crown et al. 2016. Of the cases of multiple limbs that have been described, many seem to have resulted from a Siamese twin that did not fully develop.

34. On snowflakes: Libbrecht 2017. On Thompson: W. Arthur 2006; D. Thompson 1942. Raup's studies: Raup 1966. On morphospaces in general: McGhee 2007.

35. Gould 1989, p. 46. Gould's arguments, as was usually the case, generated considerable discussion, and there is now a large literature both on his arguments about the Burgess Shale (see Erwin 2016) and about the importance of contingency in the history of life, which will be taken up again in chapter 10. On disparity: Erwin 2007a; Foote 1996; Hughes et al. 2013. On early tetrapod limbs: Dickson et al. 2021.

36. Clark et al. 2023.

37. McGhee 2011. On saber-tooths: Lautenschlager et al. 2020. On venom-spitting cobras: Kazandjian et al. 2021.

38. Cheng et al. 2023; Van Belleghem et al. 2023.

39. One attempt to encompass a greater number of forms was the development of a "space" of skeletal elements: whether the skeleton is internal or external, the type of growth, and other factors. Such a space assumes that skeletons are composed of a relatively limited number of elements that are used in many different clades: Thomas and Reif 1993; Thomas et al. 2000.

40. Bambach 1983; Bambach et al. 2002. On occupied ecospaces: Bambach et al. 2007. The comparison between dinosaur and mammalian ecosystems: McGhee 2011. Closely related to ecospaces are functional spaces—for example, of proteins: A. Dean and Thornton 2007.

41. Dennett 1995 (quote p. 143), 2017. Dennett's Library of Mendel is based on Dawkins (1986) and is only a subset of the Library of Babel. Godfrey-Smith 1998, 2009.

42. Quote Schumpeter 1935, p. 4.

43. Hidalgo and Hausmann 2009; Hidalgo et al. 2007. The product space is based on export data for 775 products and the relationships among them.

44. Hidalgo et al. 2007, p. 482.

45. The lumpy product space is described by Hidalgo et al. (2007).

46. Dediu et al. 2013; N. Evans 2013; Greenberg 1963.

47. Skirgard et al. 2023, p. 4 of 15.

48. Avena-Koenigsberger et al. 2015; Charbonneau 2015.

49. The review of N-K models applied to management is by Baumann et al. (2019, quote pp. 285–286). Baumann et al. credit the introduction of the N-K model in management to Levinthal (1997). On cultural spaces: Boyer 2018; Claidiere et al. 2014. Reviews by Gerrits and Marks (2014, 2015) discuss applications of fitness landscapes to the social sciences.

50. Mitteroecker and Huttegger 2009.

51. Block 2008. As Block describes, the library could be either a 3-torus or a 3-Klein bottle, but that is more topology than needed here.

52. The wonderful term "Euclidean intuitions" is from Mitteroecker and Huttegger (2009).

53. For two of the most perceptive treatments of the complexities of different adaptive and morphospaces: Huttegger and Mitteroecker 2011; Mitteroecker and Huttegger 2009.

54. A. Wagner 2014, p. 176.

55. On whether evolution is bounded or unbounded: De Vladar et al. 2017.

56. Simon discusses human problem solving in much of his work, but see: Newell and Simon 1972; Simon 1955.

57. Kauffman 1993, p. 118.

58. Baumann et al. 2019, pp. 285–286.

59. On wrasses: Wainwright 2007; Wainwright et al. 2005.

60. Beyond Lévy flights, Rayleigh flights are similar, but the step lengths are drawn from a normal distribution.

61. Gawne et al. 2018. The incipient development of an evolutionary cell biology (M. Lynch et al. 2014) is beginning to address this problem.

62. On pre-statable spaces and novelty: De Vladar et al. 2017; Kauffman 2014. Montévil has taken a similar approach (2018), based on Bergson's ideas of emergence. Bergson illustrated his ideas of emergence with a discussion of the space of all possible symphonies, and Montévil extends this idea by distinguishing between the generic and specific aspects of possibilities, in which the generic properties of a space define a set of "pre-possibilities," which become "explicit possibilities" "if and only if we define explicitly also how they 'work.'" (p. 7).

63. On constraints and biases in developmental trajectories: Gerber 2014; Klingenberg and Monteiro 2005; Maynard Smith et al. 1985; Schwenk and Wagner 2004; Uller et al. 2018.

64. Wainwright (2009) considers the relationships between novelties and the expansions of morphospaces.

65. A. Wagner et al. 2016.

66. Two excellent introductions are by G. West (2017) and Whitfield (2006).

67. J. Brown et al. 2004; G. West and Brown 2005; G. West et al. 1997. The optimal configuration of body plans is discussed by Kempes et al. (2019), and a synthesis of the transitions between different organisms in the ocean is by Andersen et al. (2016). On mammalian deviations from three-quarter-power scaling: Kolokotrones et al. 2010.

68. Kempes et al. 2019.

69. Bettencourt et al. 2007a, b, 2020; G. West 2017. The sublinear scaling exponent for infrastructure, supply, and transportation lines in cities is 0.85, while the supralinear exponent is 1.15. In other words, these values are mirror images of each other. Settlement scaling in the Basin of Mexico: Ortman et al. 2015.

70. G. West 2017.

71. On antifreeze proteins: L. Chen et al. 1997. On the venom of sicariid spiders: Binford and Wells 2003; Casewell et al. 2013. On venom convergence more generally: Fry et al. 2009a, b; Van Thiel et al. 2022. On the regulation of venom secreting cells: Zancolli et al. 2022.

Chapter 5

1. On the LTEE in general: Lenski 2017, 2024; Lenski and Travisano 1994. Details of the *Cit+* mutation: Blount 2016; Blount et al. 2008. On the model of novelty: Blount et al. 2012, 2018.

2. G. Müller and Wagner 2003.

3. Disparity was discussed in chapter 3, but see: Erwin 2007a; Gerber 2016; Hughes et al. 2013.

4. Monterio and Podlaha 2009; Pal and Papp 2017; Salazar-Ciudad and Jernvall 2005; Suzuki 2017; G. Wagner and Altenberg 1996.

5. Stanley 2014; Strömberg 2011. On macroevolutionary lags more generally: Jablonski and Bottjer 1990b; Stanley 2014.

6. Lozada-Chavez et al. 2018; Sebe-Pedros et al. 2016; Vanneste et al. 2014.

7. Ji et al. 2006.

8. Zabell 1992, pp. 205–206; see also Loreto et al. 2016.

9. On general-purpose technologies and innovation: Eco 1983; Jovanovic and Rousseau 2005; Landes 1976.

10. On cumulative cultural evolution in chimpanzees: Gunasekaram et al. 2024.

11. This model was developed in earlier papers: Erwin 2015a, 2017b, 2019, 2020a; Wood and Erwin 2018.

12. A subsequent study of the *Cit+* mutant revealed more of the evolutionary complexity of this event. As discussed more conceptually in chapter 4, the interaction between alleles can be critical, and this study suggested that the potentiating mutations may not always have been necessary. The genetic history of the lineage, the ecological conditions, and the history of subsequent evolutionary changes may be of equal or greater importance (Quandt et al. 2014). Mutational order effects are one aspect of a more general phenomenon in genetics known as epistasis, which involves circumstances where the expression of gene A is affected by the expression of gene B (and sometimes gene C as well). An intensive study traced the history of epistatic effects in the heat shock protein 90 (Hsp90) in the yeast *Saccharomyces cerevisiae*, showing that at least 75% of the mutational changes were contingent upon prior potentiating changes essentially constructing opportunities for other mutations to succeed—and sometimes closing other opportunities. On mutational order: Mani and Clarke 1990; Starr et al. 2018. On the order of changes in gene regulatory networks: Sorrells et al. 2015. On technology: Fink et al. 2017.

13. On snapping shrimp: Anker et al. 2006. On ants: Rajakumar et al. 2012.

14. Lister 2013a, b.

15. Agrawal 2001; Leonard et al. 1999; Reimer and Tedengren 1996.

16. True and Carroll 2002.

17. Farmer 1997; Gould and Vrba 1982.

18. Brigandt and Love 2012. They also argue that the most important issue is agreement on the structure of the problem rather than agreement on a single definition.

19. On combinations or fusions as generators of novelty: Oakley 2017.

20. G. Wagner 2014.

21. Charbonneau 2015; Henrich 2015.

22. Nakamura et al. 2015.

23. MacMillan's *War: How Conflict Shaped Us* has a nice discussion of lags and potentiation in warfare in chapter 3 (MacMillan 2020). The first draft of this note included: "Despite subsequent developments in tanks, the rapid spread of cheap antitank weapons seems likely to make the US-Iraqi War of 1991 the last great tank battles." The Russian invasion of Ukraine in 2022 further emphasized the liability of tanks in the face of operational innovations by the Ukrainian Army, although the delivery of German and US main battle tanks in 2023 may yet show that tanks still have some utility.

24. There is a robust literature in statistical mechanics (a branch of physics) on network dynamics and changes in network structure, and some of these approaches have been applied to ecological problems (R. Thompson et al. 2012). Networks based simply on the presence or absence of a species rather than on the strength of interactions among species are less likely to capture the effects of novelties on an ecosystem (Hobbs et al. 2006; Pandolfi et al. 2020; Radeloff et al. 2015, p. 2052).

25. Page 2018.

26. W. B. Arthur 2009; Hodgson and Knudsen 2010; Mokyr 1992, 2016.

27. Erwin 2008a.

Chapter 6

1. Jacob 1977.

2. Estimates for the number of genes and genome size vary, particularly as new types of gene, such as various small RNAs, are recognized. For consistency, here I use values cited in figure 1 of Sebe-Pedros et al.'s 2018 paper.

3. The 1977 paper is by Salvini-Plawen and Mayr. On kernels: Davidson 2006; Davidson and Erwin 2006.

4. DiFrisco et al. 2020; G. Wagner 2007, 2014b. Waddington described developmental canalization, albeit without any mechanistic understanding, and although much of the work is separate from concerns about novelty, there is a long history of work on canalization and developmental constraints (Maynard Smith et al. 1985; Uller et al. 2018).

5. Droser and Gehling 2015; Droser et al. 2017; Yang et al. 2021. *Dickinsonia* is described by S. Evans et al. (2017).

6. For more detail on the Ediacaran–Cambrian radiation: Erwin 2020c; Erwin and Valentine 2013; Wood et al. 2019; Yang et al. 2021. On disparity: Gould 1989; with a reconsideration by Erwin (2016). Subsequent studies of disparity: Erwin 2007a; Foote 1997; Oyston et al. 2016.

7. Mangano and Buatois 2014, 2020. A third line of evidence for Ediacaran and earlier sponges comes from characteristic fossil biomolecules; there are fairly firm reports of biomarkers from Ediacaran-aged rocks, but earlier reports have been questioned (Erwin 2020c).

8. On metazoan phylogeny: Dunn et al. 2014; Laumer et al. 2019. On molecular clock estimates: Carlisle et al. 2024; Cunningham et al. 2017; Erwin 2021; Erwin et al. 2011; Sperling and Stockey 2018. Several aspects of this phylogeny remain contentious, including the position of a clade known as ctenophores, which some analyses place below sponges, but this seems likely to be an artifact (Pisani et al. 2015); the enigmatic placozoans, which may be a sister clade of cnidarians or branch between cnidarians and bilaterians (Laumer et al. 2018); and the Xenacoelomorpha, a clade superficially similar to worms but lacking a gut (Laumer et al. 2019). Although the deuterostomes have been considered a clade for more than a century, this has recently been questioned (Kapli et al. 2021).

9. Erwin 2020c.

10. For reviews of this transition: Brunet and King 2017; Sebe-Pedros et al. 2016; Sogabe et al. 2019. Sheetlike choanoflagellates: Brunet et al. 2019.

11. On distal enhancers: Gaiti et al. 2017; Sebe-Pedros et al. 2016.

12. Erwin 2020c, 2021. Multifunctional cell types: Tarashansky et al. 2020. Flat versus hierarchically structured regulatory control: Arenas-Mena 2017; Davidson 2006; Davidson and Erwin 2006; Erwin and Davidson 2009.

13. DiFrisco et al. 2020; Erwin 2021.

14. Chipman 2010, 2020.

15. For the change in regulatory control: Arenas-Mena 2017; Davidson 2006; Davidson and Erwin 2006; Erwin and Davidson 2009; Gaiti et al. 2018. The *Amphioxus* example: Acemel et al. 2016.

16. On co-option and novelty: S. Carroll 2005; McQueen and Rebeiz 2020; True and Carroll 2002. "Plug-ins": Davidson and Erwin 2006, p. 796.

17. Davidson and I developed this model in: Davidson and Erwin 2006; Erwin and Davidson 2002, 2009. It is elaborated in: Erwin 2020c, 2021. Specifically, I envision that segmentation patterning was built off co-opted GRNs involving the genes *hairy* and *engrailed*; for the brain and nervous system *otx*, *emx*, and *six3/6*; for sensory systems Pax genes; for appendage development *distalless*; and a regionalized gut with GATA and *brachyury*. Co-option of biomineralization regulatory machinery is described by Murdock (2020).

18. Erwin 2016; Gould 1989.

19. On serial homology: G. Wagner 2014.

20. This discussion of Cambrian fossil arthropod novelties draws on: Aria 2022; Chipman and Edgecombe 2019; Daley et al. 2018; Edgecombe 2020; Jockusch 2017; Legg et al. 2012; Ortega-Hernandez et al. 2017. On appendage formation: Jockusch 2017.

21. On plant disparity: Oyston et al. 2016. On dancing to a different beat: Traverse 1988. On the impact of plants: Boyce and Lee 2017.

22. An excellent recent overview of the origins of land plants is by P. Donoghue et al. (2021), who also describe the limitations of the lower Paleozoic terrestrial record in establishing the timing of the transition to land. On the *Chara* genome: Nishiyama et al. 2018.

23. On plant novelties: Harrison 2017. Leaf morphology: Boyce and Knoll 2002.

24. On the origin of early land plants, the most recent molecular clock estimates are by J. Morris et al. (2018) and Nie et al. (2020). For an early description of the macroevolutionary lag at the base of vascular plants: Knoll et al. 1984.

25. On the pulsed increase in the complexity of vascular plants: Leslie et al. 2021. This study examined increases in the number of plant reproductive parts (sporangia, seeds, parts of flowers, etc.), following McShea's (2000) discussion of parts as a general metric of complexity, rather than defined novelties. As in the Ediacaran–Cambrian radiation, the authors found that increased functional diversity tracked increases in complexity, suggesting, at least in this analysis, that form and function were not decoupled. The low productivity in pre-angiosperm plant ecosystems is from Boyce et al. (2017).

26. On the new genes: Fang et al. 2021; Liu et al. 2022; Marchant et al. 2021. Evidence for the two bursts of new genes: Bowles et al. 2020. A broader synthesis of the developmental changes: P. Donoghue et al. 2021.

27. The architectural varieties of trees: Niklas 1997. A detailed analysis of plant architectures: Chomicki et al. 2017.

28. On angiosperm molecular clock results: Beaulieu et al. 2015; Doyle 2012; Magallon et al. 2015; Sauquet et al. 2017. The 200 million years ago date: H. Li et al. 2019. There have been many reports of Triassic and Jurassic fossil plants or pollen, but an authoritative review cast doubt on these (Herendeen et al. 2017), while not discounting the possibility of the eventual discovery of more reliable Jurassic-age plant fossils; subsequent publications have not altered these conclusions. A reanalysis of the fossil data to estimate the origin of angiosperms indicates that several families must have originated by the Jurassic, although with a marked increase in the number of taxa in the Early Cretaceous (Silvestro et al. 2021).

29. Becker 2020; Sauquet et al. 2017.

30. Van de Peer et al. 2017.

31. On seed plant and conifer genome duplications: Z. Li et al. 2015. For angiosperms: Bowles et al. 2020; P. Soltis and Soltis 2016. Dates for the whole-genome duplications are from Clark and Donoghue (2017), who also identify the purported evolutionary correlates of WGD. The difficulties with WGDs as a driver of evolutionary dynamics have been dissected by Clark and Donoghue (2017) and P. Soltis and Soltis (2016).

32. On the ecological lag: Ramirez-Barahona et al. 2020.

33. Giribet and Edgecombe 2019. The success of insects has also involved several other novelties not discussed here, including complete metamorphosis, in which the body is reconfigured during the larval stage, and the ability to fold wings over the body (compare dragonflies to butterflies) in more advanced insect clades. The fossil nymphs are described by Prokop et al. (2022).

34. Averof and Cohen 1997.

35. On the origin of insect wings: Bruce and Patel 2020; Clark-Hachtel et al. 2021; Clark-Hachtel and Tomoyasu 2020; Shiga et al. 2017. There are useful commentaries by F. Smith and Jockusch (2020) and Tomoyasu (2021).The history of investigations of this problem is discussed by Clark-Hachtel and Tomoyasu (2016).

36. J. Wang et al. 2022.

37. Hu et al. 2019; Hu and Moczek 2021; Linz and Moczek 2020; Moczek 2009. On potentiation: Zattara et al. 2016.

38. Fisher et al. 2020.

39. The cryptic persistence scenario: Bruce and Patel 2022. The developmental module scenario: Fisher et al. 2020.

40. Hu et al. 2019, p. 1007. Moczek quote: Moczek 2008, p. 436.

41. Babonis et al. 2023.

42. On the origin of new vertebrate organs: Griffith and Wagner 2017.

43. Griffith and Wagner 2017.

44. Griffith and Wagner 2017; V. Lynch et al. 2015.

45. Gallant 2019; Gallant et al. 2014. The analogy to batteries is from Gallant et al. (2014). The downregulated genes are largely transcription factors involved in muscle development. Gallant (2019) cautions that because of the difficulties of raising most of these fish in a laboratory setting, few of these studies involved early developmental stages. Until such studies are undertaken there will be some questions about the gene regulatory networks underlying electrocyte formation. On gene co-option: Arnegard et al. 2010.

46. Griffith and Wagner 2017.

47. Despite the underlying similarities among these vertebrate epidermal structures, they represent independent morphological novelties built upon a common developmental core. Any of these novelties could have been discussed. For an introduction to the literature on the evolution of teeth, scales, hair, and similar structures: Dhouailly et al. 2019; Fraser et al. 2010; Rucklin and Donoghue 2019. G. Wagner (2014) discusses scales, feathers, and mammalian hair as novelties. On feathers as novelties: Benton et al. 2019; Prum and Brush 2002. The key developmental regulators involved in placode development are Wnts, BMP, and Shh, all familiar players in other vertebrate developmental pathways. The interactions between cell layers in these cases are examples of more general interactions between epithelial cell layers and the underlying mesenchyme, which have been a frequent source of developmental opportunity in different animal clades but are also implicated in the origins of many cancers. The duplications are among keratins and corneous β-proteins. On the protein novelties: Lowe et al. 2015.

48. Benton et al. 2019; Brusatte et al. 2015.

49. Moczek 2022.

50. There is a rich literature on vertebrate limb development and the fin-to-limb transition. G. Wagner (2014) provided a useful introduction, albeit one that predates work described in the next paragraph.

51. These studies include: Bi et al. 2021 (on primitive fish); K. Wang et al. 2021 (on African lungfish); Hawkins et al. 2021 (comparing zebra fish fin development to tetrapod limbs). The deep conservation of patterning between fin rays and digits involves the *Shh-Gli* gene pathway, as described by Letelier et al. (2021).

52. Nakamura et al. 2015.

53. On pangenomes: Golicz et al. 2020; J. McInerney et al. 2017, 2020.

54. The origin of eukaryotes reflects additional horizontal gene transfers beyond those associated with the transfer of genes from mitochondria and chloroplasts to the host genome. On the origin of eukaryotes: Eme et al. 2017; L. Katz 2012; Martin et al. 2015. For alternative views: Lake 2015; Ragan et al. 2009. An excellent summary and analysis of the overabundance of theories for the origin of eukaryotes is by Martin et al. (2015).

55. Martin et al. 2015, p. 8. On gene duplication: Vosseberg et al. 2021.

56. McFadden 2014; Miao et al. 2024.

57. Booth and Doolittle 2015, quote p. 7; N. Lane and Martin 2015. The debate was sparked by N. Lane and Martin (2010). See also Blackstone 2013; M. Lynch and Marinov 2015. On the energetics of eukaryotes and mitochondria: Schavemaker and Munoz-Gomez 2022.

58. Eukaryotic cellular complexity: Eme et al. 2017; Ten Tusscher 2020; Vosseberg et al. 2021.

59. On plastid symbiotic events: Delwiche 1999; Keeling 2010. Strictly speaking, the chloroplast symbiosis was not unique. At least one lineage, represented by the amoeba *Paulinella*, expelled its primary chloroplast and later took up a replacement. The recent study of sea slugs is by Cartaxana et al. (2021).

60. On horizontal gene transfer and novelty: Fournier et al. 2009; Soucy et al. 2015; Van Etten and Bhattacharya 2020. On challenges for phylogenetics: J. McInerney et al. 2011a, 2014; Oakley 2017. On herring antifreeze proteins: Graham and Davies 2021.

61. On transposons: Fedoroff 2012. Tetrapods: Cosby et al. 2021. Primates: Trizzino et al. 2017.

62. Simpson 1944, p. 1.

Chapter 7

1. M. Lynch 2007.

2. The contrast between development-push and ecosystem-pull views of novelty and innovation is explored more fully by Erwin (2017a). See also Godfrey-Smith 2014.

3. Hochberg et al. 2017. While Hochberg and colleagues described this potential as latent innovations, that assumes that the nature of an innovation is inherent in the novelty.

4. Bambach 2006; P. Hull and Darroch 2013.

5. Ezcurra and Butler 2018; T. Simoes et al. 2020. Examining novelties after an event can lead to confirmation bias and similar problems and is not a proper test of the hypothesis.

6. On PETM vertebrates: Grossnickle et al. 2019. On the role of anoxia in novelty and innovation: Wood and Erwin 2018.

7. On birds: Brusatte et al. 2015; Jarvis et al. 2014. On ammonoids: Whalen et al. 2020.

8. On tropics as a source of evolutionary innovation: Jablonski 1993. On the importance of energy: Vermeij 2002. For caution in interpreting claims for higher rates of innovation in the tropics: Jablonski 2017a.

9. Jablonski 2005, 2017a. As noted in these papers, there is also some evidence for a similar pattern during the Paleozoic.

10. Vermeij 2018.

11. Jablonski (2005) addresses ecological (innovation) versus developmental (novelty) explanations for preferential nearshore appearance.

12. Macekura 2020.

13. Key sources on adaptive radiations: Givnish 2015; Glor 2010; Losos 2010; Schluter 2000; Stroud and Losos 2016; Yoder et al. 2010.

14. On the ubiquity of disparity: Hughes et al. 2013. A classic test of early-burst models is by Harmon et al. (2010), although the results of this work have been questioned by later studies (Landis and Schraiber 2017; Uyeda et al. 2011, 2018; see also G. Slater 2015b).

15. Novack-Gottshall et al. 2022.

16. These pulsed patterns are described as a Lèvy process or a Lèvy flight, in which long intervals of relative stasis are interrupted by bouts of rapid evolution, as discussed in chapter 4. Key references: Landis and Schraiber 2017; Uyeda et al. 2011, 2018. Periodic pulses of evolutionary change have been described in echinoids by Hopkins and Smith (2015), in canids by G. Slater (2015a) and Van Valkenburgh (1991), and in reptiles by T. Simoes et al. (2020).

17. The compilation is by Soulebeau et al. (2015), who combed through a decade's worth of papers from the *Proceedings of the National Academy of Sciences, Journal of Geography, Systematic Biology, Proceedings of the Royal Society B,* and *Molecular Phylogenetics and Evolution.* In addition to the 10 cases mentioned, there were another 7 cases that required further study. Quote from Simpson 1960, p. 162.

18. On *Anolis:* Losos 2009; Mahler et al. 2013. On nonadaptive radiations: Czekanski-Moir and Rundell 2019; Gittenberger 1991; Rundell and Price 2009. These correspond to the type III radiations of Jablonski (2017b). See also M. Simoes et al. 2016. The radiation of sigmodontine rodents: Maestri et al. 2017. And frogs: Moen et al. 2021.

19. Erwin 2015a.

20. On *Prochlorococcus*: Biller et al. 2015; Chisholm 2017. The core genome is about 1000 genes of a total genome of 1800 to 2700 genes, with a known pangenome of about 80,000 genes, allowing for extensive diversity. The actual size of the pangenome is unknown, since each newly sequenced genome adds 100–200 new genes.

21. The dissimilarity approach to novel communities is from Radeloff et al. (2015), with a useful extension by Pandolfi et al. (2020). A related approach to the problem is by Muscente et al. (2022), but they use collections in the Paleobiology Database as ecological communities, and assuming coexisting species form ecological networks is a poor assumption (Freilich et al. 2018).

22. Novack-Gottshall 2007.

23. For a recent review of food webs and other ecological networks: Delmas et al. 2018. For everything you might ever want to know about networks: Newman 2018.

24. On bears and salmon: Helfield and Naiman 2006.

25. Darwin 1881; Simpson 1953b, see pp. 182, 199, 203.

26. Key references on niche construction: Odling-Smee 2024; Odling-Smee et al. 2003, 2013. On ecosystem engineering: Cuddington et al. 2007; C. G. Jones et al. 1997; J. Wright and Jones 2006. On the macroevolutionary dimensions: Crespi 2004; Erwin 2008b, 2024.

27. On complicated ecological networks: Angulo et al. 2021; Delmas et al. 2018; Golubski et al. 2016. A compatible but different approach to expanding ecological networks is by Segar et al. (2020). The Brazilian study is by Barbosa et al. (2023).

28. On the Burgess Shale food web: Dunne et al. 2008. The study disentangling ecological and taphonomic signals is by Shaw et al. (2021).

29. For complex networks: Battiston et al. 2020; Majhi et al. 2022. A newly developed approach described as new equilibrium theory of biodiversity dynamics incorporates nonlinear relationships between species numbers, community abundance, and estimates of extinction and origination, and shows promise for illuminating aspects of innovation (Storch et al. 2022).

30. Schumpeter 1942. On general-purpose technologies: Bresnahan and Trajtenberg 1995; Jovanovic and Rousseau 2005.

31. C_3 grasses employ a three-carbon molecule in sequestration of carbon dioxide, while C_4 grasses utilize a molecule with four carbon atoms at the critical site.

32. Vermeij 2017. Vermeij does not distinguish novelty from innovation and considers a variety of novelties and innovations. Among the innovations he evaluates are plant vascular anatomy, nutrient-mining plants, high photosynthetic capacity, vertebrate herbivory, nonmicrobial farming, aerial locomotion, endothermy, eusociality, animal defense of plants, echolocation, animal-mediated dispersal of plants and other animals, and communal construction of shelters and burrows.

Chapter 8

1. This section is updated from earlier work on the Ediacaran and Cambrian, particularly that of Erwin et al. (2011), Erwin and Tweedt (2011), and Erwin and Valentine (2013). Although the four phases I describe here differ from those identified by some colleagues, the basic structure reflects Wood et al. (2019), with recognition of these phases dependent upon recent papers integrating high-resolution geochemistry, age dates, and other data producing global correlations of the patterns of appearance of different clades, particularly by Bowyer et al. (2022), Darroch et al. (2018b), Wood et al. (2019), and Yang et al. (2021).

2. On Snowball Earth: Hoffman et al. 2017. And its aftermath: Shen et al. 2008. On continental emergence and its aftermath: Peters and Gaines 2012; but see Flowers et al. 2020. On increased nutrient supply: Brasier 1991, 1992; Wood et al. 2019. And implications for productivity: Laakso and Schrag 2019; Reinhard et al. 2017.

3. Among the wealth of papers on oxygen levels and early animals: Butterfield 2018; Cole et al. 2020; Erwin and Valentine 2013; Knoll and Sperling 2014; Lenton et al. 2014; Planavsky et al. 2018; Sperling 2016. On the oxygen control hypothesis: Mills and Canfield 2014. T. Lyons et al. (2021) provide a recent comprehensive review.

4. Butterfield 2009; Lenton et al. 2014.

5. On ecosystems of this interval: Erwin et al. 2011; Erwin and Tweedt 2011; Hoffman et al. 2017. On cnidarian diversification: Quattrini et al. 2020.

6. Several assemblages of possible metazoans predate the classic Ediacaran macroscopic fossils described here as phase 2, including the Lantian and Weng'an biotas in southern China, and fossils from northwest Canada. The Lantian biota (~615–585 million years ago; Yang et al. 2022) has abundant microfossils, macroscopic algae, and a number of the enigmatic forms, possible cnidarians and one possible bilaterian (Wan et al. 2016). The Weng'an fossils (~587–574 million years ago; Yang et al. 2022) are minute larvae; some may represent holozoans (Xiao et al. 2014; Yin et al. 2020). Metazoan affinities have been proposed for several of these forms, but such assignments remain uncertain.

7. Droser et al. (2017) and Xiao and Laflamme (2008) provide excellent introductions to the Ediacaran fauna, and Droser et al. (2019) to Nilpena. Bowyer et al. (2022) and Yang et al. (2021) provide the latest temporal frameworks for the geological and biological events.

8. Droser et al. 2017; S. Evans et al. 2019; Mangano and Buatois 2020.

9. S. Evans et al.'s (2021) analysis is based in part on groupings of the Ediacaran macrofossils by an earlier postdoctoral fellow, Marc Laflamme. See Erwin et al. 2011; Laflamme et al. 2013.

10. Darroch et al. 2018a; Wood et al. 2019; Xiao et al. 2020. The Shibantan trackway is described by Z. Chen et al. (2018).

11. Mangano and Buatois 2020; Turk et al. 2022. On the increase in ecosystem engineering during this phase: Cribb et al. 2019; Erwin and Tweedt 2011; Erwin and Valentine 2013; Seilacher 1999; Turk et al. 2022.

12. Aria 2022; Daley et al. 2018. On molecular clock estimates: Howard et al. 2022. The ratio of magnesium to calcium and the levels of carbon dioxide in Cambrian seas appear to have strongly influenced whether animals began secreting skeletons that were dominantly composed of the mineral calcite, or that of aragonite.

13. On biomineralization: Murdock 2020; Wood et al. 2017.

14. Daley et al. 2018; Paterson et al. 2019. The oldest trilobites mark the base of Series 2 and Stage 3 of the Cambrian and lie about 18 million years after the beginning of the period.

15. This account of the early history of arthropods is taken from Aria (2022), Chipman and Edgecombe (2019), Daley et al. (2018), Edgecombe (2020), and Giribet and Edgecombe (2019), and arguments from Erwin and Valentine (2013).

16. Erwin and Valentine 2013; Valentine 2004.

17. Nanglu et al. 2022; Zamora and Rahman 2014. The estimate of the number of clades is from Zamora and Rahman (2014). The comparison of morphological and ecological patterns is Novack-Gottshall et al. (2022).

18. Cloud 1976; Sagan 1967.

19. On the early history of eukaryotes: Cohen and Kodner 2022; Knoll 2014; Martin et al. 2015; J. McInerney et al. 2014; Mills et al. 2022.

20. On the fossil record: Cohen and Macdonald 2015; Knoll 2014. Fossils from northern Canada: Loron et al. 2021. The billion-year-old red algae: Butterfield 2000. But see discussion of age and affinities by Cohen and Kodner (2022). Structures as old as 3200 million years ago could be eukaryotic, but these are equivocal: Butterfield 2015; Javaux 2019.

21. Betts et al. 2018; Eme et al. 2014; Parfrey et al. 2011.

22. On the origins of mitochondria: Martin 2017; Martin et al. 2020; Mentel and Martin 2008; Mills et al. 2022; M. Müller et al. 2012. On syntrophy: Lopez-Garcia and Moreira 2020; M. McInerney et al. 2009; B. Morris et al. 2013.

23. Brocks et al. 2017; Gold et al. 2017. Gold and colleagues describe the origin of sterol biosynthesis through horizontal gene transfer between bacteria and early eukaryotes. An excellent introduction to lipid biomarkers and their pitfalls is by Summons et al. (2022).

24. There is a large and often contentious literature on Proterozoic oxygen levels, fueled by the introduction of new geochemical proxies followed by debates over the reliability of these proxies. Indeed, as I write this it remains unclear whether oxygen levels were low but stable through the Proterozoic, reached Cambrian levels as early as 2 billion years ago, or varied considerably through the Proterozoic. My reading of this literature is that oxygen levels were likely ~1% to 10% of present atmospheric levels but also highly variable. For recent overviews: Cole et al. 2020; Knoll 2014; Lenton et al. 2014; T. Lyons et al. 2014, 2021.

25. Sperling et al. 2021.

26. Porter 2020. On chromium isotopes: Cole et al. 2016; Planavsky et al. 2014.

27. Laakso and Schrag 2019; Ozaki et al. 2019. Eukaryotic metabolism requires more zinc than does bacterial, so examining the record of zinc isotopes through time provides another metric for ecological dominance of eukaryotes. Such a study independently corroborated the conclusion of increased ecological importance of eukaryotes after 800 million years ago (Isson et al. 2018).

28. Grossnickle et al. 2019; Luo 2007.

29. Grossnickle et al. 2019. On disparity and character saturation: Brocklehurst et al. 2021. On brawn before brains in Cenozoic mammals: Bertrand et al. 2022.

30. M. Chen et al. 2019. Although this study did not employ the network approach I advocated in chapter 7, it is notable for emphasizing ecological structure rather than taxonomic or phylogenetic diversity.

31. Further support for this scenario comes, as Chen et al. point out, from the independent evolution of grinding with pretribosphenic teeth among a clade of mammals in South America during the Late Cretaceous; see Harper et al. (2018).

32. Sauquet and Magallon 2018.

33. On angiosperm origins I have relied upon: Barba-Montoya et al. 2018; Friis et al. 2011; Herendeen et al. 2017; O. Katz 2018; H. Li et al. 2019; Ramirez-Barahona et al. 2020; Silvestro et al. 2021; Van der Kooi and Ollerton 2020.

34. Li et al. 2019. Another study with a slightly smaller data set more thoroughly explored the limitations of both fossil data and molecular analyses but reached divergence estimates and concluded that the initial crown-group divergences occurred during the Jurassic (Barba-Montoya et al. 2018).

35. Silvestro et al. 2021.

36. The Big Cedar Ridge flora: Wing et al. 1992, 2012. The broader analysis of the delayed innovation is by Ramirez-Barahona et al. (2020), involving some 1209 taxa distributed across 435 angiosperm families to construct an angiosperm phylogeny, with divergences dated using a molecular clock calibrated with 238 dated fossils; quote from p. 1235.

37. The Cretaceous terrestrial revolution: Lloyd et al. 2008; Vermeij and Grosberg 2010.

38. Benton et al. (2022) usefully summarize the patterns of diversification, their broader context, and the importance of changes in the Paleogene; they also cite the most recent work on Cretaceous diversifications. On insect diversity: Labandeira 2014; Labandeira and Sepkoski 1993. Although whether this pattern reflects poor preservation of insects remains debated. The increased diversity of Hemiptera, Hymenoptera, Coleoptera, and Lepidoptera is detailed by Benton et al. (2022). On insect-plant coevolution: Asar et al. 2022; Van der Kooi and Ollerton 2020.

39. Onstein (2020) describes trait flexibility and reinvention as critical features of angiosperms. The estimate is from Benton et al. (2022).

40. More specifically, as used here *birds* refers to the inclusive clade including all living birds as well as related extinct taxa. Some refer to this group as Avialae, and others prefer Aves, while

Brusatte et al. use Avialae/Aves; see discussion by Brusatte et al. (2015). For key sources on early bird innovations: Brusatte et al. 2014, 2015; Field et al. 2019; Xu et al. 2014; Yu et al. 2021.

41. On miniaturization of birds: Lee et al. 2014.

42. On bird diversification: Yu et al. 2021.

43. The whole-genome analysis is by Jarvis et al. (2014) and is consistent with earlier work (e.g., Feduccia 2003); the three major radiations were dissected by Yu et al. (2021). These results differ from earlier work suggesting greater diversification of Neoaves during the Cretaceous, but are based on larger data sets and more reliable analyses. Ksepka et al. (2017) compress the duration of radiation from the estimates by Jarvis et al. Yu et al. identified a third radiation 45–40 million years ago, near the Eocene–Oligocene cooling event. The ecological implications of the Jarvis result were discussed in that paper and by Brusatte et al. (2015).

44. Lister 2013a, b. Similar decoupling of morphology and diversity was discussed in chapter 7 and has been examined in horses (Cantalapiedra et al. 2017) and in other clades (Cooney and Thomas 2020).

45. Lister 2013b. On the broader context: Cantalapiedra et al. 2021.

46. Douglas 2014; Frare et al. 2018; D. Soltis et al. 1995; Van Velzen et al. 2018; Werner et al. 2014.

47. On the origin of snakes: T. Simoes et al. 2020. On transposable elements in repatterning GRNs: Erwin 2020b. On dietary diversification: Grundler and Rabosky 2021. The latter paper is a notable example of defining and studying the evolution of an ecological space, in this case a dietary space. Although this differs from the ecological network approach advocated in chapter 7, it nonetheless demonstrates coupling phylogenetic approaches to ecological innovations.

48. The classic paper on mantle fusion is by Stanley (1968); on the ecological impact of bulldozing, Thayer (1983).

49. Allf et al. 2016.

50. Vassallo et al. 2021.

51. On plasticity-first approaches: Levis et al. 2018; Levis and Pfennig 2018; Lister 2013a; West-Eberhard 2003, 2005.

52. Reader and Laland 2003b; Zuk et al. 2014.

53. The Baldwin effect was proposed independently in 1896 by James Baldwin, Henry Fairfield Osborn, and Conwy Lloyd Morgan, but when Simpson excavated these long-forgotten papers he christened it the Baldwin effect (Simpson 1953a; Weber and Depew 2003). Some work suggests that the issue may be more complex, with highly conserved genes involved in some behaviors representing a "behavioral tool kit" similar to the suite of highly conserved developmental regulatory genes (Rittschof and Robinson 2016).

54. Klump et al. 2021.

55. E. Wilson 1975. On social learning and culture in animals: Hoppit and Laland 2013; Whiten 2019, 2021.

56. That animal culture can impact selection, shape population structure, and influence evolution via gene-culture coevolution has been established for birds, cetaceans, and primates (H. Whitehead et al. 2019); the example of chimpanzee nut cracking as cultural inheritance is from Whiten (2017), who gives relevant references.

57. Whiten 2017.

58. On the possibilities of cumulative culture among nonhuman animals: Whiten 2019, 2021. On chimpanzees: Gunasekaram et al. 2024. On the cultural ratchet: Tennie et al. 2009. The extensions to cumulative cultural evolution described by Mesoudi and Thornton (2018) in the next chapter appear to be largely confined to humans.

59. The history of the cultural drive hypothesis and the efforts of Lala (previously Laland) to test and improve the idea are covered in chapter 6 of his (2017) *Darwin's Unfinished Symphony*. Laland's work evaluated primate cognitive capacities, and thus a measure of generalized

intelligence that is related to brain volume. This work also demonstrated the importance of generalized intelligence, thus discounting the claims of evolutionary psychologists that cognitive abilities have evolved as discrete modules.

60. Rehan and Toth 2015, p. 426.

61. Romiguier et al. 2022.

Chapter 9

1. I am indebted here to Corballis (2017), particularly his dissection of Chomsky in chapter 2. Also Pinker (2007), and on Merge, Berwick and Chomsky (2016).

2. On genetic impacts of gene-culture coevolution: Laland et al. 2010.

3. Pinker 2010; Tooby and Devore 1987. Boyd et al. (2011) critique the cognitive niche hypothesis. On the social brain hypothesis: Dunbar and Shultz 2007, 2021. In largely focusing here on social brains I do not discount the significant increases in human brain size and structure, some of which included important genetic novelties. Researchers have identified a change in a single amino acid in the human *transketolase-like* (*TKTL1*) gene, which they argue increased particular neurons in the neocortex, the outer region of the cerebral cortex, which is essential for human cognition (Pinson et al. 2022).

4. C. Everett 2017; Nunez 2017a, b. On Australian languages: Bowern and Zentz 2012. There are similar results from Nunez (2017a). Rafael Nunez also has excellent videos on YouTube.

5. Introductions to cultural evolution: Boyd 2018; Boyd and Richerson 1985; Henrich 2015; Laland 2017; Laland and Brown 2011; Mesoudi 2011; Sterelny 2016.

6. On cognitive gadgets: Heyes 2018. On lowered information costs: Boyd et al. 2011.

7. The significance of cumulative cultural evolution was established by Boyd and Richerson (1996) and Tomasello (1999). Quote from Henrich 2015, p. 7. On complex recipes: Charbonneau 2015. Essential recent syntheses: Charbonneau 2015; L. Dean et al. 2014; Henrich 2015; Laland 2017; Mesoudi and Thornton 2018; O'Madagain and Tomasello 2022; Tennie et al. 2009.

8. Mesoudi and Thornton 2018. The zone of latent solutions was described by Tennie et al. (2009). A recent paper considers the open-ended nature of human cultural evolution: T. Morgan and Feldman 2024.

9. Henrich 2015. Henrich's table 5.1 has multiple examples of gene-culture coevolution. See also Laland et al. 2010.

10. This feedback pattern is borrowed from Kirby (2017).

11. A recent restatement of these views is by Berwick and Chomsky (2016).

12. Bickerton 2014; Pinker 1994, 2007.

13. Over the past decade or so there have been numerous significant books on the constructivist approach to language. I have drawn upon: Bickerton 2014; Corballis 2017; D. Everett 2017; Heyes 2018; Planer and Sterelny 2022; Tomasello 2014. On charades: Christiansen and Chater 2022.

14. Kirby 2017, p. 118; Christiansen and Chater 2008, 2022.

15. Nishimura et al. 2022.

16. Jensen 2016.

17. As Boyd (2018) documents, kin selection does not apply in such cases as empirical studies have shown that groups of hunter-gatherers are much less closely related than commonly assumed. Although I disagree with Boyd's discussion of public goods in this book, he raises valid concerns about the origins of prosocial behavior.

18. Henrich 2020.

19. If you do not recognize Ada Lovelace there is no hope, but Lonnie Johnson invented the supersoaker and Grace Hopper invented compilers to translate computer languages into machine code. She was also the first woman to achieve the rank of rear admiral in the US Navy.

20. Muthukrishna and Henrich 2016; Schimmelpfennig et al. 2022.

21. Fogarty 2018; Fogarty and Creanza 2017; see also Henrich 2015.

22. Schimmelpfennig et al. 2022.

23. O'Madagain and Tomasello 2022; Planer and Sterelny 2022; Tomasello 2014.

24. As discussed by Bergstrom et al. (2021), it is challenging and probably pointless to attempt to define a single date for the origin of modern humans, particularly as evidence accumulates for matings between early *Homo sapiens*, Neandertals, and Denisovans, and equally complex population structures. Kuhn 2021. On "language-ready" hominins: Planer and Sterelny 2022, p. xix.

25. Shea (2017) and Snyder et al. (2022) have shown that Oldowan and similar tools can readily be constructed by naïve students and require no cumulative cultural evolution. Work in Kenya recently dated Oldowan tools to between 3.032 and 2.581 million years ago, in association with *Paranthropus* (Plummer et al. 2023). While the authors of this work do not suggest that *Paranthropus* necessarily made the tools, this discovery confirms that several hominin lineages may have used Oldowan tools. Moreover, tools similar to those made by early hominins (before about 1.5 million years ago) can also be produced by macaques (Proffitt et al. 2023), emphasizing the need for caution in ascribing cumulative cultural evolution to older tool assemblages.

26. On the Upper Paleolithic nonrevolution: McBrearty and Brooks 2000. On lags: Tattersall 2008. Quote from Sterelny 2016, p. 180.

27. D. Everett 2017. Everett argues that effective languages appeared with *Homo erectus* perhaps 1.9 million years ago, far older than most other estimates.

28. The definition of innovation as invention plus social learning is from Perry et al. (2021).

29. Technically this definition applies to utility patents, not design or plant patents. See "Patent Essentials," United States Patent and Trademark Office, published April 27, 2023, last updated February 18, 2025, https://www.uspto.gov/patents/basics/essentials. Patent law and the idea that protecting the rights of inventors to receive some rewards for their discovery long predate the Industrial Revolution. Article 1, Section 8, of the US Constitution provides that the Congress has the right to "Promote the progress of science and useful Arts, by securing for limited times to the authors and inventors the exclusive right to their respective writings and discoveries." The authors of the Constitution based this system on the English concept of letters patent issued by the sovereign to inventors to protect their discoveries and to encourage further invention. Letters patent date back to the Middle Ages, but despite the long history of patent law, defining what is novel, or sufficiently novel to be an invention warranting patent protection, continues to promote acrimony.

30. Darwin 1868.

31. Purugganan 2022; Zeder 2015. Termites, ants, and other insects also have domesticated various fungal species, and there have been some promising comparative studies of domestication and agriculture in human versus nonhuman systems. Here I confine my attention to human domestication of plants and animals, largely for agricultural purposes.

32. Zeder 2012, 2015. G. Larson and Fuller (2014) expanded upon these three pathways in the context of animal domestication.

33. Evershed et al. 2022.

34. B. Smith 2012, p. 1.

35. B. Smith 2016; Zeder 2016.

36. D. Lane 2016.

37. Q. Chen et al. 2020.

38. Fuentes 2017.

39. Henrich 2020; Powers et al. 2016.

40. Currie et al. 2021; Powers et al. 2016.

41. Turchin et al. 2022.

42. Carneiro 1970. On Oaxaca: Bawaya 2022; Nichols and Feinman 2022. On collective action: Crawford and Ostrom 1995. Graeber and Wengrow (2021) provide examples of

structured democracies similar to quasi-voluntary compliance, and indeed they make the controversial claim that the development of democracy in Europe was influenced by an indigenous critique of colonialism. Thompson and colleagues identified archaeological evidence for council houses among the ancestral Muskogean of the American Southeast dating to about 500 CE (V. Thompson et al. 2022). These proposals differ from earlier discussions of egalitarian societies, because social stratification was clearly present at Monte Alban and among the ancestral Muskogean.

43. Padgett and Powell 2012, quote p. 113.

44. Acemoglu and Robinson 2012.

45. On Moore's law: Moore 1965, 1975; Waldrop 2016.

46. W. B. Arthur 2009. Arthur recounts the story of clock subassemblies, crediting Herbert Simon.

47. W. B. Arthur and Polak 2006, p. 23.

48. W. B. Arthur 2009.

49. Valverde et al. 2007; Youn et al. 2015. Product spaces are analyzed by Fink and Reeves (2019) and Fink et al. (2017).

50. Wallace 1869.

Chapter 10

1. Le Goff 2015.

2. Perceptive readers may recognize that I have not framed this argument within the context of proposals for an extended evolutionary synthesis. Such proposals for replacing or extending the Modern Synthesis have their origins in comparative developmental studies, niche construction, multilevel selection theory, and epigenetics (Laland et al. 2015; Pigliucci 2007, 2009). Several of these topics have featured in this book. The Modern Synthesis has stalwart defenders (e.g., Futuyma 2015). For a paleontologist of my generation, proposals for extended evolution are reminiscent of arguments from Stephen Jay Gould, Jim Valentine, and others for macroevolution (Gould 1980a, 1985; Stanley 1979; Valentine 2004; for recent surveys: Jablonski 2017a, b).

3. Calcott 2009.

4. Lewontin 2003.

5. Mayr 1963. On species as individuals: Godfrey-Smith 2009; D. S. Hull 1980; Lidgard and Nyhart 2017; R. Wilson 1999; R. Wilson and Barker 2019.

6. There is a large literature on individuation and "scientific kinds," with contributions from both biologists and philosophers of biology. See Brigandt (2009), Buss (1987), and papers in the works edited by Bueno et al. (2018) and by R. Wilson (1999), particularly the paper by Boyd (1999). Some of this work initially focused on whether species constituted individuals, while work on major evolutionary transitions invokes new evolutionary kinds and will be considered later in this chapter.

7. On homeostatic property clusters: Boyd 1999; Griffiths 1999. On cell types: M. Slater 2013.

8. Moczek 2008.

9. DiFrisco et al. 2020, 2022.

10. The lucky leap model: Kolodny et al. 2015. This model is similar to (but more sophisticated than) earlier models: Valentine 1980; Valentine and Walker 1986.

11. On the invasion of open oceans: Whalen and Briggs 2018.

12. Berlin 1953.

13. Dworkin 2011; Gould 1989; Tetlock 2005.

14. James and Berlin quotes in Gould 1980b, p. 116; Gould quotes from Gould 1980b, pp. 113, 116. Dobzhansky 1937; Gould 2002.

15. On generalized Darwinism: Aldrich et al. 2008; Beinhocker 2006; Campbell 1960; Hodgson and Knudsen 2010; Nelson and Winter 1982. This may seem like academic hairsplitting, but generalized evolutionary approaches are very different from the universal Darwinism associated with Richard Dawkins, which seeks to extend a gene-centric approach with memes and replicators to many other domains (Dawkins 1983).

16. On theory reduction: Dupre 1993; Laubichler et al. 2018.

17. Van Valen 1973. Recent work has raised questions about whether extinctions are truly age independent. Red Queen quote: L. Carroll 1871, p. 18.

18. Schumpeter 1942. The distinction between Smithian and Schumpeterian growth is from work by Mokyr (esp. 2005, 2016); see also Landes (1976). Smithian growth in the prehistoric southwestern United States is described by Ortman and Lobo (2020).

19. A. Whitehead 1925, p. 141. See also Cowles 2020.

20. Nordhaus 1996. On GDP more generally: Macekura 2020. Since this is a pet peeve of mine, this comparison, although basically sound, involves an egregious example of the faulty comparison of rates. Nordhaus has compared a long-term rate to a short-term rate, evidently unaware that long-term rates of almost *anything* are lower than short-term rates, simply because a longer period averages over many intervals of no net change. The appropriate comparison would be of the 19 200-year bins between ancient Babylon and the dawn of the nineteenth century to the last 200 years.

21. On evolvability: R. Brown 2014; Jablonski 2022; Love et al. 2022; Payne and Wagner 2019; Riederer et al. 2022. Other key references for this section: R. Brown 2014; Gerhart and Kirschner 2007; Pigliucci 2008a; G. Wagner and Zhang 2011. Quote Jablonski 2022, p. 2.

22. On echinoids: Hopkins and Smith 2015.

23. Clune et al. 2013; Gerhart and Kirschner 2007; Jablonski 2017b, 2022; Pereira-Leal et al. 2006; Simpson 1944; Vermeij 2015; West-Eberhard 2019.

24. Gould 1989, quote p. 57.

25. On Cambrian contingency: Beatty 2006; Erwin 2016.

26. Desjardins 2011.

27. Conway Morris 2009; McGhee 2011, 2016.

28. Erwin 2015c.

29. Ogburn and Thomas 1922, quote p. 87.

30. Pinker 2011, 2018.

31. This quote comes from a letter from Lyell to Murchison of January 15, 1829 (Lyell 1881, 234). On the development of the ideas: Rudwick 2008. Gould (1965) and Simpson (1970) distinguish actualism from uniformitarianism.

32. Quote Bonner 2019, p. 301. For his earlier discussions: Bonner 2013, 2015. The mean size of organisms has increased through the history of life, and Bonner proposed that the appearance of sexual reproduction with the origin of the eukaryotic cell increased competition among species, enhancing the efficacy of natural selection. See also Erwin 2011; Maynard Smith and Szathmáry 1995.

33. M. Lynch 2007; discussion in Erwin and Valentine 2013.

34. Colnaghi et al. 2022.

Epilogue

1. Gordon 2016; similar discussion in Phelps 2013.

2. C. I. Jones, 2023. Jones's paper was inspired by the 2019 book *Empty Planet* by Bricker and Ibbitson.

3. In the original draft of this section, I had credited Obama's speech. Boyd's book (2018) recounts the more complete story . The quote is Boyd's transcription (p. 67). This was part of

Warren's standard stump speech, and one version is on YouTube: "Before Barack Said You Didn't Build That There Was Elizabeth Warren," posted by LSUDVM, YouTube, September 5, 2012, https://www.youtube.com/watch?v=LIyIJiPQRTI. Obama speech of 13 July: Eugene Kiely, "'You Didn't Build That,' Uncut and Unedited," FactCheck.org, posted July 23, 2012, last updated July 24, 2012, https://www.factcheck.org/2012/07/you-didnt-build-that-uncut-and-unedited/.

4. Duke study: Arora et al. 2015. US Government R&D spending: "Trends in Federal R&D, FY 1976–2014," American Association for the Advancement of Science, AAAS *Research and Development* series, 2014, https://www.aaas.org/sites/default/files/DefNon.jpg.

REFERENCES

Acemel RD, Tena JJ, Irastorza-Azcarate I, Marletaz F, Gomez-Marin C, et al. 2016. A single three-dimensional chromatin compartment in amphioxus indicates a stepwise evolution of vertebrate Hox bimodal regulation. *Nature Genetics* 48: 336–41.

Acemoglu D, Robinson JA. 2012. *Why Nations Fail: The Origins of Power, Prosperity, and Poverty.* New York: Crown.

Aghion P, Antonin C, Bunel S. 2021. *The Power of Creative Destruction.* Cambridge, MA: Harvard University Press.

Agrawal AA. 2001. Phenotypic plasticity in the interactions and evolution of species. *Science* 294: 321–26.

Aguirre J, Catalan P, Cuesta JA, Manrubia S. 2018. On the networked architecture of genotype spaces and its critical effects on molecular evolution. *Open Biology* 8: 180069.

Alberch P. 1989. The logic of monsters: Evidence for internal constraint in development and evolution. *Geobios* 22: 21–57.

Albu A. 2017. Fundamentals of innovation. In *Key Issues for Management of Innovative Projects,* ed. BL Moya, MDS de Garcia, LF Mazadiego, pp. 3–24. Rijeka, Croatia: InTech.

Aldrich HE, Hodgson GM, Hull DL, Knudsen T, Mokyr J. 2008. In defense of generalized Darwinism. *Journal of Evolutionary Economics* 18: 577–96.

Allf BC, Durst PA, Pfennig DW. 2016. Behavioral plasticity and the origins of novelty: The evolution of the rattlesnake rattle. *American Naturalist* 188: 475–83.

Amundson R. 2005. *The Changing Role of the Embryo in Evolutionary Thought.* Cambridge: Cambridge University Press.

Andersen KH, Berge T, Goncalves RJ, Hartvig M, Heuschele J, et al. 2016. Characteristic sizes of life in the oceans, from bacteria to whales. *Annual Review of Marine Science* 8: 217–41.

Andersson DI, Jerlstrom-Hultqvist J, Nasvall J. 2015. Evolution of new functions de novo and from preexisting genes. *Cold Spring Harbor Perspectives in Biology* 7: a017996.

Angulo MT, Kelley A, Montejano L, Song C, Saavedra S. 2021. Coexistence holes characterize the assembly and disassembly of multispecies systems. *Nature Ecology & Evolution* 5: 1091–101.

Anker A, Ahyong ST, Noel PY, Palmer AR. 2006. Morphological phylogeny of alpheid shrimps: Parallel preadaptation and the origin of a key morphological innovation, the snapping claw. *Evolution* 60: 2507–28.

Antognazza MR. 2009. *Leibniz: An Intellectual Biography.* Cambridge: Cambridge University Press.

Appel TA. 1987. *The Cuvier-Geoffroy Debate: French Biology in the Decades before Darwin.* Oxford: Oxford University Press.

Arenas-Mena C. 2017. The origins of developmental gene regulation. *Evolution & Development* 19: 96–107.

Aria C. 2022. The origin and early evolution of arthropods. *Biological Reviews* 97: 1786–809.

Arnegard ME, Zwickl DJ, Lu Y, Zakon HH. 2010. Old gene duplication facilitates origin and diversification of an innovative communication system—twice. *Proceedings of the National Academy of Sciences of the USA* 107: 22172–77.

Arnold FH. 2011. The library of Maynard Smith: My search for meaning in the protein universe. *Microbe* 6: 316–18.

Arnold SJ, Frender ME, Jones AG. 2001. The adaptive landscape as a conceptual bridge between micro- and macroevolution. *Genetica* 112: 9–32.

Arora A, Belenzon S, Paracconi A. 2015. Killing the golden goose? The decline of science in corporate R&D. *NBER Working Paper* 20902.

Arthur W. 2006. D'Arcy Thompson and the theory of transformations. *Nature Reviews Genetics* 7: 401–7.

Arthur WB. 2009. *The Nature of Technology: What It Is and How It Evolves.* New York: Free Press.

Arthur WB, Polak W. 2006. The evolution of technology within a simple computer model. *Complexity* 11: 23–31.

Asar Y, Ho SYW, Sauquet H. 2022. Early diversifications of angiosperms and their insect pollinators: Were they unlinked? *Trends in Plant Science* 27: 858–69.

Avena-Koenigsberger A, Goni J, Sole R, Sporns O. 2015. Network morphospace. *Journal of the Royal Society Interface* 12: 20140881.

Averof M, Cohen SM. 1997. Evolutionary origin of insect wings from ancestral gills. *Nature* 385: 627–30.

Babonis LS, Enjolras C, Reft AJ, Foster BM, Hugosson F, et al. 2023. Single-cell atavism reveals an ancient mechanism of cell type diversification in a sea anemone. *Nature Communications* 14: 885.

Baedke J. 2013. The epigenetic landscape in the course of time: Conrad Hal Waddington's methodological impact on the life sciences. *Studies in the History and Philosophy of Biological and Biomedical Sciences* 44: 756–73.

Baer KE, von. 1828. *Entwickelungsgeschichte der Thiere. Beobachtung und Reflexion.* Part 1. Königsberg: Bontrager.

Bambach RK. 1983. Ecospace utilization and guilds in marine communities through the Phanerozoic. In *Biotic Interactions in Recent and Fossil Benthic Communities*, ed. MJS Tevesz, PL McCall, pp. 719–46. New York: Plenum.

Bambach RK. 2006. Phanerozoic biodiversity mass extinctions. *Annual Review of Earth and Planetary Science* 34: 127–55.

Bambach RK, Bush AM, Erwin DH. 2007. Autecology and the filling of ecospace: Key metazoan radiations. *Palaeontology* 50: 1–22.

Bambach RK, Knoll AH, Sepkoski JJ, Jr. 2002. Anatomical and ecological constraints on Phanerozoic animal diversity in the marine realm. *Proceedings of the National Academy of Sciences of the USA* 99: 6854–59.

Barba-Montoya J, Dos Reis M, Schneider H, Donoghue PCJ, Yang Z. 2018. Constraining uncertainty in the timescale of angiosperm evolution and the veracity of a Cretaceous Terrestrial Revolution. *New Phytologist* 218: 819–34.

Barbosa M, Fernandes GW, Morris RJ. 2023. Experimental evidence for a hidden network of higher-order interactions in a diverse arthropod community. *Current Biology* 33: 381–88.e4.

Barnett HG. 1953. *Innovation: The Basis of Cultural Change.* New York: McGraw-Hill.

Bataillon T, Ezard THG, Kopp M, Masel J. 2022. Genetics of adaptation and fitness landscapes: From toy models to testable quantitative predictions. *Evolution* 76: 1104–7.

Bateman RM, DiMichele WA. 1994. Saltational evolution of form in vascular plants: A neo-Goldschmidtian synthesis. In: *Shape and Form in Plants and Fungi*, pp. 61–100. London: Linnean Society.

Bateman RM, DiMichele WA. 2002. Generating and filtering major phenotypic novelties: Neo-Goldschmidtian saltation revisited. In: *Developmental Genetics and Plant Evolution*, ed. QC Cronk, RM Bateman, JA Hawkins, pp. 109–59. London: Taylor and Francis.

Battiston F, Cencetti G, Iacopini I, Latora V, Lucas M, et al. 2020. Networks beyond pairwise interactions: Structure and dynamics. *Physics Reports* 874: 1–92.

Baumann O, Schmidt J, Stieglitz N. 2019. Effective search in rugged performance landscapes: A review and outlook. *Journal of Management* 45: 285–318.

Bawaya M. 2022. Identifying good government. *American Archaeology* 26: 19–25.

Beatty J. 2006. Replaying life's tape. *Journal of Philosophy* 103: 336–62.

Beatty J. 2016. The creativity of natural selection? Part i: Darwin, Darwinism, and the mutationists. *Journal of the History of Biology* 49: 659–84.

Beatty J. 2019. The creativity of natural selection? Part ii: The synthesis and since. *Journal of the History of Biology* 52: 705–31.

Beaulieu JM, O'Meara BC, Crane PR, Donoghue MJ. 2015. Heterogeneous rates of molecular evolution and diversification could explain the Triassic age estimate for angiosperms. *Systematic Biology* 64: 869–78.

Becker A. 2020. A molecular update on the origin of the carpel. *Current Opinion in Plant Biology* 53: 15–22.

Beinhocker ED. 2006. *Origins of Wealth: Evolution, Complexity and the Radical Remaking of Economics*. Cambridge, MA: Harvard Business Review.

Benson RBJ, Hunt G, Carrano MT, Campione N. 2018. Cope's rule and the adaptive landscape of dinosaur body size evolution. *Palaeontology* 61: 13–48.

Benton MJ, Dhouailly D, Jiang B, McNamara M. 2019. The early origin of feathers. *Trends in Ecology & Evolution* 34: 856–69.

Benton MJ, Wilf P, Sauquet H. 2022. The angiosperm terrestrial revolution and the origins of modern biodiversity. *New Phytologist* 233: 2017–35.

Bergstrom A, Stringer C, Hajdinjak M, Scerri EML, Skoglund P. 2021. Origins of modern human ancestry. *Nature* 590: 229–37.

Berlin I. 1953. *The Fox and the Hedgehog*. London: Weidenfeld and Nicholsons.

Bertrand OC, Shelley SL, Williamson TE, Wible JR, Chester SGB, et al. 2022. Brawn before brains in placental mammals after the end-Cretaceous extinction. *Science* 376: 80–85.

Berwick RC, Chomsky N. 2016. *Why Only Us: Language and Its Evolution*. Cambridge, MA: MIT Press.

Bettencourt LMA. 2020. Urban growth and the emergent statistics of cities. *Science Advances* 6: eaat8812.

Bettencourt LMA, Lobo J, Helbing D, Kuhnert C, West GB. 2007a. Growth, innovation, scaling, and the pace of life in cities. *Proceedings of the National Academy of Sciences of the USA* 104: 7301–6.

Bettencourt LMA, Lobo J, Strumsky D. 2007b. Invention in the city: Increasing returns to patenting as a scaling function of metropolitan size. *Research Policy* 36: 107–20.

Betts HC, Puttick MN, Clark JW, Williams TA, Donoghue PCJ, Pisani D. 2018. Integrated genomic and fossil evidence illuminates life's early evolution and eukaryote origin. *Nature Ecology & Evolution* 2: 1556–62.

Bhattacharya S, Zhang Q, Andersen ME. 2011. A deterministic map of Waddington's epigenetic landscape for cell fate specification. *BMC Systems Biology*: 5: 1–12.

Bi X, Wang K, Yang L, Pan H, Jiang H, et al. 2021. Tracing the genetic footprints of vertebrate landing in non-teleost ray-finned fishes. *Cell* 184: 1377–91.

Bickerton D. 2014. *More than Nature Needs*. Cambridge, MA: Harvard University Press.

Biller SJ, Berube PM, Lindell D, Chisholm SW. 2015. *Prochlorococcus*: The structure and function of collective diversity. *Nature Reviews Microbiology* 13: 13–27.

Bimber B. 1990. Karl Marx and the three faces of technological determinism. *Social Studies of Science* 20: 333–51.

Binford GJ, Wells MA. 2003. The phylogenetic distribution of sphingomyelinase D activity in venoms of *Haplogyne* spiders. *Comparative Biochemistry and Physiology Part B: Biochemistry and Molecular Biology* 135: 25–33.

Blackstone NW. 2013. Why did eukaryotes evolve only once? Genetic and energetic aspects of conflict and conflict mediation. *Philosophical Transactions of the Royal Society B* 368: 20120266.

Block WG. 2008. *The Unimaginable Mathematics of Borges' Library of Babel*. Oxford: Oxford University Press.

Blount ZD. 2016. History's windings in a flask: Microbial experiments into evolutionary contingency. In: *Chance in Evolution*, ed. GR Ramsey, CH Pence, pp. 244–63. Chicago: University of Chicago Press.

Blount ZD, Barrick JE, Davidson CJ, Lenski RE. 2012. Genomic analysis of a key innovation in an experimental *Escherichia coli* population. *Nature* 489: 513–18.

Blount ZD, Borland CZ, Lenski RE. 2008. Historical contingency and the evolution of a key innovation in an experimental population of *Escherichia coli*. *Proceedings of the National Academy of Sciences of the USA* 105: 7899–906.

Blount ZD, Lenski RE, Losos JB. 2018. Contingency and determinism in evolution: Replaying life's tape. *Science* 362: aam5979.

Bogorad LD, Deem MW. 1999. A hierarchical approach to protein molecular evolution. *Proceedings of the National Academy of Sciences of the USA* 96: 2591–95.

Bolhuis JJ, Tattersall I, Chomsky N, Berwick RC. 2014. How could language have evolved? *PLoS Biology* 12: e1001934.

Bonner JT, ed. 1982. *Evolution and Development*. Berlin: SpringerVerlag.

Bonner JT. 1988. *The Evolution of Complexity*. Princeton, NJ: Princeton University Press.

Bonner JT. 2013. *Randomness in Evolution*. Princeton, NJ: Princeton University Press.

Bonner JT. 2015. The evolution of evolution: Seen through the eyes of a slime mold. *BioScience* 65: 1184–87.

Bonner JT. 2019. The evolution of evolution. *Journal of Experimental Biology (Molecular Developmental Evolution)* 332: 301–6.

Booth A, Doolittle WF. 2015. Eukaryogenesis, how special really? *Proceedings of the National Academy of Sciences of the USA* 112: 10278–85.

Borges JL. 2007. Library of Babel. In: *Labyrinths*, ed. DA Yates, JE Irby, pp. 51–58. New York: New Directions.

Bowern C, Zentz J. 2012. Diversity in the numeral systems of Australian languages. *Anthropological Linguistics* 54: 133–60.

Bowler PJ. 1992. *The Eclipse of Darwinism*. Baltimore, MD: Johns Hopkins University Press.

Bowler PJ. 1996. *Evolution: History of an Idea*. Berkeley: University of California Press.

Bowles AMC, Bechtold U, Paps J. 2020. The origin of land plants is rooted in two bursts of genomic novelty. *Current Biology* 30: 530–36.e2.

Bowyer FT, Zhuravlev AY, Wood R, Shields GA, Zhou Y, et al. 2022. Calibrating the temporal and spatial dynamics of the Ediacaran–Cambrian radiation of animals. *Earth-Science Reviews* 225: 103913.

Boyce CK, Fan Y, Zwieniecki MA. 2017. Did trees grow up to the light, up to the wind, or down to the water? How modern high productivity colors perception of early plant evolution. *New Phytologist* 215: 552–57.

Boyce CK, Knoll AH. 2002. Evolution of developmental potential and the multiple independent origins of leaves in Paleozoic vascular plants. *Paleobiology* 28: 70–100.

Boyce CK, Lee JE. 2017. Plant evolution and climate over geological timescales. *Annual Review of Earth and Planetary Science* 45: 61–87.

Boyd R. 1999. Homeostasis, species, and higher taxa. In: *Species. New Interdisciplinary Essays*, ed. RA Wilson, pp. 141–85. Cambridge, MA: MIT Press.

Boyd R. 2018. *A Different Kind of Animal*. Princeton, NJ: Princeton University Press.

Boyd R, Richerson PJ. 1985. *Culture and the Evolutionary Process*. Chicago: University of Chicago Press.

Boyd R, Richerson PJ. 1996. Why culture is common, but cultural evolution is rare. *Proceedings of the British Academy* 88: 77–93.

Boyd R, Richerson PJ, Henrich J. 2011. The cultural niche: Why social learning is essential for human adaptation. *Proceedings of the National Academy of Sciences of the USA* 108, Suppl 2: 10918–25.

Boyer P. 2018. *Minds Make Societies. How Cognition Explains the World Humans Create*. New Haven, CT: Yale University Press.

Branch G. 2015. Whence "arrival of the fittest"? *National Center for Science Education blog*, May 27, 2015. https://ncse.ngo/whence-arrival-fittest.

Brasier MD. 1991. Nutrient flux and the evolutionary explosion across the Precambrian-Cambrian boundary interval. *Historical Biology* 5: 85–93.

Brasier MD. 1992. Nutrient-enriched waters and the early skeletal fossil record. *Journal of the Geological Society* 149: 621–29.

Bresnahan TF, Trajtenberg M. 1995. General purpose technologies "engines of growth"? *Journal of Econometrics* 65: 83–108.

Brigandt I. 2009. Natural kinds in evolution and systematics: Metaphysical and epistemological considerations. *Acta Biotheoretica* 57: 77–79.

Brigandt I, Love AC. 2012. Conceptualizing evolutionary novelty: Moving beyond definitional debates. *Journal of Experimental Biology (Molecular Developmental Evolution)* 318B: 417–27.

Brocklehurst N, Panciroli E, Benevento GL, Benson RBJ. 2021. Mammaliaform extinctions as a driver of the morphological radiation of Cenozoic mammals. *Current Biology* 31: 2955–63.

Brocks JJ, Jarrett AJM, Sirantoine E, Hallmann C, Hoshino Y, Liyanage T. 2017. The rise of algae in Cryogenian oceans and the emergence of animals. *Nature* 548: 578–81.

Brown JH, Gillooly JF, Allen AP, Savage VM, West GB. 2004. Toward a metabolic theory of ecology. *Ecology* 85: 1771–89.

Brown RL. 2014. What evolvability really is. *British Journal for the Philosophy of Science* 65: 549–72.

Bruce HS, Patel NH. 2020. Knockout of crustacean leg patterning genes suggests that insect wings and body walls evolved from ancient leg segments. *Nature Ecology & Evolution* 4: 1703–12.

Bruce HS, Patel NH. 2022. The *Daphnia* carapace and other novel structures evolved via the cryptic persistence of serial homologs. *Current Biology* 32: 3792–99.e3.

Brunet T, King N. 2017. The origin of animal multicellularity and cell differentiation. *Developmental Cell* 43: 124–40.

Brunet T, Larson BT, Linden TA, Vermeij MJA, McDonald K, King N. 2019. Light-regulated collective contractility in a multicellular choanoflagellate. *Science* 366: 326–34.

Brunette TJ, Parmeggiani F, Huang PS, Bhabha G, Ekiert DC, et al. 2015. Exploring the repeat protein universe through computational protein design. *Nature* 528: 580–84.

Brusatte SL, Lloyd GT, Wang SC, Norell MA. 2014. Gradual assembly of avian body plan culminated in rapid rates of evolution across the dinosaur–bird transition. *Current Biology* 24: 2386–92.

Brusatte SL, O'Connor JK, Jarvis ED. 2015. The origin and diversification of birds. *Current Biology* 25: R888–98.

Bueno O, Chen RL, Fagan MB, eds. 2018. *Individuation, Processes, and Scientific Practices*. Oxford: Oxford University Press.

Buffon GL Leclerc de. 1778. *Histoire naturelle des époques de la nature.* Paris: Imprimerie Royale.

Burkhardt RW, Jr. (1977) 1995. *The Spirit of System: Lamark and Evolutionary Biology.* Cambridge, MA: Harvard University Press.

Buss LW. 1987. *The Evolution of Individuality.* Princeton, NJ: Princeton University Press.

Butterfield NJ. 2000. *Bangiomorpha pubescens* n. gen., n. sp.: Implications for the evolution of sex, multicellularity, and the Mesoproterozoic/Neoproterozoic radiation of eukaryotes. *Paleobiology* 26: 386–404.

Butterfield NJ. 2009. Oxygen, animals and oceanic ventilation: An alternative view. *Geobiology* 7: 1–7.

Butterfield NJ. 2015. Early evolution of the Eukaryota. *Palaeontology* 58: 5–17.

Butterfield NJ. 2018. Oxygen, animals and aquatic bioturbation: An updated account. *Geobiology* 16: 3–16.

Cain J. 1994. Ernst Mayr as community architect: Launching the Society for the Study of Evolution and the journal *Evolution. Biology & Philosophy* 9: 387–427.

Calcott B. 2009. Lineage explanations: Explaining how biological mechanisms change. *British Journal of Philosophy of Science* 60: 51–78.

Calcott B, Sterelny K, eds. 2011. *The Major Transitions in Evolution Revisited.* Cambridge, MA: MIT Press.

Campbell D. 1960. Blind variation and selective retention in creative thought as in other knowledge processes. *Psychological Review* 67: 380–400.

Canfield DE. 2005. The early history of atmospheric oxygen: Homage to Robert M. Garrels. *Annual Review of Earth and Planetary Sciences* 33: 1–36.

Canfield DE. 2014. *Oxygen: A Four Billion Year History.* Princeton, NJ: Princeton University Press.

Cantalapiedra JL, Prado JL, Hernandez Fernandez M, Alberdi MT. 2017. Decoupled ecomorphological evolution and diversification in Neogene–Quaternary horses. *Science* 355: 627–30.

Cantalapiedra JL, Sanisidro O, Zhang H, Alberdi MT, Prado JL, et al. 2021. The rise and fall of proboscidean ecological diversity. *Nature Ecology & Evolution* 5: 1266–72.

Carlisle E, Yin Z, Pisani D, Donoghue PC. 2024. Ediacaran origin and Ediacaran–Cambrian diversification of Metazoa. *Science Advances* 10: eadp7161.

Carneiro RL. 1970. A theory of the origin of the state. *Science* 169: 733–38.

Carroll L. 1871. *Through the Looking-Glass, and What Alice Found There.* London: Macmillan.

Carroll L. 1879. A new puzzle. *Vanity Fair* 21: 185–86.

Carroll SB. 2005. Evolution at two levels: On genes and form. *PLoS Biology* 3: 1159–66.

Cartaxana P, Rey F, LeKieffre C, Lopes D, Hubas C, et al. 2021. Photosynthesis from stolen chloroplasts can support sea slug reproductive fitness. *Proceedings of the Royal Society B* 288: 20211779.

Casewell NR, Wuster W, Vonk FJ, Harrison RA, Fry BG. 2013. Complex cocktails: The evolutionary novelty of venoms. *Trends in Ecology & Evolution* 28: 219–29.

Catling DC, Glein CR, Zahnle KJ, McKay CP. 2005. Why O_2 is required by complex life on habitable planets and the concept of planetary "oxygenation time." *Astrobiology* 5: 415–37.

Cavalli-Sforza LL, Feldman MW. 1981. *Cultural Transmission and Evolution: A Quantitative Approach.* Princeton, NJ: Princeton University Press.

Chamberlin TC. 1909. Diastrophism as the ultimate basis of correlation. *Journal of Geology* 17: 685–93.

Charbonneau M. 2015. All innovations are equal, but some more than others: (Re)integrating modification processes to the origins of cumulative culture. *Biological Theory* 10: 322–35.

Chatterjee K, Pavlogiannis A, Adlam B, Nowak MA. 2014. The time scale of evolutionary innovation. *PLoS Computational Biology* 10: e1003818.

Chen LB, DeVries AL, Cheng CC. 1997. Convergent evolution of antifreeze glycoproteins in Antarctic notothenioid fish and Arctic cod. *Proceedings of the National Academy of Sciences of the USA* 94: 3817–22.

Chen M, Strömberg CAE, Wilson GP. 2019. Assembly of modern mammal community structure driven by Late Cretaceous dental evolution, rise of flowering plants, and dinosaur demise. *Proceedings of the National Academy of Sciences of the USA* 116: 9931–40.

Chen Q, Samayoa LF, Yang CJ, Bradbury PJ, Olukolu BA, et al. 2020. The genetic architecture of the maize progenitor, teosinte, and how it was altered during maize domestication. *PLoS Genetics* 16: e1008791.

Chen Z, Chen X, Zhou CM, Yuan XL, Xiao SH. 2018. Late Ediacaran trackways produced by bilaterian animals with paired appendages. *Science Advances* 4: eaao6691.

Cheng J, Yao X, Li X, Yue L, Duan X, et al. 2023. Diversification of ranunculaceous petals in shape supports a generalized model for plant lateral organ morphogenesis and evolution. *Science Advances* 9: eadf8049.

Chipman AD. 2010. Parallel evolution of segmentation by co-option of ancestral gene regulatory networks. *BioEssays* 32: 60–70.

Chipman AD, ed. 2020. *Cellular Processes in Segmentation.* Boca Raton, FL: CRC.

Chipman AD, Edgecombe GD. 2019. Developing an integrated understanding of the evolution of arthropod segmentation using fossils and evo-devo. *Proceedings of the Royal Society B* 286: 20191881.

Chisholm SW. 2017. *Prochlorococcus. Current Biology* 27: R447–48.

Chomicki G, Coiro M, Renner SS. 2017. Evolution and ecology of plant architecture: Integrating insights from the fossil record, extant morphology, developmental genetics and phylogenies. *Annals of Botany* 120: 855–91.

Chorley RJ. 1963. Diastrophic background to twentieth-century geomorphological thought. *Geological Society of America Bulletin* 74: 953–70.

Christensen CM. 1997. *The Innovator's Dilemma: When New Technologies Cause Great Firms to Fail.* Cambridge, MA: Harvard Business School.

Christiansen MH, Chater N. 2008. Language as shaped by the brain. *Behavioral and Brain Sciences* 31: 489–509.

Christiansen MH, Chater N. 2022. *The Language Game.* New York: Basic Books.

Claidiere N, Scott-Phillips TC, Sperber D. 2014. How Darwinian is cultural evolution? *Philosophical Transactions of the Royal Society B* 369: 20130368.

Clark JW, Donoghue PCJ. 2017. Constraining the timing of whole genome duplication in plant evolutionary history. *Proceedings of the Royal Society B* 284: 20170912.

Clark JW, Hetherington AJ, Morris JL, Pressel S, Duckett JG, et al. 2023. Evolution of phenotypic disparity in the plant kingdom. *Nature Plants* 9: 1618–26.

Clark-Hachtel C, Fernandez-Nicolas A, Belles X, Tomoyasu Y. 2021. Tergal and pleural wing-related tissues in the German cockroach and their implication to the evolutionary origin of insect wings. *Evolution & Development* 23: 100–116.

Clark-Hachtel C, Tomoyasu Y. 2016. Exploring the origin of insect wings from an evo-devo perspective. *Current Opinion in Insect Science* 13: 77–85.

Clark-Hachtel C, Tomoyasu Y. 2020. Two sets of candidate crustacean wing homologues and their implication for the origin of insect wings. *Nature Ecology & Evolution* 4: 1694–702.

Clements T, Dolocan A, Martin P, Purnell MA, Vinther J, Gabbott SE. 2016. The eyes of *Tullimonstrum* reveal a vertebrate affinity. *Nature* 532: 500–503.

Cloud PE. 1948. Some problems and patterns of evolution exemplified by fossil invertebrates. *Evolution* 2: 322–50.

Cloud PE. 1976. Beginnings of biospheric evolution and their biogeochemical consequences. *Paleobiology* 2: 351–87.

Clune J, Mouret JB, Lipson H. 2013. The evolutionary origins of modularity. *Proceedings of the Royal Society B* 280: 20122863.

Coddington JA. 1988. Cladistic tests of adaptational hypotheses. *Cladistics* 4: 3–22.

Cohen PA, Kodner RB. 2022. The earliest history of eukaryotic life: Uncovering an evolutionary story through the integration of biological and geological data. *Trends in Ecology & Evolution* 37: 246–56.

Cohen PA, Macdonald FA. 2015. The Proterozoic record of eukaryotes. *Paleobiology* 41: 610–32.

Cole DB, Mills DB, Erwin DH, Sperling EA, Porter SM, et al. 2020. On the co-evolution of surface oxygen levels and animals. *Geobiology* 18: 260–81.

Cole DB, Reinhard CT, Wang X, Gueguen B, Halverson GP, et al. 2016. A shale-hosted Cr isotope record of low atmospheric oxygen during the Proterozoic. *Geology* 44: 555–58.

Colnaghi M, Lane N, Pomiankowski A. 2022. Repeat sequences limit the effectiveness of lateral gene transfer and favored the evolution of meiotic sex in early eukaryotes. *Proceedings of the National Academy of Sciences of the USA* 119: e2205041119.

Conway Morris S. 2009. The predictability of evolution: Glimpses into a post-Darwinian world. *Naturwissenschaften* 96: 1313–37.

Cooney CR, Thomas GH. 2020. Heterogeneous relationships between rates of speciation and body size evolution across vertebrate clades. *Nature Ecology & Evolution* 5: 101–10.

Corballis MC. 2017. *The Truth about Language: What It Is and Where It Came From.* Chicago: University of Chicago Press.

Cornes R, Sandler T. 1986. *The Theory of Externalities, Public Goods and Club Goods.* Cambridge: Cambridge University Press.

Cosby RL, Judd J, Zhang R, Zhong A, Garry N, et al. 2021. Recurrent evolution of vertebrate transcription factors by transposase capture. *Science* 371: eabc6405.

Cowles HI. 2020. *The Scientific Method: An Evolution of Thinking from Darwin to Dewey.* Cambridge, MA: Harvard University Press.

Cracraft J. 1985. Biological diversification and its causes. *Annals of the Missouri Botanical Garden* 72: 794–822.

Cracraft J. 1990. The origin of evolutionary novelties: Pattern and process at different hierarchical levels. In: *Evolutionary Innovations*, ed. MH Nitecki, pp. 21–44. Chicago: University of Chicago Press.

Crawford SES, Ostrom E. 1995. A grammar of institutions. *American Political Science Review* 89: 582–600.

Crespi BJ. 2004. Vicious circles: Positive feedback in major evolutionary and ecological transitions. *Trends in Ecology & Evolution* 19: 627–33.

Cribb AT, Kenchington CG, Koester B, Gibson BM, Boag TH, et al. 2019. Increase in metazoan ecosystem engineering prior to the Ediacaran-Cambrian boundary in the Nama Group, Namibia. *Royal Society Open Science* 6: 190548.

Crown PL, Marden K, Mattson HV. 2016. Foot notes: The social implications of polydactyly and foot-related imagery at Pueblo Bonito, Chaco Canyon. *American Antiquity* 81: 426–48.

Cuddington K, Byers JE, Wilson WG, Hastings A, eds. 2007. *Ecosystem Engineers: Plants to Protists.* London: Academic.

Cunningham JA, Liu AG, Bengtson S, Donoghue PCJ. 2017. The origin of animals: Can molecular clocks and the fossil record be reconciled? *BioEssays* 39: 1–12.

Currie TE, Campenni M, Flitton A, Njagi T, Ontiri E, et al. 2021. The cultural evolution and ecology of institutions. *Philosophical Transactions of the Royal Society B* 376: 20200047.

Currie TE, Mace R. 2011. Mode and tempo in the evolution of socio-political organization: Reconciling "Darwinian" and "Spencerian" evolutionary approaches in anthropology. *Philosophical Transactions of the Royal Society B* 366: 1108–17.

Czekanski-Moir JE, Rundell RJ. 2019. The ecology of nonecological speciation and nonadaptive radiations. *Trends in Ecology & Evolution* 34: 400–415.

Daley AC, Antcliffe JB, Drage HB, Pates S. 2018. Early fossil record of Euarthropoda and the Cambrian Explosion. *Proceedings of the National Academy of Sciences of the USA* 115: 5323–31.

Darroch SAF, Laflamme M, Wagner PJ. 2018a. High ecological complexity in benthic Ediacaran communities. *Nature Ecology & Evolution* 2: 1541–47.

Darroch SAF, Smith EF, Laflamme M, Erwin DH. 2018b. Ediacaran Extinction and Cambrian Explosion. *Trends in Ecology & Evolution* 33: 653–63.

Darwin C. 1859. *On the Origin of Species by means of Natural Selection.* London: John Murray.

Darwin C. 1868. *The Variation of Animals and Plants under Domestication.* London: John Murray.

Darwin C. 1872. *On the Origin of Species by means of Natural Selection.* 6th ed. London: John Murray.

Darwin C. 1881. *The Formation of Vegetable Mould, through the Action of Worms.* London: John Murray.

Davidson EH. 2006. *The Regulatory Genome.* San Diego, CA: Academic.

Davidson EH, Erwin DH. 2006. Gene regulatory networks and the evolution of animal body plans. *Science* 311: 796–800.

Davidson EH, Erwin DH. 2010. Evolutionary innovation and stability in animal gene networks. *Journal of Experimental Zoology Part B (Molecular Developmental Evolution)* 312: 182–86.

Dawkins R. 1983. Universal Darwinism. In *Evolution from Molecules to Men,* ed. DS Bendall, pp. 403–25. Cambridge: Cambridge University Press.

Dawkins R. 1986. *The Blind Watchmaker.* New York: Norton.

Dawkins R. 1989. *The Selfish Gene.* Oxford: Oxford University Press.

Dean AM, Thornton JW. 2007. Mechanistic approaches to the study of evolution: The functional synthesis. *Nature Reviews Genetics* 8: 675–88.

Dean LG, Vale GL, Laland KN, Flynn E, Kendal RL. 2014. Human cumulative culture: A comparative perspective. *Biological Reviews* 89: 284–301.

Dediu D, Cysouw M, Levinson SC, Baronchelli A, Christiansen MH, et al. 2013. Cultural evolution of language. In *Cultural Evolution: Society, Technology, Language, and Religion,* ed. PJ Richerson, MH Christiansen, pp. 303–32. Cambridge, MA: MIT Press.

Degler CN. 1991. *In Search of Human Nature.* Oxford: Oxford University Press.

Delmas E, Besson M, Brice MH, Burkle LA, Dalla Riva GV, et al. 2018. Analysing ecological networks of species interactions. *Biological Reviews* 94: 16–36.

Delwiche CF. 1999. Tracing the thread of plastid diversity through the tapestry of life. *American Naturalist* 154, Suppl: S164–77.

Dennett DC. 1995. *Darwin's Dangerous Idea.* New York: Simon and Schuster.

Dennett DC. 2017. *From Bacteria to Bach and Back: The Evolution of Minds.* New York: Norton.

Desjardins EC. 2011. Historicity and experimental evolution. *Biology & Philosophy* 26: 339–64.

De Visser JAGM, Krug J. 2014. Empirical fitness landscapes and the predictability of evolution. *Nature Reviews Genetics* 15: 480–90.

De Vladar HP, Santos M, Szathmary E. 2017. Grand views of evolution. *Trends in Ecology & Evolution* 32: 324–34.

De Vries H. 1906. *Species and Varieties: Their Origin by Mutation.* Chicago: Open Court.

Dhouailly D, Godefroit P, Martin T, Nonchev S, Caraguel F, Oftedal O. 2019. Getting to the root of scales, feather and hair: As deep as odontodes? *Experimental Dermatology* 28: 503–8.

Dickson BV, Clack JA, Smithson TR, Pierce SE. 2021. Functional adaptive landscapes predict terrestrial capacity at the origin of limbs. *Nature* 589: 242–45.

Dietrich MR. 2003. Richard Goldschmidt: Hopeful monsters and other "heresies." *Nature Reviews Genetics* 4: 68–74.

Dietrich MR, Skipper RA, Jr. 2012. A shifting terrain: A brief history of the adaptive landscape. In *The Adaptive Landscape in Evolutionary Biology*, ed. EI Svenson, R Calsbeek, pp. 3–25. Oxford: Oxford University Press.

DiFrisco J, Love AC, Wagner GP. 2020. Character identity mechanisms: A conceptual model for comparative-mechanistic biology. *Biology & Philosophy* 35: 44.

DiFrisco J, Wagner GP, Love AC. 2022. Reframing research on evolutionary novelty and co-option: Character identity mechanisms versus deep homology. *Seminars in Cell & Developmental Biology* 145: 3–12.

Di Gregorio M. 1984. *T. H. Huxley's Place in Natural Science*. New Haven, CT: Yale University Press.

Dobzhansky T. 1937. *Genetics and the Origin of Species*. New York: Columbia University Press.

Dochtermann NA, Klock B, Roff DA, Royaute R. 2023. Drift on holey landscapes as a dominant evolutionary process. *Proceedings of the National Academy of Sciences of the USA* 120: e2313282120.

Dohrn A. 1875. *Princip des Functionswechsels*. Leipzig: Englemann.

Donoghue MJ. 1992. Homology. In *Keywords in Evolutionary Biology*, ed. EF Keller, EA Lloyd, pp. 170–79. Cambridge, MA: Harvard University Press.

Donoghue MJ, Sanderson MJ. 2015. Confluence, synnovation, and depauperons in plant diversification. *New Phytologist* 207: 260–74.

Donoghue PCJ, Harrison CJ, Paps J, Schneider H. 2021. The evolutionary emergence of land plants. *Current Biology* 31: R1281–98.

Douglas AE. 2014. Symbiosis as a general principle in eukaryotic evolution. *Cold Spring Harbor Perspectives in Biology* 6: a016113.

Doyle JA. 2012. Molecular and fossil evidence on the origin of angiosperms. *Annual Review of Earth and Planetary Science* 40: 301–26.

Droser ML, Gehling JG. 2015. The advent of animals: The view from the Ediacaran. *Proceedings of the National Academy of Sciences of the USA* 112: 4865–70.

Droser ML, Gehling JG, Tarhan LG, Evans SD, Hall CMS, et al. 2019. Piecing together the puzzle of the Ediacara biota: Excavation and reconstruction at the Ediacara National Heritage site Nilpena. *Palaeogeography, Palaeoclimatology, Palaeoecology* 513: 132–45.

Droser ML, Tarhan LG, Gehling JG. 2017. The rise of animals in a changing environment: Global ecological innovation in the late Ediacaran. *Annual Review of Earth and Planetary Sciences* 45: 593–617.

Dunbar RI, Shultz S. 2007. Evolution in the social brain. *Science* 317: 1344–47.

Dunbar RIM, Shultz S. 2021. Social complexity and the fractal structure of group size in primate social evolution. *Biological Reviews* 96: 1889–906.

Dunn CW, Giribet G, Edgecombe GD, Hejnol A. 2014. Animal phylogeny and its evolutionary implications. *Annual Review of Ecology, Evolution and Systematics* 45: 371–95.

Dunne JA, Williams RJ, Martinez ND, Wood RA, Erwin DH. 2008. Compilation and network analysis of Cambrian food webs. *PLoS Biology* 6: e102.

Dunnell RC. 1980. Evolutionary theory and archaeology. *Advances in Archaeological Method and Theory* 3: 35–99.

Dupre J. 1993. *The Disorder of Things: Metaphysical Foundations of the Disunity of Science*. Cambridge, MA: Harvard University Press.

Dworkin R. 2011. *Justice for Hedgehogs*. Cambridge, MA: Harvard University Press.

Eco U. 1983. *The Name of the Rose*. New York: Harcourt Brace Jovanovich.

Edgecombe GD. 2020. Arthropod origins: Integrating paleontological and molecular evidence. *Annual Review of Ecology, Evolution and Systematics* 51: 1–25.

Edwards EJ, Osborne CP, Stromberg CA, Smith SA, Bond WJ, et al. 2010. The origins of C4 grasslands: Integrating evolutionary and ecosystem science. *Science* 328: 587–91.

Eldredge N, Gould SJ. 1972. Punctuated equilibria: An alternative to phyletic gradualism. In *Models in Paleobiology*, ed. TJM Schopf, pp. 82–115. San Francisco: Freeman.

Eme L, Sharpe SC, Brown MW, Roger AJ. 2014. On the age of eukaryotes: Evaluating evidence from fossils and molecular clocks. *Cold Spring Harbor Perspectives in Biology* 6: a016139.

Eme L, Spang A, Lombard J, Stairs CW, Ettema TJG. 2017. Archaea and the origin of eukaryotes. *Nature Reviews Microbiology* 15: 711–23.

Erwin DH. 1992. A preliminary classification of radiations. *Historical Biology* 6: 133–47.

Erwin DH. 2007a. Disparity: Morphological pattern and developmental context. *Palaeontology* 50: 57–73.

Erwin DH. 2007b. Innovation without imagination [review of R.G.B. Reid's *Biological Emergences*]. *Trends in Ecology & Evolution* 22: 567–68.

Erwin DH. 2008a. Extinction as the loss of evolutionary history. *Proceedings of the National Academy of Sciences of the USA* 105: 11520–27.

Erwin DH. 2008b. Macroevolution of ecosystem engineering, niche construction and diversity. *Trends in Ecology & Evolution* 23: 304–10.

Erwin DH. 2011. Evolutionary uniformitarianism. *Developmental Biology* 357: 27–34.

Erwin DH. 2015a. Novelty and innovation in the history of life. *Current Biology* 25: R930–40.

Erwin DH. 2015b. A public goods approach to major evolutionary transitions. *Geobiology* 13: 308–15.

Erwin DH. 2015c. Was the Ediacaran-Cambrian radiation a unique evolutionary event? *Paleobiology* 41: 1–15.

Erwin DH. 2016. *Wonderful Life* revisited: Chance and contingency in the Ediacaran-Cambrian radiation. In *Chance in Evolution*, ed. G Ramsay, CH Pence, pp. 279–98. Chicago: University of Chicago Press.

Erwin DH. 2017a. Developmental push or ecological pull? The causes of macroevolutionary dynamics. *History and Philosophy of Life Science* 39: 36.

Erwin DH. 2017b. The topology of evolutionary novelty and innovation in macroevolution. *Philosophical Transactions of the Royal Society B* 372: 20160422.

Erwin DH. 2019. Prospects for a general theory of evolutionary novelty. *Journal of Computer Biology* 26: 735–44.

Erwin DH. 2020a. A conceptual framework of evolutionary novelty and innovation. *Biological Reviews* 96: 1–15.

Erwin DH. 2020b. Evolutionary dynamics of gene regulation. *Current Topics in Developmental Biology* 139: 407–31.

Erwin DH. 2020c. Origin of animal bodyplans: A view from the regulatory genome. *Development* 147: dev182899.

Erwin DH. 2021. Developmental capacity and the early evolution of animals. *Journal of the Geological Society* 178: jgs2020-245.

Erwin, DH. 2024. Macroevolutionary dynamics of ecosystem-engineering and niche construction. *Palaeontology* 67: e12718.

Erwin DH, Davidson EH. 2002. The last common bilaterian ancestor. *Development* 129: 3021–32.

Erwin DH, Davidson EH. 2009. The evolution of hierarchical gene regulatory networks. *Nature Reviews Genetics* 10: 141–48.

Erwin DH, Laflamme M, Tweedt SM, Sperling EA, Pisani D, Peterson KJ. 2011. The Cambrian conundrum: Early divergence and later ecological success in the early history of animals. *Science* 334: 1091–97.

Erwin DH, Tweedt SM. 2011. Ecosystem engineering and the Ediacaran-Ordovician diversification of Metazoa. *Evolutionary Ecology* 26: 417–33.

Erwin DH, Valentine JW. 2013. *The Cambrian Explosion: The Construction of Animal Biodiversity.*
Greenwood, CO: Roberts.

Erwin DH, Valentine JW, Sepkoski JJ, Jr. 1987. A comparative study of diversification events:
The Early Paleozoic vs. the Mesozoic. *Evolution* 41: 1177–86.

Evans N. 2013. Language diversity as a resource for understanding cultural evolution. In *Cultural
Evolution: Society, Technology, Language, and Religion,* ed. PJ Richerson, MH Christiansen,
pp. 233–68. Cambridge, MA: MIT Press.

Evans SD, Droser ML, Erwin DH. 2021. Developmental processes in Ediacaran macrofossils.
Proceedings of the Royal Society B 288: 20203055.

Evans SD, Droser ML, Gehling JG. 2017. Highly regulated growth and development of the Ediacara macrofossil *Dickinsonia costata. Plos One* 12: e0176874.

Evans SD, Gehling JG, Droser ML. 2019. Slime travelers: Early evidence of animal mobility and
feeding in an organic mat world. *Geobiology* 17: 490–509.

Everett C. 2017. *Numbers and the Making of Us.* Cambridge, MA: Harvard University Press.

Everett DL. 2017. *How Language Began.* New York: Liveright.

Evershed RP, Davey Smith G, Roffet-Salque M, Timpson A, Diekmann Y, et al. 2022. Dairying,
diseases and the evolution of lactase persistence in Europe. *Nature* 608: 336–45.

Ezcurra MD, Butler RJ. 2018. The rise of the ruling reptiles and ecosystem recovery from the
Permo-Triassic mass extinction. *Proceedings of the Royal Society B* 285: 20180361.

Fang Y, Qin X, Liao Q, Du R, Luo X, et al. 2021. The genome of homosporous maiden-
hair fern sheds light on the euphyllophyte evolution and defenses. *Nature Plants* 8:
1024–37.

Farber PL. 1976. The type-concept in zoology during the first half of the nineteenth century.
Journal of the History of Biology 9: 93–119.

Farmer C. 1997. Did lungs and the intracardiac shunt evolve to oxygenate the heart in vertebrates? *Paleobiology* 23: 358–72.

Fedoroff NV. 2012. Transposable elements, epigenetics, and genome evolution. *Science* 338:
758–67.

Feduccia A. 2003. "Big bang" for Tertiary birds? *Trends in Ecology & Evolution* 18: 172–76.

Felix MA. 2012. Evolution in developmental phenotype space. *Current Opinion in Genetics &
Development* 22: 593–99.

Field DJ, Berv JS, Hsiang AJ, Lanfear R, Landis MJ, Dornburg A. 2019. Timing the extant avian
radiation: The rise of modern birds, and the importance of modeling molecular rate variation. *PeerJ* 7: e27521v1.

Filipchenko JP. 1927. *Variabilitat und Variation.* Berlin: Gebruder Bortraeger.

Fink TMA, Reeves M. 2019. How much can we influence the rate of innovation? *Science Advances* 5: eaat6107.

Fink TMA, Reeves M, Palma R, Farr RS. 2017. Serendipity and strategy in rapid innovation.
Nature Communications 8: 2002.

Fishburn G. 2004. *Natura non facit saltum* in Alfred Marshall (and Charles Darwin). *History of
Economics Review* 40: 59–69.

Fisher CR, Wegrzyn JL, Jockusch EL. 2020. Co-option of wing-patterning genes underlies the
evolution of the treehopper helmet. *Nature Ecology & Evolution* 4: 250–60.

Flowers RM, Macdonald FA, Siddoway CS, Havranek R. 2020. Diachronous development of
great unconformities before Neoproterozoic Snowball Earth. *Proceedings of the National
Academy of Sciences of the USA* 117: 10172–80.

Fogarty L. 2018. Cultural complexity and evolution in fluctuating environments. *Philosophical
Transactions of the Royal Society B* 373: 20170063.

Fogarty L, Creanza N. 2017. The niche construction of cultural complexity: Interactions between innovations, population size and the environment. *Philosophical Transactions of the Royal Society B* 372: 20160428.

Fogarty L, Creanza N, Feldman MW. 2015. Cultural evolutionary perspectives on creativity and human innovation. *Trends in Ecology & Evolution* 30: 736–54.

Fontana W. 2006. Topology of the possible. In *Understanding Change: Models, Methodologies and Metaphors*, ed. A Wimmer, R Kossler, pp. 67–84. Houndmills, UK: Palgrave Macmillan UK

Fontana W, Buss LW. 1994. What would be conserved if "the tape were played twice"? *Proceedings of the National Academy of Sciences of the USA* 91: 757–61.

Fontana W, Schuster P. 1998. Continuity in evolution: On the nature of transitions. *Science* 280: 1451–55.

Foote M. 1996. Ecological controls on the evolutionary recovery of post-Paleozoic crinoids. *Science* 274: 1492–95.

Foote M. 1997. Evolution of morphological diversity. *Annual Review of Ecology and Systematics* 28: 129–52.

Fournier GP, Huang J, Gogarten JP. 2009. Horizontal gene transfer from extinct and extant lineages: Biological innovation and the coral of life. *Philosophical Transactions of the Royal Society B* 364: 2229–39.

Fragata I, Blanckaert A, Dias Louro MA, Liberles DA, Bank C. 2019. Evolution in the light of fitness landscape theory. *Trends in Ecology & Evolution* 34: 69–82.

Frare R, Ayub N, Alleva K, Soto G. 2018. The ammonium channel NOD26 is the evolutionary innovation that drives the emergence, consolidation, and dissemination of nitrogen-fixing symbiosis in angiosperms. *Journal of Molecular Evolution* 86: 554–65.

Fraser GJ, Cerny R, Soukup V, Bronner-Fraser M, Streelman JT. 2010. The odontode explosion: The origin of tooth-like structures in vertebrates. *BioEssays* 32: 808–17.

Freilich MA, Wieters E, Broitman BR, Marquet PA, Navarrete SA. 2018. Species co-occurrence networks: Can they reveal trophic and non-trophic interactions in ecological communities? *Ecology* 99: 690–99.

Friis EM, Crane PR, Pedersen KR. 2011. *Early Flowers and Angiosperm Evolution*. Cambridge: Cambridge University Press.

Frison G. 1988. Technical and technological innovation in Marx. *History and Technology* 6: 299–324.

Fry BG, Roelants K, Champagne DE, Scheib H, Tyndall JD, et al. 2009a. The toxicogenomic multiverse: Convergent recruitment of proteins into animal venoms. *Annual Review of Genomics and Human Genetics* 10: 483–511.

Fry BG, Roelants K, Norman JA. 2009b. Tentacles of venom: Toxic protein convergence in the kingdom Animalia. *Journal of Molecular Evolution* 68: 311–21.

Fuentes A. 2017. *The Creative Spark*. New York: Dutton.

Fusco G, Carrer R, Serrelli E. 2014. The landscape metaphor in development. In *Towards a Theory of Development*, ed. A Minelli, T Pradeu, pp. 114–28. Oxford: Oxford University Press.

Fusco G, Minelli A. 2010. Phenotypic plasticity in development and evolution: Facts and concepts. Introduction. *Philosophical Transactions of the Royal Society B* 365: 547–56.

Futuyma D. 2015. Can modern evolutionary theory explain macroevolution? In *Macroevolution*, ed. E Serreli, N Grontier, pp. 29–85. Cham, Switzerland: Springer Nature.

Gaiti F, Degnan BM, Tanurdzic M. 2018. Long non-coding regulatory RNAs in sponges and insights into the origin of animal multicellularity. *RNA Biology* 15: 696–702.

Gaiti F, Jindrich K, Fernandez-Valverde SL, Roper KE, Degnan BM, Tanurdzic M. 2017. Landscape of histone modifications in a sponge reveals the origin of animal cis-regulatory complexity. *eLife* 6: e22194.

Galis F. 2001. Key innovations and radiations. In *The Character Concept in Evolutionary Biology*, ed. GP Wagner, pp. 581–605. San Diego: Academic.

Gallant JR. 2019. The evolution and development of electric organs. In *Electroreception: Fundamental Insights from Comparative Approaches*, ed. BA Carlson, JA Sisneros, AN Popper, RR Fay, pp. 91–123. Cham, Switzerland: Springer Nature.

Gallant JR, Traeger LL, Volkening JD, Moffett H, Chen PH, et al. 2014. Genomic basis for the convergent evolution of electric organs. *Science* 344: 1522–25.

Garcia-Galindo P, Ahnert SE, Martin NS. 2023. The non-deterministic genotype-phenotype map of RNA secondary structure. *Interface* 20: 20230132.

Gavrilets S. 2010. *Fitness Landscapes and the Origin of Species*. Princeton, NJ: Princeton University Press.

Gawne R, McKenna KZ, Nijhout, HF. 2018. Unmodern synthesis: Developmental hierarchies and the origin of phenotypes. *BioEssays* 40: 1600265

Gerber S. 2014. Not all roads can be taken: Development induces anisotropic accessibility in morphospace. *Evolution & Development* 16: 373–81.

Gerber S. 2016. The geometry of morphospaces: Lessons from the classic Raup shell coiling model. *Biological Reviews* 92: 1142–55.

Gerhart J, Kirschner M. 2007. The theory of facilitated variation. *Proceedings of the National Academy of Sciences of the USA* 104, Suppl 1: 8582–89.

Gerrits L, Marks P. 2014. How fitness landscapes help further the social and behavioral sciences. *Emergence: Complexity and Organization* 16: 1–17.

Gerrits L, Marks P. 2015. The evolution of Wright's (1932) adaptive field to contemporary interpretations and uses of fitness landscapes in the social sciences. *Biology & Philosophy* 30: 459–79.

Giribet G, Edgecombe GD. 2019. The phylogeny and evolutionary history of arthropods. *Current Biology* 29: R592–602.

Gissis SB, Jablonka E, eds. 2011. *Transformations of Lamarckism*. Cambridge, MA: MIT Press.

Gittenberger E. 1991. What about non-adaptive radiation? *Biological Journal of the Linnean Society* 43: 263–72.

Givnish TJ. 2015. Adaptive radiation versus "radiation" and "explosive diversification": Why conceptual distinctions are fundamental to understanding evolution. *New Phytologist* 207: 297–303.

Gleick J. 2011. *The Information*. New York: Pantheon Books.

Glor RE. 2010. Phylogenetic insights on adaptive radiation. *Annual Review of Ecology, Evolution and Systematics* 41: 251–70.

Godfrey-Smith P. 1998. Darwin's dangerous idea: Evolution and the meanings of life. *Philosophy of Science* 65: 709–20.

Godfrey-Smith P. 2009. *Darwinian Populations and Natural Selection*. Oxford: Oxford University Press.

Godfrey-Smith, P. 2014. *Philosophy of Biology*. Princeton, NJ: Princeton University Press.

Godin B. 2017. *Models of Innovation: The History of an Idea*. Cambridge, MA: MIT Press.

Gold DA, Caron A, Fournier GP, Summons RE. 2017. Paleoproterozoic sterol biosynthesis and the rise of oxygen. *Nature* 543: 420.

Goldschmidt R. 1940. *The Material Basis of Evolution*. New Haven, CT: Yale University Press.

Golicz AA, Bayer PE, Bhalla PL, Batley J, Edwards D. 2020. Pangenomics comes of age: From bacteria to plant and animal applications. *Trends in Genetics* 36: 132–45.

Golubski AJ, Westlund EE, Vandermeer J, Pascual M. 2016. Ecological networks over the edge: Hypergraph trait-mediated indirect interaction (TMII) structure. *Trends in Ecology & Evolution* 31: 344–54.

Gordon RJ. 2016. *The Rise and Fall of American Growth: The U.S. Standard of Living since the Civil War*. Princeton, NJ: Princeton University Press.

Gould SJ. 1965. Is uniformitarianism necessary? *American Journal of Science* 263: 223–28.

Gould SJ. 1977. *Ontogeny and Phylogeny*. Cambridge, MA: Belknap Press of Harvard University Press.

Gould SJ. 1980a. Is a new and general theory of evolution emerging? *Paleobiology* 6: 119–30.

Gould SJ. 1980b. The promise of paleobiology as a nonothetic, evolutionary discipline. *Paleobiology* 6: 96–118.

Gould SJ. 1982. The uses of heresy: An introduction to Richard Goldschmidt's *The Material Basis of Evolution*. In *The Material Basis of Evolution*, pp. xii–xlii. Chicago: University of Chicago Press.

Gould SJ. 1985. The paradox of the first tier: An agenda for paleobiology. *Paleobiology* 11: 2–12.

Gould SJ. 1989. *Wonderful Life*. New York: Norton.

Gould SJ. 2002. *The Structure of Evolutionary Theory*. Cambridge, MA: Harvard University Press.

Gould SJ, Eldredge N. 1993. Punctuated equilibrium comes of age. *Nature* 366: 223–27.

Gould SJ, Vrba E. 1982. Exaptation: A missing term in the science of form. *Paleobiology* 8: 4–15.

Graeber D, Wengrow D. 2021. *The Dawn of Everything*. New York: Farrar, Straus and Giroux.

Graham LA, Davies PL. 2021. Horizontal gene transfer in vertebrates: A fishy tale. *Trends in Genetics* 37: 501–3.

Greenberg JH. 1963. Some universals of grammar with particular reference to the order of meaningful elements. In *Universals of Language*, ed. JH Greenberg, pp. 58–99. Cambridge, MA: MIT Press.

Greenbury SF, Louis AA, Ahnert SE. 2022. The structure of genotype-phenotype maps makes fitness landscapes navigable. *Nature Ecology & Evolution* 6: 1742–52.

Griffith OW, Wagner GP. 2017. The placenta as a model for understanding the origin and evolution of vertebrate organs. *Nature Ecology & Evolution* 1: 0072.

Griffiths PE. 1999. Squaring the circle: Natural kinds with historical essences. In *Species: New Interdisciplinary Essays*, ed. RA Wilson, pp. 209–28. Cambridge, MA: MIT Press.

Grossnickle DM, Smith SM, Wilson GP. 2019. Untangling the multiple ecological radiations of early mammals. *Trends in Ecology & Evolution* 34: 936–49.

Grundler MC, Rabosky DL. 2021. Rapid increase in snake dietary diversity and complexity following the end-Cretaceous mass extinction. *PLoS Biology* 19: e3001414.

Gunasekaram C, Battiston F, Sadekar O, Padilla-Iglesias C, van Noordwijk MA, et al. 2024. Population connectivity shapes the distribution and complexity of chimpanzee cumulative culture. *Science* 386: 920–25.

Hall BK. 1996. Bauplane, phylotypic stages and constraint. *Evolutionary Biology* 29: 215–61.

Hall BK. 1999. *Evolutionary Developmental Biology*. Dordrecht, Netherlands: Kleuwer.

Hallgrimsson B, Jamniczky HA, Young NM, Rolian C, Schmidt-Ott U, Marcucio RS. 2012. The generation of variation and the developmental basis for evolutionary novelty. *Journal of Experimental Zoology Part B (Molecular and Developmental Evolution)* 318: 501–17.

Hardin G. 1968. The tragedy of the commons. *Science* 162: 1243–48.

Harmon LJ, Losos JB, Jonathan Davies T, Gillespie RG, Gittleman JL, et al. 2010. Early bursts of body size and shape evolution are rare in comparative data. *Evolution* 64: 2385–96.

Harper T, Parras A, Rougier GW. 2018. *Reigitherium* (Meridiolestida, Mesungulatoidea): An enigmatic Late Cretaceous animal from Patagonia, Argentina. *Journal of Mammalian Evolution* 8: 1–32.

Harris M. 1968. *The Rise of Anthropological Theory*. New York: Thomas Crowell.

Harrison CJ. 2017. Development and genetics in the evolution of land plant body plans. *Philosophical Transactions of the Royal Society B* 372: 20150490.

Hawkins MB, Henke K, Harris MP. 2021. Latent developmental potential to form limb-like skeletal structures in zebrafish. *Cell* 184: 899–911.

Heard SB, Hauser DL. 1995. Key evolutionary innovations and their ecological mechanism. *Historical Biology* 10: 151–73.

Helfield JM, Naiman RJ. 2006. Keystone interactions: Salmon and bear in riparian forests of Alaska. *Ecosystems* 9: 167–80.

Henbest LG, Marble JP, Cooper GA, Williams A, Moore RC, et al. 1952. Distributions of evolutionary explosions in evolutionary time: A symposium. *Journal of Paleontology* 26: 297–394.

Henrich J. 2020. *The WEIRDest People in the World*. New York: Farrar, Straus and Giroux.

Henrich JA. 2015. *The Secret of Our Success: How Culture Is Driving Human Evolution, Domesticating Our Species, and Making Us Smarter*. Princeton, NJ: Princeton University Press.

Herendeen PS, Friis EM, Pedersen KR, Crane PR. 2017. Palaeobotanical redux: Revisiting the age of the angiosperms. *Nature Plants* 3: 17015.

Heyes C. 2018. *Cognitive Gadgets*. Cambridge, MA: Harvard University Press.

Hidalgo CA, Hausmann R. 2009. The building blocks of economic complexity. *Proceedings of the National Academy of Sciences of the USA* 106: 10570–75.

Hidalgo CA, Klinger B, Barabasi AL, Hausmann R. 2007. The product space conditions the development of nations. *Science* 317: 482–87.

Hilgers L, Roth O, Nolte AW, Schuller A, Spanke T, et al. 2022. Inflammation and convergent placenta gene co-option contributed to a novel reproductive tissue. *Current Biology* 32: 715–24.e4.

Hobbs RJ, Arico S, Aronson J, Baron JS, Bridgewater P, et al. 2006. Novel ecosystems: Theoretical and management aspects of the new ecological world order. *Global Ecology and Biogeography* 15: 1–7.

Hochberg ME, Marquet PA, Boyd R, Wagner A. 2017. Innovation: An emerging focus from cells to societies. *Philosophical Transactions of the Royal Society B* 372: 20160414.

Hodgson GM, Knudsen T. 2010. *Darwin's Conjecture: The Search for General Principles of Social and Economic Evolution*. Chicago: University of Chicago Press.

Hoffman PF, Abbot DS, Ashkenazy Y, Benn DI, Brocks JJ, et al. 2017. Snowball Earth climate dynamics and Cryogenian geology-geobiology. *Science Advances* 3: e1600983.

Hopkins MJ, Smith AB. 2015. Dynamic evolutionary change in post-Paleozoic echinoids and the importance of scale when interpreting changes in rates of evolution. *Proceedings of the National Academy of Sciences of the USA* 112: 3758–63.

Hoppit W, Laland KN. 2013. *Social Learning*. Princeton, NJ: Princeton University Press.

Howard RJ, Giacomelli M, Lozano-Fernandez J, Edgecombe GD, Fleming JF, et al. 2022. The Ediacaran origin of Ecdysozoa: Integrating fossil and phylogenomic data. *Journal of the Geological Society* 179: gs2021-107.

Hu Y, Linz DM, Moczek AP. 2019. Beetle horns evolved from wing serial homologs. *Science* 366: 1004–7.

Hu Y, Moczek AP. 2021. Wing serial homologues and the diversification of insect outgrowths: Insights from the pupae of scarab beetles. *Proceedings of the Royal Society B* 288: 20202828.

Huang S. 2012. The molecular and mathematical basis of Waddington's epigenetic landscape: A framework for post-Darwinian biology? *BioEssays* 34: 149–57.

Hughes M, Gerber S, Wills MA. 2013. Clades reach highest morphologic disparity early in their evolution. *Proceedings of the National Academy of Sciences of the USA* 110: 13875–79.

Hull DL. 1973. *Darwin and His Critics*. Chicago: University of Chicago Press.

Hull DS. 1980. Individuality and selection. *Annual Review of Ecology and Systematics* 11: 311–32.

Hull PM, Darroch SA. 2013. Mass extinction and the structure and function of ecosystems. In *Ecosystem Paleobiology and Geobiology: Paleontological Society Short Course*, ed. AM Bush, SB Pruss, JL Payne, pp. 1–42. N.p.: Paleontological Society.

Hunter JP. 1998. Key innovations and the ecology of macroevolution. *Transactions of the Royal Society of Edinburgh* 13: 31–36.

Huttegger SM, Mitteroecker P. 2011. Invariance and meaningfulness in phenotypic spaces. *Evolutionary Biology* 38: 335–51.

Hutton J. 1788. Theory of the Earth; or An investigation of the laws observable in the composition, dissolution, and restoration of land upon the Globe. *Transactions of the Royal Society of Edinburgh* 1, part 2: 209–304.

Huxley JS. 1942. *Evolution: The Modern Synthesis*. New York: Harper.

Huxley TH. 1894. *Darwiniana: Essays*. New York: D. Appleton.

Huxley TH, Huxley L. 1900. *Life and Letters of Thomas Henry Huxley*. London: Macmillan.

Hyde D, Raffman D. 2018. Sorites paradox. In *The Stanford Encyclopedia of Philosophy*, Summer 2018 ed., ed. EN Zalta. https://plato.stanford.edu/archives/sum2018/entries/sorites-paradox/.

Isson TT, Love GD, Dupont CL, Reinhard CT, Zumberge AJ, et al. 2018. Tracking the rise of eukaryotes to ecological dominance with zinc isotopes. *Geobiology* 16: 341–52.

Jablonski D. 1993. The tropics as a source of evolutionary novelty through geologic time. *Nature* 364: 142–44.

Jablonski D. 2005. Evolutionary innovations in the fossil record: The intersection of ecology, development, and macroevolution. *Journal of Experimental Zoology Part B (Molecular and Developmental Evolution)* 304: 504–19.

Jablonski D. 2008. Species selection: Theory and data. *Annual Review of Ecology, Evolution and Systematics* 39: 501–24.

Jablonski D. 2017a. Approaches to macroevolution: 1. General concepts and origin of variation. *Evolutionary Biology* 44: 427–50.

Jablonski D. 2017b. Approaches to macroevolution: 2. Sorting of variation, some overarching issues, and general conclusions. *Evolutionary Biology* 44: 451–75.

Jablonski D. 2022. Evolvability and macroevolution: Overview and synthesis. *Evolutionary Biology* 49: 265–91.

Jablonski D, Bottjer DJ. 1990a. The ecology of evolutionary innovation: The fossil record. In *Evolutionary Innovations*, ed. MH Nitecki, pp. 253–88. Chicago: University of Chicago Press.

Jablonski D, Bottjer DJ. 1990b. The origin and diversification of major groups: Environmental patterns and macroevolutionary lags. In *Major Evolutionary Radiations*, ed. PD Taylor, GP Larwood, pp. 17–57. Oxford: Clarendon.

Jablonski D, Bottjer DJ. 1991. Environmental patterns in the origins of higher taxa: The post-Paleozoic fossil record. *Science* 252: 1831–33.

Jablonski D, Lidgard S, Taylor PD. 1997. Comparative ecology of bryozoan radiations: Origin of novelties in cyclostomes and cheilostomes. *Palaios* 12: 505–12.

Jablonski D, Sepkoski JJ, Bottjer DJ, Sheehan PM. 1983. Onshore-offshore patterns in the evolution of phanerozoic shelf communities. *Science* 222: 1123–25.

Jacob F. 1977. Evolution and tinkering. *Science* 196: 1161–66.

Jaekel O. 1902. *Über verschiedene Wege phylogenetischer Entwicklung*. Jena, Germany: Gustav Fischer.

Jarvis ED, Mirarab S, Aberer AJ, Li B, Houde P, et al. 2014. Whole-genome analyses resolve early branches in the tree of life of modern birds. *Science* 346: 1320–31.

Javaux EJ. 2019. Challenges in evidencing the earliest traces of life. *Nature* 572: 451–60.

Jenner RA. 2022. *Ancestors in Evolutionary Biology*. Cambridge: Cambridge University Press.

Jensen K. 2016. Prosociality. *Current Biology* 26: R748–52.

Ji Q, Luo ZX, Yuan CX, Tabrum AR. 2006. A swimming mammaliform from the Middle Jurassic and ecomorphological diversification of early mammals. *Science* 311: 1123–26.

Jockusch EL. 2017. Developmental and evolutionary perspectives on the origin and diversification of arthropod appendages. *Integrative and Comparative Biology* 57: 533–45.

Johnston DT, Wolfe-Simon F, Pearson A, Knoll AH. 2009. Anoxygenic photosynthesis modulated Proterozoic oxygen and sustained Earth's middle age. *Proceedings of the National Academy of Sciences of the USA* 106: 16925–29.

Jones CG, Lawton JH, Shachak M. 1997. Positive and negative effects of organisms as physical ecosystem engineers. *Ecology* 78: 1946–57.

Jones CI. 2002. *Introduction to Economic Growth*. New York: Norton.

Jones CI. 2019. Paul Romer: Ideas, nonrivalry, and endogenous growth. *Scandinavian Journal of Economics* 121: 859–83.

Jones CI. 2023. Recipes and economic growth: A combinatorial march down an exponential tail. *Journal of Political Economy* 131: 1994–2031.

Jovanovic B, Rousseau PL. 2005. General purpose technologies. In *Handbook of Economic Growth*, ed. P Aghion, SN Durlaur, pp. 1181–224. Amsterdam: Elsevier.

Kapli P, Natsidis P, Leite DJ, Fursman M, Jeffrie N, et al. 2021. Lack of support for Deuterostomia prompts reinterpretation of the first Bilateria. *Science Advances* 7: eabe2741.

Kassen R. 2019. Experimental evolution of innovation and novelty. *Trends in Ecology & Evolution* 34: 712–22.

Katz LA. 2012. Origin and diversification of eukaryotes. *Annual Review of Microbiology* 66: 411–27.

Katz O. 2018. Extending the scope of Darwin's "abominable mystery": Integrative approaches to understanding angiosperm origins and species richness. *Annals of Botany* 121: 1–8.

Kauffman SA. 1993. *The Origins of Order*. Oxford: Oxford University Press.

Kauffman SA. 2014. Foreword: Stable and non-prestatable fitness landscapes. In *Recent Advances in the Theory and Application of Fitness Landscapes*, ed. H Richter, A Engelbrecht, pp. vii–xvi. Berlin: Springer-Verlag.

Kauffman SA, Levin S. 1987. Towards a general theory of adaptive walks on rugged landscapes. *Journal of Theoretical Biology* 128: 11–45.

Kauffman SA, Weinberger ED. 1988. Origins of order in evolution: Self organization and selection. In *Biomathematics and Related Computational Problems*, ed. LM Ricciardi, pp. 311–30. Dordrecht: Kluwer Academic.

Kazandjian TD, Petras D, Robinson SD, Van Thiel J, Greene HW, et al. 2021. Convergent evolution of pain-inducing defensive venom components in spitting cobras. *Science* 371: 386–90.

Keeling PJ. 2010. The endosymbiotic origin, diversification and fate of plastids. *Philosophical Transactions of the Royal Society B* 365: 729–48.

Kemp TS. 2007a. The concept of correlated progression as the basis of a model for the evolutionary origin of major new taxa. *Proceedings of the Royal Society B* 274: 1667–73.

Kemp TS. 2007b. The origin of higher taxa: Macroevolutionary processes, and the case of the mammals. *Acta zoologica* 88: 3–22.

Kemp TS. 2016. *The Origin of Higher Taxa*. Oxford: Oxford University Press.

Kempes CP, Koehl MA, West GB. 2019. The scales that limit: The physical boundaries of evolution. *Frontiers in Ecology and the Environment* 7: 242.

Kirby S. 2017. Culture and biology in the origins of linguistic structure. *Psychonomic Bulletin and Review* 24: 118–37.

Kirch PV. 2010. *How Chiefs Became Kings*. Berkeley: University of California Press.

Kline SJ. 1985. Innovation is not a linear process. *Research Management* 28: 36–45.

Klingenberg CP, Monteiro LR. 2005. Distances and directions in multidimensional shape spaces: Implications for morphometric applications. *Systematic Biology* 54: 678–88.

Klump BC, Martin JM, Wild S, Horsch JK, Major RE, Aplin LM. 2021. Innovation and geographic spread of a complex foraging culture in an urban parrot. *Science* 373: 456–60.

Knight KE. 1967. A descriptive model of the intra-firm innovation process. *Journal of Business* 40: 478–96.

Knoll AH. 2014. Paleobiological perspectives on early eukaryotic evolution. *Cold Spring Harbor Perspectives in Biology* 6: a016121.

Knoll AH, Bambach RK. 2000. Directionality in the history of life: Diffusion from the left wall or repeated scaling of the right? In *Deep Time: Paleobiology's Perspective*, ed. DH Erwin, SL Wing, pp. 2–14. Lawrence, KS: Paleontological Society.

Knoll AH, Niklas KJ, Gensel PG, Tiffney BH. 1984. Character diversification and patterns of evolution in early vascular plants. *Paleobiology* 10: 34–47.

Knoll AH, Sperling EA. 2014. Oxygen and animals in Earth history. *Proceedings of the National Academy of Sciences of the USA* 111: 3907–8.

Kokko H, Chaturvedi A, Croll D, Fischer MC, Guillaume F, et al. 2017. Can evolution supply what ecology demands? *Trends in Ecology & Evolution* 32: 187–97.

Kolodny O, Creanza N, Feldman MW. 2015. Evolution in leaps: The punctuated accumulation and loss of cultural innovations. *Proceedings of the National Academy of Sciences of the USA* 112: e6762–69.

Kolokotrones T, Van S, Deeds EJ, Fontana W. 2010. Curvature in metabolic scaling. *Nature* 464: 753–56.

Ksepka DT, Stidham TA, Williamson TE. 2017. Early Paleocene landbird supports rapid phylogenetic and morphological diversification of crown birds after the K-Pg mass extinction. *Proceedings of the National Academy of Sciences of the USA* 114: 8047–52.

Kuhn SL. 2021. *The Evolution of Paleolithic Technologies*. New York: Routledge.

Laakso TA, Schrag DP. 2019. A small marine biosphere in the Proterozoic. *Geobiology* 17: 161–71.

Labandeira CC. 2014. Why did terrestrial insect diversity not increase during the angiosperm radiation? Mid-Mesozoic, plant-associated insect lineages harbor clues. In *Evolutionary Biology: Genome Evolution, Speciation, Coevolution and Origin of Life*, ed. P Pontarotti, pp. 261–99. Basel, Switzerland: Springer.

Labandeira CC, Sepkoski JJ, Jr. 1993. Insect diversity in the fossil record. *Science* 261: 310–15.

Laflamme M, Darroch SA, Tweedt SM, Peterson KJ, Erwin DH. 2013. The end of the Ediacara biota: Extinction, replacement or Cheshire cat? *Gondwana Research* 23: 558–73.

Lake JA. 2015. Eukaryotic origins. *Philosophical Transactions of the Royal Society B* 370: 20140321.

Laland KN. 2017. *Darwin's Unfinished Symphony*. Princeton, NJ: Princeton University Press.

Laland KN, Brown GR. 2011. *Sense and Nonsense: Evolutionary Perspectives on Human Behavior*. Oxford: Oxford University Press.

Laland KN, Odling-Smee FJ, Myles S. 2010. How culture shaped the human genome: Bringing genetics and the human sciences back together. *Nature Reviews Genetics* 11: 137–48.

Laland KN, Uller T, Feldman MW, Sterelny K, Muller GB, et al. 2015. The extended evolutionary synthesis: Its structure, assumptions and predictions. *Proceedings of the Royal Society B* 282: 20151019.

Lamarck JB. 1809. *Zoological Philosophy*. Chicago: University of Chicago Press.

Landes DS. 1976. *Unbound Prometheus: Technological Change and Industrial Development in Western Europe from 1750 to the Present*. Cambridge: Cambridge University Press.

Landis MJ, Schraiber JG. 2017. Pulsed evolution shaped modern vertebrate body sizes. *Proceedings of the National Academy of Sciences of the USA* 114: 13224–29.

Lane DA. 2016. Innovation cascades: Artefacts, organization and attributions. *Philosophical Transactions of the Royal Society B* 371: 20150194.

Lane N, Martin W. 2010. The energetics of genome complexity. *Nature* 467: 929–34.

Lane N, Martin WF. 2015. Eukaryotes really are special, and mitochondria are why. *Proceedings of the National Academy of Sciences of the USA* 112: e4823.

Larson A. 2014. Concepts in character macroevolution: Adaptation, homology and evolvability. In *Princeton Guide to Evolution*, ed. JB Losos, pp. 87–99. Princeton, NJ: Princeton University Press.

Larson G, Fuller DQ. 2014. The evolution of animal domestication. *Annual Review of Ecology, Evolution and Systematics* 45: 115–36.

Laubichler MD, Niklas KJ. 2009. The morphological tradition in German palaeontology: Otto Jaekel, Walter Zimmermann, and Otto Schindewolf. In *The Paleobiological Revolution*, ed. D Sepkoski, M Ruse, pp. 279–300. Chicago: University of Chicago Press.

Laubichler MD, Prohaska SJ, Stadler PF. 2018. Toward a mechanistic explanation of phenotypic evolution: The need for a theory of theory integration. *Journal of Experimental Zoology Part B (Molecular and Developmental Evolution)* 330: 5–14.

Lauder GV. 1981. Form and function: Structural analysis in evolutionary biology. *Paleobiology* 7: 430–42.

Laumer CE, Fernandez R, Lemer S, Combosch D, Kocot KM, et al. 2019. Revisiting metazoan phylogeny with genomic sampling of all phyla. *Proceedings of the Royal Society B* 286: 20190831.

Laumer CE, Gruber-Vodicka H, Hadfield MG, Pearse VB, Riesgo A, et al. 2018. Support for a clade of Placozoa and Cnidaria in genes with minimal compositional bias. *Elife* 7: e36278.

Lautenschlager S, Figueirido B, Cashmore DD, Bendel EM, Stubbs TL. 2020. Morphological convergence obscures functional diversity in sabre-toothed carnivores. *Proceedings of the Royal Society B* 287: 20201818.

Lee MS, Cau A, Naish D, Dyke GJ. 2014. Sustained miniaturization and anatomical innovation in the dinosaurian ancestors of birds. *Science* 345: 562–66.

Legg DA, Sutton MD, Edgecombe GD, Caron JB. 2012. Cambrian bivalved arthropod reveals origin of arthropodization. *Proceedings of the Royal Society B* 279: 4699–704.

Le Goff J. 2015. *Must We Divide History into Periods?* New York: Columbia University Press.

Le Guyader H. 2004. *Geoffroy Saint-Hilaire. A Visionary Naturalist.* Chicago: University of Chicago Press.

Leibniz GW. (1666) 1923. *Dissertatio de arte combinatoria.* Berlin: Akademie Verlag.

Leibniz GW. (1715) 1991. *De l'horizon de la doctrine humaine.* Paris: Vrin.

Lenski RE. 2017. Experimental evolution and the dynamics of adaptation and genome evolution in microbial populations. *ISME Journal* 11: 2181–94.

Lenski RE. 2024. The *E. coli* long-term Experimental Evolution project site. https://lenski.mmg.msu.edu/ecoli/index.html.

Lenski RE, Travisano M. 1994. Dynamics of adaptation and diversification: A 10,000-generation experiment with bacterial populations. *Proceedings of the National Academy of Sciences of the USA* 91: 6808–14.

Lenton TM, Boyle RA, Poulton SW, Shields-Zhou G, Butterfield NJ. 2014. Co-evolution of eukaryotes and ocean oxygenation in the Neoproterozoic era. *Nature Geoscience* 7: 257–65.

Lenton TM, Watson A. 2011. *Revolutions That Made the Earth.* Oxford: Oxford University Press.

Leonard GH, Bertness MD, Yund PO. 1999. Crab predation, waterborne cues, and inducible defenses in the blue mussel, *Mytilus edulis. Ecology* 80: 1–14.

Lepore J. 2014. The disruption machine. *New Yorker,* June 23, 2014.

Leslie AB, Simpson C, Mander L. 2021. Reproductive innovations and pulsed rise in plant complexity. *Science* 373: 1368–72.

Letelier J, Naranjo S, Sospedra-Arrufat I, Martinez-Morales JR, Lopez-Rios J, et al. 2021. The Shh/Gli3 gene regulatory network precedes the origin of paired fins and reveals the deep homology between distal fins and digits. *Proceedings of the National Academy of Sciences of the USA* 118: e2100575118.

Levinthal, DA. 1997. Adaptation on rugged landscapes. *Management Science* 43: 934–50.

Levis NA, Isdaner AJ, Pfennig DW. 2018. Morphological novelty emerges from pre-existing phenotypic plasticity. *Nature Ecology & Evolution* 2: 1289–97.

Levis NA, Pfennig DW. 2018. Phenotypic plasticity, canalization, and the origins of novelty: Evidence and mechanisms from amphibians. *Seminars in Cell & Developmental Biology* 88: 80–90.

Levit GS, Hossfeld U, Olsson L. 2014. The Darwinian revolution in Germany: From evolutionary morphology to the modern synthesis. *Endeavour* 38: 268–79.

Levit GS, Meister K. 2006. The history of essentialism vs. Ernst Mayr's "Essentialism Story": A case study of German idealistic morphology. *Theory in Biosciences* 124: 281–307.

Levit GS, Olsson L. 2006. "Evolution on rails": Mechanisms and levels of orthogenesis. *Annals for the History and Philosophy of Biology* 11: 99–138.

Lewens T. 2015. *Cultural Evolution*. Oxford: Oxford University Press.

Lewis CM. 2000. *The Dating Game: One Man's Search for the Age of the Earth*. Cambridge: Cambridge University Press.

Lewontin RC. 1978. Adaptation. *Scientific American* 239: 212–29.

Lewontin RC. 1983. Gene, organism and environment. In *Evolution from Molecules to Men*, ed. DS Bendall, pp. 273–85. Cambridge: Cambridge University Press.

Lewontin RC. 2003. Four complications in understanding the evolutionary process. *SFI Bulletin* 18: 17–23.

Li HT, Yi TS, Gao LM, Ma PF, Zhang T, et al. 2019. Origin of angiosperms and the puzzle of the Jurassic gap. *Nature Plants* 5: 461–70.

Li Z, Baniaga AE, Sessa EB, Scascitelli M, Graham SW, et al. 2015. Early genome duplications in conifers and other seed plants. *Science Advances* 1: e1501084.

Libbrecht KG. 2017. Physical dynamics of ice crystal growth. *Annual Review of Materials Research* 47: 271–95.

Lidgard S, Nyhart LK, eds. 2017. *Biological Individuality: Integrating Scientific, Philosophical, and Historical Perspectives*. Chicago: University of Chicago Press.

Lieberman P. 2015. Language did not spring forth 100,000 years ago. *PLoS Biology* 13: e1002064.

Liem KF. 1990. Key evolutionary innovations, differential diversity, and symecomorphosis. In *Evolutionary Innovations*, ed. MH Nitecki, pp. 147–70. Chicago: University of Chicago Press.

Linton R. 1936. *The Study of Man*. New York: Appleton-Century-Crofts.

Linz DM, Moczek AP. 2020. Integrating evolutionarily novel horns within the deeply conserved insect head. *BMC Biology* 18: 41.

Lister AM. 2013a. Behavioral leads in evolution: Evidence from the fossil record. *Biological Journal of the Linnean Society* 112: 315–31.

Lister AM. 2013b. The role of behaviour in adaptive morphological evolution of African proboscideans. *Nature* 500: 331–34.

Liu Y, Wang S, Li L, Yang T, Dong S, et al. 2022. The *Cycas* genome and the early evolution of seed plants. *Nature Plants* 8: 389–401.

Lloyd GT, Davis KE, Pisani D, Tarver JE, Ruta M, et al. 2008. Dinosaurs and the Cretaceous Terrestrial Revolution. *Proceedings of the Royal Society B* 275: 2483–90.

Lopez-Garcia P, Moreira D. 2020. The syntrophy hypothesis for the origin of eukaryotes revisited. *Nature Microbiology* 5: 655–67.

Loreto V, Servedio VDP, Strogatz SH, Tria F. 2016. Dynamics on expanding spaces: Modeling the emergence of novelties. In *Creativity and Universality in Language: Lecture Notes in Morphogenesis*, ed. M Degli Esposito, E Altmann, F Pachet, pp. 59–83. Cham, Switzerland: Springer International.

Loron CC, Halverson GP, Rainbird RH, Skulski T, Turner EC, Javaux EJ. 2021. Shale-hosted biota from the Dismal Lakes Group in Arctic Canada supports an early Mesoproterozoic diversification of eukaryotes. *Journal of Paleontology* 95: 1113–37.

Losos JB. 2009. *Lizards in an Evolutionary Tree: Ecology and Adaptive Radiation of Anolis*. Berkeley: University of California Press.

Losos JB. 2010. Adaptive radiation, ecological opportunity, and evolutionary determinism. *American Naturalist* 175: 623–39.

Louis AA. 2016. Contingency, convergence and hyper-astronomical numbers in biological evolution. *Studies in the History and Philosophy of Biological and Biomedical Sciences* 58: 107–16.

Love AC. 2003. Evolutionary morphology, innovation and the synthesis of evolutionary and developmental biology. *Biology & Philosophy* 18: 309–45.

Love AC. 2006. Evolutionary morphology and evo-devo: Hierarchy and novelty. *Theory in Biosciences* 124: 317–33.

Love AC. 2008. Explaining evolutionary innovations and novelties: Criteria of explanatory adequacy and epistemological prerequisites. *Philosophy of Science* 75: 874–86.

Love AC, ed. 2015. *Conceptual Change in Biology: Scientific and Philosophical Perspectives on Evolution and Development*. Dordrecht, Netherlands: Springer.

Love AC, Grabowski M, Houle D, Liow LH, Porto A, et al. 2022. Evolvability in the fossil record. *Paleobiology* 48: 186–209.

Lovejoy AO. 1936. *Great Chain of Being*. Cambridge, MA: Harvard University Press.

Lowe CB, Clarke JA, Baker AJ, Haussler D, Edwards SV. 2015. Feather development genes and associated regulatory innovation predate the origin of Dinosauria. *Molecular Biology and Evolution* 32: 23–28.

Lozada-Chavez I, Stadler PF, Prohaska SJ. 2018. Genome-wide features of introns are evolutionary decoupled among themselves and from genome size throughout Eukarya. *BioRxiv*: 283549.

Luo ZX. 2007. Transformation and diversification in early mammal evolution. *Nature* 450: 1011–19.

Lyell C. 1881. *Life, Letters and Journals of Sir Charles Lyell, Bart*. Ed. Mrs. [KM] Lyell. Vol. 1. London: John Murray.

Lynch M. 2007. *The Origins of Genome Architecture*. Sunderland, MA: Sinauer Associates.

Lynch M, Field MC, Goodson HV, Malik HS, Pereira-Leal JB, et al. 2014. Evolutionary cell biology: Two origins, one objective. *Proceedings of the National Academy of Sciences of the USA* 111: 16990–94.

Lynch M, Marinov GK. 2015. The bioenergetic costs of a gene. *Proceedings of the National Academy of Sciences of the USA* 112: 15690–95.

Lynch VJ, Nnamani MC, Kapusta A, Brayer K, Plaza SL, et al. 2015. Ancient transposable elements transformed the uterine regulatory landscape and transcriptome during the evolution of mammalian pregnancy. *Cell Reports* 10: 551–61.

Lyons SL. 1995. The origins of T. H. Huxley's saltationism: History in Darwin's shadow. *Journal of the History of Biology* 28: 463–94.

Lyons TW, Diamond CW, Planavsky NJ, Reinhard CT, Li C. 2021. Oxygenation, life, and the planetary system during earth's middle history: An overview. *Astrobiology* 21: 906–23.

Lyons TW, Fike DA, Zerkle A. 2015. Emerging biogeochemical views of earth's ancient microbial worlds. *Elements* 11: 415–21.

Lyons TW, Reinhard CT, Planavsky NJ. 2014. The rise of oxygen in Earth's early ocean and atmosphere. *Nature* 506: 307–15.

Macekura SJ. 2020. *The Mismeasure of Progress*. Chicago: University of Chicago Press.

Maclaurin WR. 1950. The process of technological innovation: The launching of a new scientific industry. *American Economic Review* 40: 90–112.

Maclaurin WR. 1953. The sequence from invention to innovation and its relation to economic growth. *Quarterly Journal of Economics* 67: 97–111.

MacMillan M. 2020. *War: How Conflict Shaped Us*. New York: Random House.

Maestri R, Monteiro LR, Fornel R, Upham NS, Patterson BD, de Freitas TR. 2017. The ecology of a continental evolutionary radiation: Is the radiation of sigmodontine rodents adaptive? *Evolution* 71: 610–32.

Magallon S, Gomez-Acevedo S, Sanchez-Reyes LL, Hernandez-Hernandez T. 2015. A metacalibrated time-tree documents the early rise of flowering plant phylogenetic diversity. *New Phytologist* 207: 437–53.

Mahler DL, Ingram T, Revell LJ, Losos JB. 2013. Exceptional convergence on the macroevolutionary landscape in island lizard radiations. *Science* 341: 292–95.

Majhi S, Perc M, Ghosh D. 2022. Dynamics on higher-order networks: A review. *Journal of the Royal Society Interface* 19: 20220043.

Malthus TR. (1798) 1926. *An Essay on the Principle of Population, as It Affects the Future Improvement of Society with Remarks on the Speculations of Mr: Godwin, M: Condorcet, and Other Writers.* London: Macmillan.

Mangano MG, Buatois LA. 2014. Decoupling of body-plan diversification and ecological structuring during the Ediacaran-Cambrian transition: Evolutionary and geobiological feedbacks. *Proceedings of the Royal Society B* 281: 20140038.

Mangano MG, Buatois LA. 2020. The rise and early evolution of animals: Where do we stand from a trace-fossil perspective? *Interface Focus* 10: 20190103.

Mani GS, Clarke BC. 1990. Mutational order: A major stochastic-process in evolution. *Proceedings of the Royal Society B* 240: 29–37.

Manrubia S, Cuesta JA, Aguirre J, Ahnert SE, Altenberg L, et al. 2021. From genotypes to organisms: State-of-the-art and perspectives of a cornerstone in evolutionary dynamics. *Physics of Life Reviews* 38: 55–106.

Marchant DB, Chen G, Cai S, Chen F, Schafran P, et al. 2021. Dynamic genome evolution in a model fern. *Nature Plants* 8: 1038–51.

Martin WF. 2017. Physiology, anaerobes, and the origin of mitosing cells 50 years on. *Journal of Theoretical Biology* 434: 2–10.

Martin WF, Garg S, Zimorski V. 2015. Endosymbiotic theories for eukaryote origin. *Philosophical Transactions of the Royal Society B* 370: 20140330.

Martin WF, Tielens AG, Mentel M. 2020. *Mitochondria and Anaerobic Energy Metabolism in Eukaryotes: Biochemistry and Evolution.* Berlin: Walter de Gruyter.

Mason OT. 1902. *The Origins of Invention: A Study of Industry among Primitive Peoples.* New York: Scribner's Sons.

Maynard Smith J. 1970. Natural selection and the concept of a protein space. *Nature* 225: 563–64.

Maynard Smith J, Burian R, Kauffman S, Alberch P, Campbell J, et al. 1985. Developmental constraints and evolution. *Quarterly Review of Biology* 60: 265–87.

Maynard Smith J, Szathmáry E. 1995. *The Major Transitions in Evolution.* New York: W. H. Freeman.

Mayr E. 1942. *Systematics and the Origin of Species.* New York: Columbia University Press.

Mayr E. 1954. Change of genetic environment and evolution. In *Evolution as a Process*, ed. JS Huxley, AC Hardy, EB Ford, pp. 157–80. London: George Allen and Unwin.

Mayr E. 1960. The emergence of novelty. In *The Evolution of Life*, ed. S Tax, pp. 349–80. Chicago: University of Chicago Press.

Mayr E. 1963. *Animal Species and Evolution.* Cambridge, MA: Belknap Press of Harvard University Press.

Mayr E. 1976. *Evolution and the Diversity of Life.* Cambridge, MA: Harvard University Press.

Mayr E. 1982. *The Growth of Biological Thought.* Cambridge, MA: Harvard University Press.

McBrearty S, Brooks AS. 2000. The revolution that wasn't: A new interpretation of the origin of modern human behavior. *Journal of Human Behavior* 39: 453–563.

McCoy VE, Saupe EE, Lamsdell JC, Tarhan LG, McMahon S, et al. 2016. The "Tully monster" is a vertebrate. *Nature* 532: 496–99.

McCoy VE, Wiemann J, Lamsdell JC, Whalen CD, Lidgard S, et al. 2020. Chemical signatures of soft tissues distinguish between vertebrates and invertebrates from the Carboniferous Mazon Creek Lagerstatte of Illinois. *Geobiology* 18: 560–65.

McCraw TK. 2007. *Prophet of Innovation: Joseph Schumpeter and Creative Destruction.* Cambridge, MA: Harvard University Press.

McFadden GI. 2014. Origin and evolution of plastids and photosynthesis in eukaryotes. *Cold Spring Harbor Perspectives in Biology* 6: A016105.

McGhee GR. 2007. *The Geometry of Evolution.* Cambridge: Cambridge University Press.

McGhee GR, Jr. 2011. *Convergent Evolution*. Cambridge, MA: MIT Press.

McGhee GR, Jr. 2016. Can evolution be directional without being teleological? *Studies in the History and Philosophy of Biological and Biomedical Sciences* 58: 93–99.

McInerney JO, Cummins C, Haggerty L. 2011a. Goods-thinking vs. tree-thinking: Finding a place for mobile genetic elements. *Mobile Genetic Elements* 1: 304–8.

McInerney JO, Erwin DH. 2016. Public goods and the early evolution of life. *Philosophical Transactions of the Royal Society B* 375: 20160359.

McInerney JO, Erwin DH. 2017. The role of public goods in planetary evolution. *Philosophical Transactions of the Royal Society A* 375: 20160359.

McInerney JO, McNally A, O'Connell MJ. 2017. Why prokaryotes have pangenomes. *Nature Microbiology* 2: 1–5.

McInerney JO, O'Connell MJ, Pisani D. 2014. The hybrid nature of the Eukaryota and a consilient view of life on Earth. *Nature Reviews Microbiology* 12: 449–55.

McInerney JO, Pisani D, Bapteste E, O'Connell MJ. 2011b. The public goods hypothesis for the evolution of life on Earth. *Biology Direct* 6: 1–17.

McInerney JO, Whelan FJ, Domingo-Sananes MR, McNally A, O'Connell MJ. 2020. Pangenomes and selection: The public goods hypothesis. In *The Pangenome: Diversity, Dynamics and Evolution of Genomes*, ed. H Tettelin, D Medini, pp. 151–67. Cham, Switzerland: Springer.

McInerney MJ, Sieber JR, Gunsalus RP. 2009. Syntrophy in anaerobic global carbon cycle. *Current Opinion in Biotechnology* 20: 623–32.

McQueen E, Rebeiz M. 2020. On the specificity of gene regulatory networks: How does network co-option affect subsequent evolution? *Current Topics in Developmental Biology* 139: 375–405.

McShea DW. 1998. Possible largest-scale trends in organismal evolution: Eight "live hypotheses." *Annual Review of Ecology and Systematics* 29: 293–318.

McShea DW. 2000. Functional complexity in organisms: Parts as proxies. *Biology & Philosophy* 15: 641–68.

Mentel M, Martin W. 2008. Energy metabolism among eukaryotic anaerobes in light of Proterozoic ocean chemistry. *Philosophical Transactions of the Royal Society B* 363: 2717–29.

Merton RK. 1961. Singletons and multiples in scientific discovery: A chapter in the sociology of science. *Proceedings of the American Philosophical Society* 105: 470–86.

Mesoudi A. 2011. *Cultural Evolution: How Darwinian Theory Can Explain Human Culture and Synthesize the Social Sciences*. Chicago: University of Chicago Press.

Mesoudi A, Thornton A. 2018. What is cumulative cultural evolution? *Proceedings of the Royal Society B* 285: 20180712.

Miao L, Yin Z, Knoll AH, Qu Y, Zhu M. 2024. 1.63-billion-year-old multicellular eukaryotes from the Chuanlinggou Formation in North China. *Science Advances* 10: eadk3208.

Michod RE. 1999. *Darwinian Dynamics*. Princeton, NJ: Princeton University Press.

Michod RE, Roze D. 2000. Cooperation and conflict in the evolution of multicellularity. *Heredity* 86: 1–7.

Miles DB, Ricklefs RE, Losos JB. 2023. How exceptional are the classic adaptive radiations of passerine birds? *Proceedings of the National Academy of Sciences of the USA* 120: e1813976120.

Miller AI, Sepkoski JJ, Jr. 1988. Modeling bivalve diversification: The effect of interaction on a macroevolutionary system. *Paleobiology* 14: 364–69.

Mills DB, Boyle RA, Daines SJ, Sperling EA, Pisani D, et al. 2022. Eukaryogenesis and oxygen in Earth history. *Nature Ecology & Evolution* 6: 520–32.

Mills DB, Canfield DE. 2014. Oxygen and animal evolution: Did a rise of atmospheric oxygen "trigger" the origin of animals? *BioEssays* 36: 1145–55.

Minsky M. 2006. *The Emotion Machine*. New York: Simon and Schuster.

Mitteroecker P, Huttegger SM. 2009. The concept of morphospaces in evolutionary and developmental biology: Mathematics and metaphors. *Biological Theory* 4: 54–67.

Mivart SGJ. 1871. *On the Genesis of Species*. London: Macmillan.

Moczek AP. 2008. On the origins of novelty in development and evolution. *BioEssays* 30: 432–47.

Moczek AP. 2009. The origin and diversification of complex traits through micro- and macroevolution of development: Insights from horned beetles. *Current Topics in Developmental Biology* 86: 137–62.

Moczek AP. 2022. When the end modifies its means: The origins of novelty and the evolution of innovation. *Biological Journal of the Linnean Society* 139: 433–40.

Moczek AP, Sultan S, Foster S, Ledon-Rettig C, Dworkin I, et al. 2011. The role of developmental plasticity in evolutionary innovation. *Proceedings of the Royal Society B* 278: 2705–13.

Moen DS, Ravelojaona RN, Hutter CR, Wiens JJ. 2021. Testing for adaptive radiation: A new approach applied to Madagascar frogs. *Evolution* 75: 3008–25.

Mokyr J. 1992. *The Lever of Riches: Technological Creativity and Economic Progress*. Oxford: Oxford University Press.

Mokyr J. 2005. Long-term economic growth and the history of technology. In *Handbook of Economic Growth*, ed. P Aghion, SN Durlauf, pp. 1114–80. Amsterdam: Elsevier.

Mokyr J. 2016. *Culture of Growth: The Origins of the Modern Economy*. Princeton, NJ: Princeton University Press.

Monod J. 1972. *Chance and Necessity*. New York: Vintage Books.

Monterio A, Podlaha O. 2009. Wings, horns, and butterfly eyespots: How do complex traits evolve? *PLoS Biology* 7: 209–16.

Montévil M. 2018. Possibility spaces and the notion of novelty: From music to biology. *Synthese* 196: 4555–81.

Moore GE. 1965. Cramming more components onto integrated circuits. *Electronics* 38: 114–17.

Moore GE. 1975. Progress in digital integrated electronics. *IEEE, IEDM Technical Digest (1975)*: 11–13.

Morgan CL. 1923. *Emergent Evolution*. New York: Henry Holt.

Morgan LH. 1877. *Ancient Society*. New York: Henry Holt.

Morgan TJH, Feldman MW. 2024. Human culture is uniquely open-ended rather than uniquely cumulative. *Nature Human Behavior*, https://doi.org/10.1038/s41562-024-02035-y.

Morris BEL, Henneberger R, Moissl-Eichinger C. 2013. Microbial syntrophy: Interaction for the common good. *FEBS Microbiology Review* 37: 384–406.

Morris JL, Puttick MN, Clark JW, Edwards D, Kenrick P, et al. 2018. The timescale of early land plant evolution. *Proceedings of the National Academy of Sciences of the USA* 115: e2274–83.

Müller GB. 1990. Developmental mechanisms at the origin of morphologic novelty. In *Evolutionary Innovations*, ed. MH Nitecki, pp. 99–130. Chicago: University of Chicago Press.

Müller GB, Wagner GP. 1991. Novelty in evolution: Restructuring the concept. *Annual Review of Ecology and Systematics* 22: 229–56.

Müller GB, Wagner GP. 2003. Innovation. In *Keywords and Concepts in Evolutionary Developmental Biology*, ed. BK Hall, WM Olson, pp. 218–27. Cambridge, MA: Harvard University Press.

Müller M, Mentel M, Van Hellemond JJ, Henze K, Woehle C, et al. 2012. Biochemistry and evolution of anaerobic energy metabolism in eukaryotes. *Microbiology and Molecular Biology Reviews* 76: 444–95.

Murdock DJE. 2020. The "biomineralization toolkit" and the origin of animal skeletons. *Biological Reviews* 95: 1372–92.

Muscente AD, Martindale RC, Prabhu A, Ma XG, Fox P, et al. 2022. Appearance and disappearance rates of Phanerozoic marine animal paleocommunities. *Geology* 50: 341–45.

Muthukrishna M, Henrich J. 2016. Innovation in the collective brain. *Philosophical Transactions of the Royal Society B* 371: 20150192.

Nakamura T, Klomp J, Pieretti J, Schneider I, Gehrke AR, Shubin NH. 2015. Molecular mechanisms underlying the exceptional adaptations of batoid fins. *Proceedings of the National Academy of Sciences of the USA* 112: 15940–45.

Nanglu K, Cole SR, Wright DF, Souto C. 2022. Worms and gills, plates and spines: The evolutionary origins and incredible disparity of deuterostomes revealed by fossils, genes, and development. *Biological Reviews* 98: 316–51.

Nasvall J, Sun L, Roth JR, Andersson DI. 2012. Real-time evolution of new genes by innovation, amplification, and divergence. *Science* 338: 384–87.

Nedelcu AM, Michod RE. 2020. Stress responses co-opted for specialized cell types during the early evolution of multicellularity: The role of stress in the evolution of cell types can be traced back to the early evolution of multicellularity. *BioEssays* 42: e2000029.

Nelson RR, Winter SG. 1982. *An Evolutionary Theory of Economic Change*. Cambridge, MA: Harvard University Press.

Newell A, Simon HA. 1972. *Human Problem Solving*. Englewood Cliffs, NJ: Prentice-Hall.

Newman M. 2018. *Networks*. Oxford: Oxford University Press.

Newton I. 1999. *The Principia: Mathematical Principles of Natural Philosophy*. Berkeley: University of California Press.

Nichols LM, Feinman GM. 2022. The foundation of Monte Alban, intensification, and growth: Coactive processes and joint production. *Frontiers in Political Science* 4: 805407.

Nie Y, Foster CSP, Zhu T, Yao R, Duchene DA, et al. 2020. Accounting for uncertainty in the evolutionary timescale of green plants through clock-partitioning and fossil calibration strategies. *Systematic Biology* 69: 1–16.

Niklas KJ. 1997. *The Evolutionary Biology of Plants*. Chicago: University of Chicago Press.

Nishimura T, Tokuda IT, Miyachi S, Dunn JC, Herbst CT, et al. 2022. Evolutionary loss of complexity in human vocal anatomy as an adaptation for speech. *Science* 377: 760–63.

Nishiyama T, Sakayama H, de Vries J, Buschmann H, Saint-Marcoux D, et al. 2018. The *Chara* genome: Secondary complexity and implications for plant terrestrialization. *Cell* 174: 448–64.e24.

Nitecki MH. 1990. The plurality of evolutionary innovations. In *Evolutionary Innovations*, ed. MH Nitecki, pp. 3–18. Chicago: University of Chicago Press.

Nolan C. 2014. *Interstellar*. Burbank, CA: Warner Bros.

Nordhaus WD. 1996. Do real output and real wage measures capture reality? The history of lighting suggests not. In *The Economics of New Goods*, ed. TF Bresnahan, RJ Gordon, pp. 27–69. Chicago: University of Chicago Press.

North M. 2013. *Novelty: A History of the New*. Chicago: University of Chicago Press.

Novack-Gottshall PM. 2007. Using a theoretical ecospace to quantify the ecological diversity of Paleozoic and modern marine biotas. *Paleobiology* 33: 273–94.

Novack-Gottshall PM, Sultan A, Smith NS, Purcell J, Hanson KE, et al. 2022. Morphological volatility precedes ecological innovation in early echinoderms. *Nature Ecology & Evolution* 6: 263–72.

Nunez RE. 2017a. Is there really an evolved capacity for number? *Trends in Cognitive Science* 21: 409–24.

Nunez RE. 2017b. Number: Biological enculturation beyond natural selection. *Trends in Cognitive Science* 21: 404–5.

Oakley TH. 2017. Furcation and fusion: The phylogenetics of evolutionary novelty. *Developmental Biology* 431: 69–76.

Oakley TH, Speiser DI. 2015. How complexity originates: The evolution of animal eyes. *Annual Review of Ecology, Evolution and Systematics* 46: 237–60.

Odling-Smee FJ. 2024. *Niche Construction: How Life Contributes to Its Own Evolution*. Cambridge, MA: MIT Press.

Odling-Smee FJ, Erwin DH, Palkovacs EP, Feldman MW, Laland KN. 2013. Niche construction theory: A practical guide for ecologists. *Quarterly Review of Biology* 88: 4–28.

Odling-Smee FJ, Laland KN, Feldman MW. 2003. *Niche Construction: The Neglected Process in Evolution*. Princeton, NJ: Princeton University Press.

Ogburn WF. 1922. *Social Change with respect to Culture and Original Nature*. New York: Viking.

Ogburn WF, Thomas D. 1922. Are inventions inevitable? A note on social evolution. *Political Science Quarterly* 37: 83–98.

O'Madagain C, Tomasello M. 2022. Shared intentionality, reason-giving and the evolution of human culture. *Philosophical Transactions of the Royal Society B* 377: 20200320.

O'Malley MA, Powell R. 2016. Major problems in evolutionary transitions: How a metabolic perspective can enrich our understanding of macroevolution. *Biology & Philosophy* 31: 159–89.

Onstein RE. 2020. Darwin's second "abominable mystery": Trait flexibility as the innovation leading to angiosperm diversity. *New Phytologist* 228: 1741–47.

Ortega-Hernandez J, Janssen R, Budd GE. 2017. Origin and evolution of the panarthropod head: A palaeobiological and developmental perspective. *Arthropod Structure & Development* 46: 354–79.

Ortman SG, Cabaniss AHF, Strum JO, Bettencourt LMA. 2015. Settlement scaling and increasing returns in an ancient society. *Science Advances* 1: e1400066.

Ortman SG, Lobo J. 2020. Smithian growth in a nonindustrial society. *Science Advances* 6: eaba5694.

Osborn HF. 1897. Organic selection. *Science* 6: 583–87.

Osborn HF. 1933. Aristogenesis, the observed order of biomechanical evolution. *Proceedings of the National Academy of Sciences of the USA* 19: 699–703.

Ospovat D. 1981. *The Development of Darwin's Theory*. Chicago: University of Chicago Press.

Oyston JW, Hughes M, Gerber S, Wills MA. 2016. Why should we investigate the morphological disparity of plant clades? *Annals of Botany* 117: 859–79.

Ozaki K, Reinhard CT, Tajika E. 2019. A sluggish mid-Proterozoic biosphere and its effect on Earth's redox balance. *Geobiology* 17: 3–11.

Padgett JF, Powell WW. 2012. *The Emergence of Organizations and Markets*. Princeton, NJ: Princeton University Press.

Pagani M, Caldeira K, Berner R, Beerling DJ. 2009. The role of terrestrial plants in limiting atmospheric CO_2 decline over the past 24 million years. *Nature* 460: 85–88.

Page SE. 2018. *The Model Thinker*. New York: Basic Books.

Pal C, Papp B. 2017. Evolution of complex adaptations in molecular systems. *Nature Ecology & Evolution* 1: 1084–92.

Panchen AL. 1992. *Classification, Evolution and the Nature of Biology*. Cambridge: Cambridge University Press.

Pandolfi JM, Staples TL, Kiessling W. 2020. Increased extinction in the emergence of novel ecological communities. *Science* 370: 220–22.

Parfrey LW, Lahr DJG, Knoll AH, Katz LA. 2011. Estimating the timing of early eukaryotic diversification with multigene molecular clocks. *Proceedings of the National Academy of Sciences of the USA* 108: 13624–29.

Parker W. 1974. *Europe, America, and the Wider World*. Cambridge: Cambridge University Press.

Paterson JR, Edgecombe GD, Lee MSY. 2019. Trilobite evolutionary rates constrain the duration of the Cambrian Explosion. *Proceedings of the National Academy of Sciences of the USA* 116: 4394–99.

Payne JL, Wagner A. 2019. The causes of evolvability and their evolution. *Nature Reviews Genetics* 20: 24–38.

Pearce T. 2020. *Pragmatism's Evolution: Organism and Environment in American Philosophy*. Chicago: University of Chicago Press.

Pearl R, Reed LJ. 1920. On the rate of growth of the population of the United States since 1790 and its mathematical representation. *Proceedings of the National Academy of Sciences of the USA* 6: 275–88.

Pereira-Leal JB, Levy ED, Teichmann SA. 2006. The origins and evolution of functional modules: Lessons from protein complexes. *Philosophical Transactions of the Royal Society B* 361: 507–17.

Perry S, Carter A, Smolla M, Akcay E, Nobel S, et al. 2021. Not by transmission alone: The role of invention in cultural evolution. *Philosophical Transactions of the Royal Society B* 376: 20200049.

Peters SE, Gaines RR. 2012. Formation of the "Great Unconformity" as a trigger for the Cambrian Explosion. *Nature* 484: 363–66.

Peterson T, Müller GB. 2013. What is evolutionary novelty? Process versus character-based definitions. *Journal of Experimental Zoology Part B (Molecular and Developmental Evolution)* 320: 345–50.

Peterson T, Müller GB. 2016. Phenotypic novelty in EvoDevo: The distinction between continuous and discontinuous variation and its importance in evolutionary theory. *Evolutionary Biology* 43: 314–35.

Phelps E. 2013. *Mass Flourishing*. Princeton, NJ: Princeton University Press.

Phillips J. 1841. *Figures and Descriptions of the Palaeozoic Fossils of Cornwall, Devon and West Somerset; Observed in the Course of the Ordnance Survey of That District*. London: Longman, Brown, Green and Longmans.

Phillips J. 1860. *Life on the Earth: Its Origin and Succession*. Cambridge, UK: Macmillan.

Pictet FJ. 1860. *On the Origin of Species* by Charles Darwin. In *Darwin and His Critics*, ed. DL Hull, pp. 142–54. Chicago: University of Chicago Press.

Pigliucci M. 2007. Do we need an extended evolutionary synthesis? *Evolution* 61: 2743–49.

Pigliucci M. 2008a. Is evolvability evolvable? *Nature Reviews Genetics* 9: 75–82.

Pigliucci M. 2008b. What, if anything, is an evolutionary novelty? *Philosophy of Science* 75: 887–98.

Pigliucci M. 2009. An extended synthesis for evolutionary biology. *Annals of the New York Academy of Sciences* 1168: 218–28.

Pigliucci M, Kaplan J. 2006. *Making Sense of Evolution*. Chicago: University of Chicago Press.

Pinker S. 1994. *The Language Instinct*. New York: William Morrow.

Pinker S. 2007. *The Stuff of Thought: Language as a Window into Human Nature*. New York: Viking.

Pinker S. 2010. The cognitive niche: Coevolution of intelligence, sociality, and language. *Proceedings of the National Academy of Sciences of the USA* 107, Suppl 2: 8993–99.

Pinker S. 2011. *The Better Angels of Our Nature*. New York: Viking.

Pinker S. 2018. *Enlightenment Now*. New York: Viking.

Pinker S, Bloom P. 1990. Natural selection and natural language. *Behavioral and Brain Sciences* 13: 707–84.

Pinson A, Xing L, Namba T, Kalebic N, Peters J, et al. 2022. Human TKTL1 implies greater neurogenesis in frontal neocortex of modern humans than Neanderthals. *Science* 377: eabl6422.

Pisani D, Pett W, Dohrmann M, Feuda R, Rota-Stabelli O, et al. 2015. Genomic data do not support comb jellies as the sister group to all other animals. *Proceedings of the National Academy of Sciences of the USA* 112: 15402–7.

Pitt-Rivers AL-F. 1906. *The Evolution of Culture and Other Essays*. Oxford: Oxford University Press.

Planavsky NJ, Cole DB, Isson TT, Reinhard CT, Crockford PW, et al. 2018. A case for low atmospheric oxygen levels during Earth's middle history. *Emerging Topics in Life Sciences* 2: 149–59.

Planavsky NJ, Reinhard CT, Wang XL, Thomson D, McGoldrick P, et al. 2014. Low mid-Proterozoic atmospheric oxygen levels and the delayed rise of animals. *Science* 346: 635–38.

Planer RJ, Sterelny K. 2022. *From Signal to Symbol.* Cambridge, MA: MIT Press.

Plummer TW, Oliver JS, Finestone EM, Ditchfield PW, Bishop LC, et al. 2023. Expanded geographic distribution and dietary strategies of the earliest Oldowan hominins and *Paranthropus. Science* 379: 561–66.

Pomeranz K. 2000. *The Great Divergence.* Princeton, NJ: Princeton University Press.

Porter SM. 2020. Insights into eukaryogenesis from the fossil record. *Interface Focus* 10: 20190105.

Powers ST, Van Schaik CP, Lehmann L. 2016. How institutions shaped the last major evolutionary transition to large-scale human societies. *Philosophical Transactions of the Royal Society B* 371: 20150098.

Proffitt T, Reeves JS, Braun DR, Malaivijitnond S, Luncz LV. 2023. Wild macaques challenge the origin of intentional tool production. *Science Advances* 9: eade8159.

Prokop J, Rosova K, Krzeminska E, Krzeminski W, Nel A, Engel MS. 2022. Abdominal serial homologues of wings in Paleozoic insects. *Current Biology* 32: 3414–22.e1.

Provine WB. 1986. *Sewall Wright and Evolutionary Biology.* Chicago: University of Chicago Press.

Provine WB. 2005. Ernst Mayr, a retrospective. *Trends in Ecology & Evolution* 20: 411–13.

Prum RO. 2005. Evolution of the morphological innovations of feathers. *Journal of Experimental Zoology Part B (Molecular and Developmental Evolution)* 304: 570–79.

Prum RO, Brush AH. 2002. The evolutionary origin and diversification of feathers. *Quarterly Review of Biology* 77: 261–95.

Purugganan MD. 2022. What is domestication? *Trends in Ecology & Evolution* 37: 663–71.

Quandt EM, Deatherage DE, Ellington AD, Georgiou G, Barrick JE. 2014. Recursive genome-wide recombination and sequencing reveals a key refinement step in the evolution of a metabolic innovation in *Escherichia coli. Proceedings of the National Academy of Sciences of the USA* 111: 2217–22.

Quattrini AM, Rodriguez E, Faircloth BC, Cowman PF, Brugler MR, et al. 2020. Palaeoclimate ocean conditions shaped the evolution of corals and their skeletons through deep time. *Nature Ecology & Evolution* 4: 1531–38.

Queller DC. 1997. Cooperators since life began. *Quarterly Review of Biology* 72: 184–88.

Rabosky DL. 2014. Automatic detection of key innovations, rate shifts, and diversity-dependence on phylogenetic trees. *PloS One* 9: e89543.

Rabosky DL. 2017. Phylogenetic tests for evolutionary innovation: The problematic link between key innovations and exceptional diversification. *Philosophical Transactions of the Royal Society B* 372: 20160417.

Radeloff VC, Williams JW, Bateman BL, Burke KD, Carter SK, et al. 2015. The rise of novelty in ecosystems. *Ecological Applications* 25: 2051–68.

Raff RA. 1996. *The Shape of Life.* Chicago: University of Chicago Press.

Raff RA, Kaufman TC. 1983. *Embryos, Genes, and Evolution.* New York: Macmillan.

Raff RA, Sly BJ. 2000. Modularity and dissociation in the evolution of gene expression territories in development. *Evolution & Development* 2: 102–13.

Ragan MA, McInerney JO, Lake JA. 2009. The network of life: Genome beginnings and evolution. Introduction. *Philosophical Transactions of the Royal Society B* 364: 2169–75.

Rajakumar R, San Mauro D, Dijkstra MB, Huang MH, Wheeler DE, et al. 2012. Ancestral developmental potential facilitates parallel evolution in ants. *Science* 335: 79–82.

Ramirez-Barahona S, Sauquet H, Magallon S. 2020. The delayed and geographically heterogeneous diversification of flowering plant families. *Nature Ecology & Evolution* 4: 1232–38.

Raup DM. 1966. Geometric analysis of shell coiling: General problems. *Journal of Paleontology* 40: 1178–90.

Raymond J, Segre D. 2006. The effect of oxygen on biochemical networks and the evolution of complex life. *Science* 311: 1764–67.

Reader SM, Laland KN. 2003a. *Animal Innovation*. Oxford: Oxford University Press.

Reader SM, Laland KN. 2003b. Animal innovation: An introduction. In *Animal Innovation*, ed. SL Reader, KN Laland, pp. 3–35. Oxford: Oxford University Press.

Rehan SM, Toth AL. 2015. Climbing the social ladder: The molecular evolution of sociality. *Trends in Ecology & Evolution* 30: 426–33.

Reid RGB. 2007. *Biological Emergences: Evolution by Natural Experiment*. Cambridge, MA: MIT Press.

Reif W-E. 1983. Evolutionary theory in German paleontology. In *Dimensions of Darwinism*, ed. M Grene, pp. 173–203. Cambridge: Cambridge University Press.

Reif W-E. 1986. The search for a macroevolutionary theory in German paleontology. *Journal of the History of Biology* 19: 79–130.

Reif W-E, Junker T, Hoßfeld U. 2000. The synthetic theory of evolution: General problems and the German contribution to the synthesis. *Theory in Biosciences* 119: 41–91.

Reimer O, Tedengren M. 1996. Phenotypical improvement of morphological defences in the mussel *Mytilus edulis* induced by exposure to the predator *Asterias rubens*. *Oikos* 75: 383–90.

Reinhard CT, Planavsky NJ, Gill BC, Ozaki K, Robbins LJ, et al. 2017. Evolution of the global phosphorus cycle. *Nature* 541: 386–89.

Rensch B. 1959. *Evolution above the Species Level*. New York: Columbia University Press.

Rensch B. 1980. Historical development of the present synthetic neo-Darwinism in Germany. In *The Evolutionary Synthesis: Perspectives on the Unification of Biology*, ed. E Mayr, W Provine, pp. 284–303. Cambridge, MA: Harvard University Press.

Richards RJ. 1992. *The Meaning of Evolution*. Chicago: University of Chicago Press.

Riederer JM, Tiso S, Van Eldijk TJB, Weissing FJ. 2022. Capturing the facets of evolvability in a mechanistic framework. *Trends in Ecology & Evolution* 37: 430–39.

Riedl R. 1977. A systems-analytical approach to macro-evolutionary phenomena. *Quarterly Review of Biology* 52: 351–70.

Riedl R. 1978. *Order in Living Organisms*. Chichester, UK: John Wiley.

Riegner MF. 2013. Ancestor of the new archetypal biology: Goethe's dynamic typology as a model for contemporary evolutionary developmental biology. *Studies in the History and Philosophy of Biological and Biomedical Sciences* 44: 735–44.

Rieppel O. 2017. *Turtles as Hopeful Monsters*. Bloomington: Indiana University Press.

Rittschof CC, Robinson GE. 2016. Behavioral genetic toolkits: Toward the evolutionary origins of complex phenotypes. *Current Topics in Developmental Biology* 119: 157–204.

Rogers CS, Astrop TI, Webb SM, Ito S, Wakamatsu K, McNamara ME. 2019. Synchrotron X-ray absorption spectroscopy of melanosomes in vertebrates and cephalopods: Implications for the affinity of *Tullimonstrum*. *Proceedings of the Royal Society B* 286: 20191649.

Rogers EM. 1978. Re-invention during the innovation process. In *The Diffusion of Innovations: An Assessment*, ed. M Radnor, I Feller, E Rogers, pp. 343–61. Evanston, IL: Center for the Interdisciplinary Study of Science and Technology, Northwestern University.

Rogers EM. 1995. *Diffusion of Innovations*. New York: Free Press.

Romanes GJ. 1895. *Darwin, and after Darwin: An Exposition of the Darwinian Theory and a Discussion of Post-Darwinian Questions*. 3 vols. Chicago: Open Court.

Romer PM. 1986. Increasing returns and long-run growth. *Journal of Political Economy* 94: 1002–37.

Romer PM. 1990. Endogenous technological change. *Journal of Political Economy* 98: S71–102.

Romero PA, Arnold FH. 2009. Exploring protein fitness landscapes by directed evolution. *Nature Reviews Molecular Cell Biology* 10: 866–76.

Romiguier J, Borowiec ML, Weyna A, Helleu Q, Loire E, et al. 2022. Ant phylogenomics reveals a natural selection hotspot preceding the origin of complex eusociality. *Current Biology* 32: 2942–47.

Rosenberg N. 1982. *Inside the Black Box: Technology and Economics*. Cambridge: Cambridge University Press.

Rosenzweig ML, McCord RD. 1991. Incumbent replacement: Evidence for long-term evolutionary progress. *Paleobiology* 17: 202–13.

Rothwell R. 1992. Successful industrial innovation: Critical factors for the 1990s. *R&D Management* 22: 221–40.

Rucklin M, Donoghue PCJ. 2019. Evolutionary origin of teeth. *Encyclopedia of Life Sciences*, https://doi.org/10.1002/9780470015902.a0026408.

Rudwick MJS. 2008. *Worlds before Adam: The Reconstruction of Geohistory in the Age of Reform*. Chicago: University of Chicago Press.

Rundell RJ, Price TD. 2009. Adaptive radiation, nonadaptive radiation, ecological speciation and nonecological speciation. *Trends in Ecology & Evolution* 24: 394–99.

Ruse M. 1996. *Monad to Man*. Cambridge, MA: Harvard University Press.

Russell ES. 1916. *Form and Function*. London: John Murray.

Russell ES. 1936. Form and function: A historical note. *Folia biotheoretica* 1: 1–12.

Sagan L. 1967. On the origin of mitosing cells. *Journal of Theoretical Biology* 14: 255–74.

Sahlins MD, Service ER, eds. 1960. *Evolution and Culture*. Ann Arbor: University of Michigan Press.

Salazar-Ciudad I. 2006. On the origins of morphological disparity and its diverse developmental bases. *BioEssays* 28: 1112–22.

Salazar-Ciudad I, Jernvall J. 2005. Graduality and innovation in the evolution of complex phenotypes: Insights from development. *Journal of Experimental Zoology Part B (Molecular and Developmental Evolution)* 304: 619–31.

Sallan L, Giles S, Sansom RS, Clarke JT, Johanson Z, et al. 2017. The "Tully Monster" is not a vertebrate: Characters, convergence and taphonomy in Palaeozoic problematic animals. *Palaeontology* 60: 149–57.

Salvini-Plawen LV, Mayr E. 1977. On the evolution of photoreceptors and eyes. *Evolutionary Biology* 10: 207–63.

Sauquet H, Magallon S. 2018. Key questions and challenges in angiosperm macroevolution. *New Phytologist* 219: 1170–87.

Sauquet H, Von Balthazar M, Magallon S, Doyle JA, Endress PK, et al. 2017. The ancestral flower of angiosperms and its early diversification. *Nature Communications* 8: 16047.

Schaffer B, Hecht MK. 1965. Symposium: The origin of higher levels of organization. *Systematic Biology* 14: 245–48.

Schavemaker PE, Munoz-Gomez SA. 2022. The role of mitochondrial energetics in the origin and diversification of eukaryotes. *Nature Ecology & Evolution* 6: 1307–17.

Schimmelpfennig R, Razek L, Schnell E, Muthukrishna M. 2022. Paradox of diversity in the collective brain. *Philosophical Transactions of the Royal Society B* 377: 20200316.

Schindewolf OH. 1936. *Paläontologie, Entwicklungslehre und Genetik*. Berlin: Borntraeger.

Schindewolf OH. (1950) 1994. *Basic Questions in Paleontology: Geologic Time, Organic Evolution, and Biological Systematics*. Chicago: University of Chicago Press.

Schlichting CD, Pigliucci M. 1998. *Phenotypic Evolution: A Reaction Norm Perspective*. Sunderland, MA: Sinauer Associates.

Schluter D. 2000. *The Ecology of Adaptive Radiation*. Oxford: Oxford University Press.

Schmookler J. 1966. *Invention and Economic Growth*. Cambridge, MA: Harvard University Press.

Schoch RR. 2010. Riedl's burden and the body plan: Selection, constraint, and deep time. *Journal of Experimental Zoology Part B (Molecular and Developmental Evolution)* 314: 1–10.

Schumpeter JA. (1912) 1954. *Economic Doctrine and Method: An Historical Sketch*. New York: Oxford University Press.

Schumpeter JA. 1935. The analysis of economic change. *Review of Economics and Statistics* 17: 2–10.

Schumpeter JA. 1942. *Capitalism, Socialism and Democracy.* New York: Harper and Collins.

Schwenk K, Wagner GP. 2004. The relativism of constraints on phenotypic integration. In *Phenotypic Integration*, ed. M Pigliucci, K Preston, pp. 390–408. New York: Oxford University Press.

Scoville WC. 1951. Minority migrations and the diffusion of technology. *Journal of Economic History* 11: 347–60.

Sebe-Pedros A, Ballare C, Parra-Acero H, Chiva C, Tena JJ, et al. 2016. The dynamic regulatory genome of *Capsaspora* and the origin of animal multicellularity. *Cell* 165: 1224–37.

Sebe-Pedros A, Chomsky E, Pang K, Lara-Astiaso D, Gaiti F, et al. 2018. Early metazoan cell type diversity and the evolution of multicellular gene regulation. *Nature Ecology & Evolution* 2: 1176–88.

Segar ST, Fayle TM, Srivastava DS, Lewinsohn TM, Lewis OT, et al. 2020. The role of evolution in shaping ecological networks. *Trends in Ecology & Evolution* 35: 454–66.

Seilacher A. 1999. Biomat-related lifestyles in the Precambrian. *Palaios* 14: 86–93.

Sepkoski JJ, Jr. 1982. A compendium of fossil marine families. *Milwaukee Public Museum Contributions in Biology and Geology* 51: 1–125.

Sepkoski JJ, Jr., Miller AI. 1985. Evolutionary marine faunas and the distribution of Paleozoic benthic communities in space and time. In *Phanerozoic Diversity Patterns*, ed. JW Valentine, pp. 153–89. Princeton, NJ: Princeton University Press.

Serreli E. 2015. Visualizing macroevolution: From adaptive landscapes to compositions of multiple spaces. In *Macroevolution*, ed. E Serreli, N Gontier, pp. 113–62. Cham, Switzerland: Springer.

Shaw JO, Coco E, Wootton K, Daems D, Gillreath-Brown A, et al. 2021. Disentangling ecological and taphonomic signals in ancient food webs. *Paleobiology* 47: 385–401.

Shea JJ. 2017. Occasional, obligatory, and habitual stone tool use in hominin evolution. *Evolutionary Anthropology* 26: 200–217.

Shen Y, Zhang TG, Hoffman PF. 2008. On the coevolution of Ediacaran oceans and animals. *Proceedings of the National Academy of Sciences of the USA* 105: 7376–81.

Shiga Y, Kato K, Aragane-Nomura Y, Haraguchi T, Saridaki T, et al. 2017. Repeated co-option of a conserved gene regulatory module underpins the evolution of the crustacean carapace, insect wings and other flat outgrowths. *BioRxiv*: 160010.

Shubin N, Tabin C, Carroll S. 2009. Deep homology and the origins of evolutionary novelty. *Nature* 457: 818–23.

Silvestro D, Bacon CD, Ding W, Zhang Q, Donoghue PCJ, et al. 2021. Fossil data support a pre-Cretaceous origin of flowering plants. *Nature Ecology & Evolution* 5: 449–57.

Simakov O, Kawashima T. 2017. Independent evolution of genomic characters during major metazoan transitions. *Developmental Biology* 427: 179–82.

Simoes M, Breitkreuz L, Alvarado M, Baca S, Cooper JC, et al. 2016. The evolving theory of evolutionary radiations. *Trends in Ecology & Evolution* 31: 27–34.

Simoes TR, Vernygora O, Caldwell MW, Pierce SE. 2020. Megaevolutionary dynamics and the timing of evolutionary innovation in reptiles. *Nature Communications* 11: 3322.

Simon HA. 1955. A behavioral model of rational choice. *Quarterly Journal of Economics* 69: 99–118.

Simonton DK. 1999. *Origins of Genius: Darwinian Perspectives on Creativity.* New York: Oxford University Press.

Simonton DK. 2003. Human creativity: Two Darwinian analyses. In *Animal Innovation*, ed. SL Reader, KN Laland, pp. 309–25. Oxford: Oxford University Press.

Simpson GG. 1944. *Tempo and Mode in Evolution.* New York: Columbia University Press.

Simpson GG. 1953a. The Baldwin effect. *Evolution* 7: 110–17.

Simpson GG. 1953b. *The Major Features of Evolution*. New York: Columbia University Press.

Simpson GG. 1960. The history of life. In *The Evolution of Life*, ed. S Tax, pp. 117–80. Chicago: University of Chicago Press.

Simpson GG. 1970. Uniformitarianism: An inquiry into principle, theory and method in geohistory and biohistory. In *Essays in Evolution and Genetics in Honor of Theodosius Dobzhansky*, ed. MK Hecht, WC Steere, pp. 43–96. New York: Appleton.

Siskin C. 2016. *System: The Shaping of Modern Knowledge*. Cambridge, MA: MIT Press.

Skirgard H, Haynie HJ, Blasi DE, Hammarstrom H, Collins J, et al. 2023. Grambank reveals the importance of genealogical constraints on linguistic diversity and highlights the impact of language loss. *Science Advances* 9: eadg6175.

Slater GJ. 2015a. Iterative adaptive radiations of fossil canids show no evidence for diversity-dependent trait evolution. *Proceedings of the National Academy of Sciences of the USA* 112: 4897–902.

Slater GJ. 2015b. Not-so-early burst and the dynamic nature of morphological diversification. *Proceedings of the National Academy of Sciences of the USA* 112: 3595–96.

Slater MH. 2013. Cell types as natural kinds. *Biological Theory* 7: 170–79.

Smil V. 2023. *Invention and Innovation: A Brief History of Hype and Failure*. Cambridge, MA: MIT Press.

Smith A. 1776. *An Inquiry into the Nature and Causes of the Wealth of Nations*. London: Strahan and Cadell.

Smith BD. 2012. A cultural niche construction theory of initial domestication. *Biological Theory* 6: 260–71.

Smith BD. 2016. Neo-Darwinism, niche construction theory, and the initial domestication of plants and animals. *Evolutionary Ecology* 30: 307–24.

Smith FW, Jockusch EL. 2020. Into the body wall and back out again. *Nature Ecology & Evolution* 4: 1580–81.

Snyder WD, Reeves JS, Tennie C. 2022. Early knapping techniques do not necessitate cultural transmission. *Science Advances* 8: eabo2894.

Sogabe S, Hatleberg WL, Kocot KM, Say TE, Stoupin D, et al. 2019. Pluripotency and the origin of animal multicellularity. *Nature* 570: 519–22.

Soltis DE, Soltis PS, Morgan DR, Swensen SM, Mullin BC, et al. 1995. Chloroplast gene sequence data suggest a single origin of the predisposition for symbiotic nitrogen-fixation in angiosperms. *Proceedings of the National Academy of Sciences of the USA* 92: 2647–51.

Soltis PS, Soltis DE. 2016. Ancient WGD events as drivers of key innovations in angiosperms. *Current Opinion in Plant Biology* 30: 159–65.

Sorrells TR, Booth LN, Tuch BB, Johnson AD. 2015. Intersecting transcription networks constrain gene regulatory evolution. *Nature* 523: 361–65.

Soucy SM, Huang J, Gogarten JP. 2015. Horizontal gene transfer: Building the web of life. *Nature Reviews Genetics* 16: 472–82.

Soulebeau A, Aubriot X, Gaudeul M, Rouhan G, Hennequin S, et al. 2015. The hypothesis of adaptive radiation in evolutionary biology: Hard facts about a hazy concept. *Organisms Diversity & Evolution* 15: 747–61.

Spencer CS. 1997. Evolutionary approaches in archaeology. *Journal of Archaeological Research* 5: 209–64.

Sperling EA. 2016. The ecological physiology of the Cambrian "explosion" and Earth's second oxygen revolution. *Integrative and Comparative Biology* 56: e209.

Sperling EA, Melchin MJ, Fraser T, Stockey RG, Farrell UC, et al. 2021. A long-term record of early to mid-Paleozoic marine redox change. *Science Advances* 7: eabf4382.

Sperling EA, Stockey RG. 2018. The temporal and environmental context of early animal evolution: Considering all the ingredients of an "explosion." *Integrative and Comparative Biology* 58: 605–22.

Standen EM, Du TY, Larsson HC. 2014. Developmental plasticity and the origin of tetrapods. *Nature* 513: 54–58.

Stanley, SM. 1968. Post-Paleozoic adaptive radiation of infaunal bivalve molluscs: A consequence of mantle fusion and siphon formation. *Journal of Paleontology* 42: 214–29.

Stanley SM. 1975. A theory of evolution above the species level. *Proceedings of the National Academy of Sciences of the USA* 72: 646–50.

Stanley SM. 1979. *Macroevolution*. San Francisco: W. H. Freeman.

Stanley SM. 2014. Evolutionary radiation of shallow-water Lucinidae (Bivalvia with endosymbionts) as a result of the rise of seagrasses and mangroves. *Geology* 42: 803–6.

Starr TN, Flynn JM, Mishra P, Bolon DNA, Thornton JW. 2018. Pervasive contingency and entrenchment in a billion years of Hsp90 evolution. *Proceedings of the National Academy of Sciences of the USA* 115: 4453–58.

Stebbins GL. 1969. *The Basis of Progressive Evolution*. Chapel Hill: University of North Carolina Press.

Stebbins GL. 1974. Adaptive shifts and evolutionary novelty: A compositionist approach. In *Studies in the Philosophy of Biology*, ed. FJ Ayala, T Dobzhansky, pp. 285–306. Berkeley: University of California Press.

Sterelny K. 2014. A Paleolithic reciprocation crisis: Symbols, signals and norms. *Biological Theory* 9: 65–77.

Sterelny K. 2016. Cumulative cultural evolution and the origins of language. *Biological Theory* 11: 173–86.

Storch D, Simova I, Smycka J, Bohdalkova E, Toszogyova A, Okie JG. 2022. Biodiversity dynamics in the Anthropocene: How human activities change equilibria of species richness. *Ecogeography*: e05778.

Strömberg CAE. 2005. Decoupled taxonomic radiation and ecological expansion of open-habitat grasses in the Cenozoic of North America. *Proceedings of the National Academy of Sciences of the USA* 102: 11980–84.

Strömberg CAE. 2011. Evolution of grasses and grassland ecosystems. *Annual Review of Earth and Planetary Science* 39: 517–44.

Stroud LT, Losos JB. 2016. Ecological opportunity and adaptive radiation. *Annual Review of Ecology, Evolution and Systematics* 47: 507–32.

Suess E. 1904. *The Face of the Earth*. Oxford: Clarendon.

Summons RE, Welander PV, Gold DA. 2022. Lipid biomarkers: Molecular tools for illuminating the history of microbial life. *Nature Reviews Microbiology* 20: 174–85.

Suzuki TK. 2017. On the origin of complex adaptive traits: Progress since the Darwin versus Mivart debate. *Journal of Experimental Zoology Part B (Molecular and Developmental Evolution)* 328: 304–20.

Svensson EI, Calsbeek R, eds. 2012. *The Adaptive Landscape in Evolutionary Biology*. Oxford: Oxford University Press.

Swedberg R. 1991. *Schumpeter: A Biography*. Princeton, NJ: Princeton University Press.

Szathmáry E. 2015. Toward major evolutionary transitions theory 2.0. *Proceedings of the National Academy of Sciences of the USA* 112: 10104–11.

Szathmáry E, Maynard Smith J. 1995. The major evolutionary transitions. *Nature* 374: 227–32.

Tarashansky AJ, Musser JM, Khariton M, Li PY, Arendt D, et al. 2020. Mapping single-cell atlases throughout Metazoa unravels cell type evolution. *Elife* 10: e66747.

Tattersall I. 2008. An evolutionary framework for the acquisition of symbolic cognition by *Homo sapiens*. *Comparative Cognition and Behavior Reviews* 3: 99–114.

Tattersall I. 2018. Language origins: An evolutionary framework. *Topoi* 37: 289–96.

Tennie C, Call J, Tomasello M. 2009. Ratcheting up the ratchet: On the evolution of cumulative culture. *Philosophical Transactions of the Royal Society B* 364: 2405–15.

Ten Tusscher K. 2020. Of mice and plants: Comparative developmental systems biology. *Developmental Biology* 460: 32–39.

Tetlock PE. 2005. *Expert Political Judgment*. Princeton, NJ: Princeton University Press.

Thayer CW. 1983. Sediment-mediated biological disturbances and the evolution of the marine benthos. In *Biotic Interactions in Recent and Fossil Benthic Communities*, ed. MJS Tevesz, PL McCall, pp. 480–625. New York: Plenum.

Thomas RDK, Reif WE. 1993. The skeleton space: A finite set of organic designs. *Evolution* 47: 341–60.

Thomas RDK, Shearman RM, Stewart GW. 2000. Evolutionary exploitation of design options by the first animals with hard skeletons. *Science* 288: 1239–42.

Thompson DA. 1942. *On Growth and Form*. Cambridge: Cambridge University Press.

Thompson RM, Brose U, Dunne JA, Hall RO, Hladyz S, et al. 2012. Food webs: Reconciling the structure and function of biodiversity. *Trends in Ecology & Evolution* 27: 689–97.

Thompson VD, Holland-Lulewicz J, Butler RA, Hunt TW, Wendt L, et al. 2022. The early materialization of democratic institutions among the ancestral Muskogean of the American Southeast. *American Antiquity* 87: 703–23.

Tomasello M. 1999. *The Cultural Origins of Human Cognition*. Cambridge, MA: Harvard University Press.

Tomasello M. 2014. *A Natural History of Human Thinking*. Cambridge, MA: Harvard University Press.

Tomasello M, Melis A, Tennie C, Wyman E, Herrmann E. 2012. Two key steps in the evolution of human cooperation: The interdependence hypothesis. *Current Anthropology* 53: 673–92.

Tomoyasu Y. 2021. What crustaceans can tell us about the evolution of insect wings and other morphologically novel structures. *Current Opinion in Genetics & Development* 69: 48–55.

Tooby J, Devore I. 1987. The reconstruction of hominid behavioral evolution through strategic modeling. In *The Evolution of Human Behavior: Primate Models*, ed. WG Kirby, pp. 183–237. Albany: State University of New York Press.

Traverse A. 1988. Plant evolution dances to a different beat: Plant and animal evolutionary mechanisms compared. *Historical Biology* 1: 277–302.

Travis JM, Münkemüller T, Burton OJ. 2010. Mutation surfing and the evolution of dispersal during range expansions. *Journal of Evolutionary Biology* 23: 2656–67.

Tria F, Loreto V, Servedio VDP. 2019. Zipf's, Heaps' and Taylor's laws are determined by the expansion into the adjacent possible. *Entropy* 20: 752.

Trigger BG. 1998. *Sociocultural Evolution*. Oxford: Blackwell.

Trizzino M, Park Y, Holsbach-Beltrame M, Aracena K, Mika K, et al. 2017. Transposable elements are the primary source of novelty in primate gene regulation. *Genome Research* 27: 1623–33.

True JR, Carroll SB. 2002. Gene co-option in physiological and morphological evolution. *Annual Review of Cell and Developmental Biology* 18: 53–80.

Turchin P, Whitehouse H, Gavrilets S, Hoyer D, Francois P, et al. 2022. Disentangling the evolutionary drivers of social complexity: A comprehensive test of hypotheses. *Science Advances* 8: eabn3517.

Turk KA, Maloney KM, Laflamme M, Darroch SAF. 2022. Paleontology and ichnology of the late Ediacaran Nasep–Huns transition (Nama Group, southern Namibia). *Journal of Paleontology* 96: 753–69.

Turner JRG. 1983. "The hypothesis that explains mimetic resemblance explains evolution": The gradualist-saltationist schism. In *Dimensions of Darwinism*, ed. M Grene, pp. 130–69. Cambridge: Cambridge University Press.

Uller T, Moczek AP, Watson RA, Brakefield PM, Laland KN. 2018. Developmental bias and evolution: A regulatory network perspective. *Genetics* 209: 949–66.

Usselman SW. 1995. Determining a middle landscape: Competing narratives in the history of technology. *Reviews in American History* 23: 370–77.

Uyeda JC, Hansen TF, Arnold SJ, Pienaar J. 2011. The million-year wait for macroevolutionary bursts. *Proceedings of the National Academy of Sciences of the USA* 108: 15908–13.

Uyeda JC, Zenil-Ferguson R, Pennell MW. 2018. Rethinking phylogenetic comparative methods. *Systematic Biology* 67: 1091–109.

Valentine JW. 1973. *Evolutionary Paleoecology of the Marine Biosphere*. Englewood Cliffs, NJ: Prentice-Hall.

Valentine JW. 1980. Determinants of diversity in higher taxonomic categories. *Paleobiology* 6: 444–50.

Valentine JW. 1989. Phanerozoic marine faunas and the stability of the earth system. *Palaeogeography, Palaeoclimatology, Palaeoecology* 75: 137–55.

Valentine JW. 2004. *On the Origin of Phyla*. Chicago: University of Chicago Press.

Valentine JW, Walker TD. 1986. Diversity trends within a model taxonomic hierarchy and evolution. *Physica* 22D: 31–42.

Valverde S, Sole RV, Bedau MA, Packard NH. 2007. Topology and evolution of technology innovation networks. *Physical Review E* 76: 056118.

Van Belleghem SM, Ruggieri AA, Concha C, Livraghi L, Hebberecht L, et al. 2023. High level of novelty under the hood of convergent evolution. *Science* 379: 1043–49.

Van de Peer Y, Mizrachi E, Marchal K. 2017. The evolutionary significance of polyploidy. *Nature Reviews Genetics* 18: 411–24.

Van der Kooi CJ, Ollerton J. 2020. The origins of flowering plants and pollinators. *Science* 368: 1306–8.

Van Etten J, Bhattacharya D. 2020. Horizontal gene transfer in eukaryotes: Not if, but how much? *Trends in Genetics* 36: 915–25.

Vanneste K, Maere S, Van de Peer Y. 2014. Tangled up in two: A burst of genome duplications at the end of the Cretaceous and the consequences for plant evolution. *Philosophical Transactions of the Royal Society B* 369: 20130353.

Van Thiel J, Khan MA, Wouters RM, Harris RJ, Casewell NR, et al. 2022. Convergent evolution of toxin resistance in animals. *Biological Reviews* 97: 1823–43.

Van Valen L. 1973. A new evolutionary law. *Evolutionary Theory* 1: 1–30.

Van Valkenburgh B. 1991. Iterative evolution of hypercarnivory in canids (Mammalia: Carnivora): Evolutionary interactions among sympatric predators. *Paleobiology* 17: 340–62.

Van Velzen R, Holmer R, Bu FJ, Rutten L, Van Zeijl A, et al. 2018. Comparative genomics of the nonlegume *Parasponia* reveals insights into evolution of nitrogen-fixing rhizobium symbioses. *Proceedings of the National Academy of Sciences of the USA* 115: e4700–e4709.

Vassallo AI, Manzano A, Abdala V, Muzio RN. 2021. Can anyone climb? The skills of a nonspecialized toad and its bearing on the evolution of new niches. *Evolutionary Biology* 48: 293–311.

Vermeij GJ. 1998. Fossils and the social future of science. *Science* 281: 1444–45.

Vermeij GJ. 2002. The geography of evolutionary opportunity: Hypothesis and two cases in gastropods. *Integrative and Comparative Biology* 42: 935–40.

Vermeij GJ. 2015. Forbidden phenotypes and the limits of evolution. *Interface Focus* 5: 20150028.

Vermeij GJ. 2017. How the land became the locus of major evolutionary innovations. *Current Biology* 27: 3178–82.

Vermeij GJ. 2018. Comparative biogeography: Innovations and the rise to dominance of the North Pacific biota. *Proceedings of the Royal Society B* 285: 20182027.

Vermeij GJ, Grosberg RK. 2010. The Great Divergence: When did diversity on land exceed that in the sea? *Integrative and Comparative Biology* 50: 675–82.

Vosseberg J, Van Hooff JJE, Marcet-Houben M, Van Vlimmeren A, Van Wijk LM, et al. 2021. Timing the origin of eukaryotic cellular complexity with ancient duplications. *Nature Ecology & Evolution* 5: 92–100.

Waddington CH. 1940. *Organizers and Genes*. Cambridge: Cambridge University Press.

Waddington CH. 1957. *Strategy of the Genes*. London: George Allen and Unwin.

Wagner A. 2011. *The Origins of Evolutionary Innovations*. Oxford: Oxford University Press.

Wagner A. 2014. *Arrival of the Fittest*. New York: Penguin Books.

Wagner A, Ortman S, Maxfield R. 2016. From the primordial soup to self-driving cars: Standards and their role in natural and technological innovation. *Journal of the Royal Society Interface* 13: 20151086.

Wagner GP. 2007. The developmental genetics of homology. *Nature Reviews Genetics* 8: 473–79.

Wagner GP. 2014. *Homology, Genes, and Evolutionary Innovation*. Princeton, NJ: Princeton University Press.

Wagner GP, Altenberg L. 1996. Complex adaptations and the evolution of evolvability. *Evolution* 50: 967–76.

Wagner GP, Erkenbrack EM, Love AC. 2019. Stress-induced evolutionary innovation: A mechanism for the origin of cell types. *BioEssays* 41: e1800188.

Wagner GP, Laubichler MD. 2004. Rupert Riedl and the re-synthesis of evolutionary and developmental biology: Body plans and evolvability. *Journal of Experimental Zoology Part B (Molecular and Developmental Evolution)* 302: 92–102.

Wagner GP, Lynch VJ. 2010. Evolutionary novelties. *Current Biology* 20: R48–52.

Wagner GP, Zhang JZ. 2011. The pleiotropic structure of the genotype-phenotype map: The evolvability of complex organisms. *Nature Reviews Genetics* 12: 204–13.

Wainwright PC. 2007. Functional versus morphological diversity in macroevolution. *Annual Review of Ecology, Evolution and Systematics* 38: 381–401.

Wainwright PC. 2009. Innovation and diversity in functional morphology. In *Form and Function in Developmental Evolution*, ed. MD Laubichler, J Maienschein, pp. 132–52. Cambridge: Cambridge University Press.

Wainwright PC, Alfaro ME, Bolnick DI, Hulsey CD. 2005. Many-to-one mapping of form to function: A general principle in organismal design. *Integrative and Comparative Biology* 45: 256–62.

Wainwright PC, Price SA. 2016. The impact of organismal innovation on functional and ecological diversification. *Integrative and Comparative Biology* 56: 479–88.

Wake DB. 2003. Homology and homoplasy. In *Keywords & Concepts in Evolutionary Developmental Biology*, ed. BK Hall, WM Olson, pp. 191–201. Cambridge, MA: Harvard University Press.

Walcott CD. 1910. Abrupt appearance of the Cambrian fauna on the North American continent. *Smithsonian Miscellaneous Collections* 57: 1–16.

Waldrop MM. 2016. The chips are down for Moore's law. *Nature* 530: 144–47.

Wallace AR. 1869. Sir Charles Lyell on geological climates and the origin of species. *Quarterly Review* 126: 359–94.

Wan B, Yuan XL, Chen Z, Guan CG, Pang K, et al. 2016. Systematic description of putative animal fossils from the early Ediacaran Lantian Formation of South China. *Palaeontology* 59: 515–32.

Wang JJ, Zhang WT, Engel MS, Sheng XS, Shih CK, Ren D. 2022. Early evolution of wing scales prior to the rise of moths and butterflies. *Current Biology* 32: 3808–14.

Wang K, Wang J, Zhu C, Yang L, Ren Y, et al. 2021. African lungfish genome sheds light on the vertebrate water-to-land transition. *Cell* 184: 1362–76.

Wang M, Stidham TA, Zhou Z. 2018. A new clade of basal Early Cretaceous pygostylian birds and developmental plasticity of the avian shoulder girdle. *Proceedings of the National Academy of Sciences of the USA* 115: 10708–13.

Warsh D. 2006. *Knowledge and the Wealth of Nations*. New York: W. W. Norton.

Weber BH, Depew DJ, eds. 2003. *Evolution and Learning: The Baldwin Effect Reconsidered*. Cambridge, MA: MIT Press.

Werner GDA, Cornwell WK, Sprent JI, Kattge J, Kiers ET. 2014. A single evolutionary innovation drives the deep evolution of symbiotic N-2–fixation in angiosperms. *Nature Communications* 5: 4087.

West GB. 2017. *Scale: The Universal Laws of Growth, Innovation, Sustainability, and the Pace of Life in Organisms, Cities, Economies, and Companies*. New York: Penguin Books.

West GB, Brown JH. 2005. The origin of allometric scaling laws in biology from genomes to ecosystems: Towards a quantitative unifying theory of biological structure and organization. *Journal of Experimental Biology* 208: 1575–92.

West GB, Brown JH, Enquist BJ. 1997. A general model for the origin of allometric scaling laws in biology. *Science* 276: 122–26.

West SA, Fisher RM, Gardner A, Kiers ET. 2015. Major evolutionary transitions in individuality. *Proceedings of the National Academy of Sciences of the USA* 112: 10112–19.

West-Eberhard MJ. 2003. *Developmental Plasticity and Evolution*. Oxford: Oxford University Press.

West-Eberhard MJ. 2005. Phenotypic accommodation: Adaptive innovation due to developmental plasticity. *Journal of Experimental Zoology Part B (Molecular and Developmental Evolution)* 304: 610–18.

West-Eberhard MJ. 2008. Toward a modern revival of Darwin's theory of evolutionary novelty. *Philosophy of Science* 75: 899–908.

West-Eberhard MJ. 2019. Modularity as a universal emergent property of biological systems. *Journal of Experimental Zoology Part B (Molecular and Developmental Evolution)* 332: 356–64.

Whalen CD, Briggs DEG. 2018. The Palaeozoic colonization of the water column and the rise of global nekton. *Proceedings of the Royal Society B* 285: 20180883.

Whalen CD, Hull PM, Briggs DEG. 2020. Paleozoic ammonoid ecomorphometrics test ecospace availability as a driver of morphological diversification. *Science Advances* 6: eabc2365.

White LA. 1959. *The Evolution of Culture: The Development of Civilization to the Fall of Rome*. New York: McGraw-Hill.

Whitehead AN. 1925. *Science and the Modern World*. New York: Macmillan.

Whitehead H, Laland KN, Rendell L, Thorogood R, Whiten A. 2019. The reach of gene-culture coevolution in animals. *Nature Communications* 10: 2405.

Whiten A. 2017. A second inheritance system: The extension of biology through culture. *Interface Focus* 7: 20160142.

Whiten A. 2019. Cultural evolution in animals. *Annual Review of Ecology, Evolution and Systematics* 50: 27–48.

Whiten A. 2021. The burgeoning reach of animal culture. *Science* 372: eabe6514.

Whitfield J. 2006. *In the Beat of a Heart*. Washington, DC: Joseph Henry.

Wilson EO. 1975. *Sociobiology: The New Synthesis*. Cambridge, MA: Harvard University Press.

Wilson RA, ed. 1999. *Species: New Interdisciplinary Essays*. Cambridge, MA: MIT Press.

Wilson RA, Barker MJ. 2019. Biological individuals. In *The Stanford Encyclopedia of Philosophy*, Fall 2019 ed., ed. EN Zalta. https://plato.stanford.edu/archives/fall2019/entries/biology-individual/.

Wing SL, Hickey LJ, Swicher CC. 1992. Implications of an exceptional fossil flora for Late Cretaceous vegetation. *Nature* 363: 342–44.

Wing SL, Stromberg CA, Hickey LJ, Tiver F, Willis B, et al. 2012. Floral and environmental gradients on a Late Cretaceous landscape. *Ecological Monographs* 82: 23–47.

Winsor MP. 2006. The creation of the essentialism story: An exercise in metahistory. *History and Philosophy of Life Science* 28: 149–74.

Wood R, Erwin DH. 2018. Innovation not recovery: Dynamic redox promotes metazoan radiations. *Biological Reviews* 93: 863–73.

Wood R, Ivantsov AY, Zhuravlev AY. 2017. First macrobiota biomineralization was environmentally triggered. *Proceedings of the Royal Society B* 284: 20170059.

Wood R, Liu AG, Bowyer F, Wilby PR, Dunn FS, et al. 2019. Integrated records of environmental change and evolution challenge the Cambrian Explosion. *Nature Ecology & Evolution* 3: 528–38.

Wooton D. 2015. *The Invention of Science: A New History of the Scientific Revolution*. New York: HarperCollins.

Wright JP, Jones CG. 2006. The concept of organisms as ecosystem engineers ten years on: Progress, limitations, and challenges. *BioScience* 56: 203–9.

Wright S. 1932. The roles of mutation, inbreeding, crossbreeding and selection in evolution. In *Proceedings of the Sixth Annual Congress of Genetics*, Ithaca, NY, 1932, vol. 1, ed. DF Jones, pp. 356–66. Menasha, WI: Brooklyn Botanic Garden.

Wright S. 1982a. Character change, speciation and higher taxa. *Evolution* 36: 427–43.

Wright S. 1982b. The shifting balance theory and macroevolution. *Annual Review of Genetics* 16: 1–19.

Xiao SH, Chen Z, Pang K, Zhou CM, Yuan XL. 2020. The Shibantan Lagerstätte: Insights into the Proterozoic–Phanerozoic transition. *Journal of the Geological Society* 178: jgs2020-135.

Xiao SH, Laflamme M. 2008. On the eve of animal radiation: Phylogeny, ecology and evolution of the Ediacara Biota. *Trends in Ecology & Evolution* 24: 31–40.

Xiao SH, Muscente AD, Chen L, Zhou CM, Schiffbauer JD, et al. 2014. The Weng'an biota and the Ediacaran radiation of multicellular eukaryotes. *National Science Review* 1: 498–520.

Xu X, Zhou Z, Dudley R, Mackem S, Chuong CM, et al. 2014. An integrative approach to understanding bird origins. *Science* 346: 1253293.

Yang C, Li Y, Selby D, Wan B, Guan CG, et al. 2022. Implications for Ediacaran biological evolution from the ca. 602 Ma Lantian biota in China. *Geology* 50: 562–66.

Yang C, Rooney AD, Condon DJ, Li XH, Grazhdankin DV, et al. 2021. The tempo of Ediacaran evolution. *Science Advances* 7: eabi9643.

Yin Z, Sun W, Liu P, Zhu M, Donoghue PCJ. 2020. Developmental biology of *Helicoforamina* reveals holozoan affinity, cryptic diversity, and adaptation to heterogeneous environments in the Early Ediacaran Weng'an biota (Doushantuo Formation, South China). *Science Advances* 6: eabb0083.

Yochelson EL. 2006. The Lipalian interval: A forgotten, novel concept in the geologic column. *Earth Sciences History* 25: 251–69.

Yoder JB, Clancey E, des Roches S, Eastman JM, Gentry L, et al. 2010. Ecological opportunity and the origin of adaptive radiations. *Journal of Evolutionary Biology* 23: 1581–96.

Youn HJ, Strumsky D, Bettencourt LMA, Lobo J. 2015. Invention as a combinatorial process: Evidence from US patents. *Journal of the Royal Society Interface* 12: 20150272.

Yu Y, Zhang C, Xu X. 2021. Deep time diversity and the early radiations of birds. *Proceedings of the National Academy of Sciences of the USA* 118: e2019865118.

Zabell SL. 1992. Predicting the unpredictable. *Synthese* 90: 205–32.

Zamora S, Rahman IA. 2014. Deciphering the early evolution of echinoderms with Cambrian fossils. *Palaeontology* 57: 1105–19.

Zancolli G, Reijnders M, Waterhouse RM, Robinson-Rechavi M. 2022. Convergent evolution of venom gland transcriptomes across Metazoa. *Proceedings of the National Academy of Sciences of the USA* 119: e2111392119.

Zangerl R. 1948. The methods of comparative anatomy and its contribution to the study of evolution. *Evolution* 2: 351–74.

Zattara EE, Busey HA, Linz DM, Tomoyasu Y, Moczek AP. 2016. Neofunctionalization of embryonic head patterning genes facilitates the positioning of novel traits on the dorsal head of adult beetles. *Proceedings of the Royal Society B* 283: 20160824.

Zeder MA. 2012. The domestication of animals. *Journal of Anthropological Research* 68: 161–90.

Zeder MA. 2015. Core questions in domestication research. *Proceedings of the National Academy of Sciences of the USA* 112: 3191–98.

Zeder MA. 2016. Domestication as a model system for niche construction. *Evolutionary Ecology* 30: 325–48.

Zuk M, Bastiaans E, Langkilde T, Swanger E. 2014. The role of behavior in establishment of novel traits. *Animal Behavior* 92: 333–44.

INDEX

A page number in italics refers to a figure.

speciation, Eldredge and Gould on, 46
species, number of, 239
species diversity, 186
species selection, 46
Spencer, Herbert, 51, 54–55, 56
stabilizing selection, 49, 191
standards, 157, 313
Stanley, Steve, 7
state formation, 315
Stebbins, G. Ledyard, 35, 38–39, 239
steranes, 268
Sterelny, Kim, 298, 306
stress response, 91–92, 370n42
Strömberg, Caroline, 1
structure-function distinction, 74–76, 78
  in defining novelty, 111
subcircuits of genes, 80–81, 191, 324
symbiosis, and eukaryotic cell,
  221–226, 265
syntrophies, microbial, 267, 270
systematics
  evolutionary, 71
  Linnaeus as father of, 26
  Linnean hierarchy of, 35, 36, 38, 71
  phylogenetic approaches to, 76–77
  type specimen in, 19
Systematics and the Origin of Species
  (Mayr), 37
Systematic Zoology journal, 46
systemic mutations, 41, 290, 325
Szathmáry, Eörs, 7–8, 95–98, 169, 173, 291, 351

tanks in warfare, 180–181, 376n23
tardigrades, 201, 263
technological change, 9–10
  by combinations, 94
  discussions of, 59
  economic growth and, 53, 106–110
  linear model of, 62, 63, 65
  modular and hierarchical, 318
  vs. widespread conformity, 50–51
technological innovation, 5–6
  Mokyr on, 61
  Morgan on, 65
  Romer on, 73, 108
  Schumpeter on, 6, 59–61, 65, 67, 141, 248
  types of, 63–64
technology
  adaptive/evolutionary spaces and, 133
  built by culture, 306–321

collective/individual creativity and, 105,
  243, 301
combinations and, 117
complexities of innovation, 2
correlation with social complexity, 54
cumulative activities of, 293
culture's role in building, 306–308
definition, 306–307
demand-pull model in, 25
domestication and agriculture, 308–312
drivers of change, 233
early growth models, 170
evolution/evolutionary innovations, 8, 13
general-purpose, 160, 170, 172, 2224
hierarchical structured recipes of, 294
human behavior, culture, and, 99–101
invention and social complexity, 55–57
lags between invention and impact, 1–2
language as a primary technology,
  296, 304
macroevolutionary patterns of novelty, 14
military countertechnology, 181
novelty and innovation in, 13–14, 50–53,
  61–65, 153, 305–306
opportunity drivers of, 25
radar and patents, 316–318
role of innovation in economic growth,
  106–107
spaces for, 117, 142–144, 249, 318–321
standards, 157
technological determinism, 54
teeth
  of elephants, 279–280
  of mammals, 271, 272–273
Tempo and Mode in Evolution (Simpson), 35
Tetlock, Philip, 335
theory reduction, 339
Thomas, Dorothy, 348–349
Thompson, D'Arcy, 136
Thornton, Joseph, 134
tissues, 191, 192, 199, 214
Tomasello, Michael, 298, 303–304, 351
topology
  algebraic topology, 293
  problem of, 145–156
  topological spaces, 145–147, 147
  of evolutionary spaces, 11, 16, 146–149,
  148 161, 164
trace fossils, 193–194, 260, 261–262, 265,
  282–283

A NOTE ON THE TYPE

This book has been composed in Arno, an Old-style serif typeface in the classic Venetian tradition, designed by Robert Slimbach at Adobe.